BIOGENIC NANOMATERIALS

Structural Properties and Functional Applications

Innovations in Biotechnology

BIOGENIC NANOMATERIALS

Structural Properties and Functional Applications

Edited by

Devarajan Thangadurai, PhD

Saher Islam, PhD

Jeyabalan Sangeetha, PhD

Natália Cruz-Martins, PhD

AAP | APPLE ACADEMIC PRESS

First edition published 2023

Apple Academic Press Inc.
1265 Goldenrod Circle, NE,
Palm Bay, FL 32905 USA

4164 Lakeshore Road, Burlington,
ON, L7L 1A4 Canada

CRC Press
6000 Broken Sound Parkway NW,
Suite 300, Boca Raton, FL 33487-2742 USA

4 Park Square, Milton Park,
Abingdon, Oxon, OX14 4RN UK

Library and Archives Canada Cataloguing in Publication

CIP data on file with Canada Library and Archives

Library of Congress Cataloging-in-Publication Data

CIP data on file with US Library of Congress

ISBN: 978-1-77463-838-5 (hbk)
ISBN: 978-1-77463-839-2 (pbk)
ISBN: 978-1-00327-714-9 (ebk)

INNOVATIONS IN BIOTECHNOLOGY BOOK SERIES

SERIES EDITOR

Devarajan Thangadurai, PhD

Associate Professor, Karnatak University, Dharwad, South India

BOOKS IN THE SERIES

Fundamentals of Molecular Mycology

Devarajan Thangadurai, PhD, Jeyabalan Sangeetha, PhD, and Muniswamy David, PhD

Biotechnology of Microorganisms: Diversity, Improvement, and Application of Microbes for Food Processing, Healthcare, Environmental Safety, and Agriculture

Jeyabalan Sangeetha, PhD, Devarajan Thangadurai, PhD, Somboon Tanasupawat, PhD, and Pradnya Pralhad Kanekar, PhD

Phycobiotechnology: Biodiversity and Biotechnology of Algae and Algal Products for Food, Feed, and Fuel

Jeyabalan Sangeetha, PhD, Devarajan Thangadurai, PhD, Sanyasi Elumalai, PhD, and Shivasharana Chandrabanda Thimmappa, PhD

Biogenic Nanomaterials: Structural Properties and Functional Applications

Devarajan Thangadurai, PhD, Saher Islam, PhD, Jeyabalan Sangeetha, PhD, Natália Cruz-Martins, PhD

ABOUT THE EDITORS

Devarajan Thangadurai, PhD

Associate Professor, Karnatak University, Dharwad, South India, and Editor-in-Chief, Biotechnology, Bioinformatics and Bioengineering, and Acta Biologica Indica

Devarajan Thangadurai, PhD, is an Associate Professor at Karnatak University in South India and Editor-in-Chief of the international journals *Biotechnology, Bioinformatics and Bioengineering,* and *Acta Biologica Indica.* He has authored/edited over 30 books with national and international publishers and has visited 24 countries in Asia, Europe, Africa, and the Middle East for academic visits, scientific meetings, and international collaborations. He received his PhD in Botany from Sri Krishnadevaraya University in South India as a CSIR Senior Research Fellow with funding from the Ministry of Science and Technology, Government of India. He served as a Postdoctoral Fellow at the University of Madeira, Portugal; University of Delhi, India; and ICAR National Research Centre for Banana, India. He is the recipient of a Best Young Scientist Award with a Gold Medal from Acharya Nagarjuna University, India, and the VGST-SMYSR Young Scientist Award of the Government of Karnataka, Republic of India.

Saher Islam, PhD

Visiting Lecturer, Lahore College for Women University, Pakistan

Saher Islam, PhD, is Visiting Lecturer at the Department of Biotechnology of Lahore College for Women University, Pakistan. She received her PhD in Molecular Biology and Biotechnology from the University of Veterinary and Animal Sciences, Lahore, Pakistan. She was an IRSIP Scholar at Cornell University, New York and Visiting Scholar at West Virginia State University, West Virginia in USA. She has keen research interests in genetics, molecular biology, biotechnology, and bioinformatics and has ample hands-on experience in molecular marker analysis, whole genome sequencing and RNA sequencing. She has visited the USA, UK, Singapore, Germany, Italy and Russia for academic and scientific trainings, courses, and meetings. She is the recipient of 2016 Boehringer Ingelheim Fonds Travel Grant from European Molecular Biology Laboratory, Germany. She is an author/coauthor of 60 publications including journal articles, book chapters, books and conference presentations.

Jeyabalan Sangeetha, PhD

Assistant Professor, Central University of Kerala, Kasaragod, South India

Jeyabalan Sangeetha, PhD, is an Assistant Professor at Central University of Kerala, Kasaragod, South India. She has edited/co-edited several books in her research areas, which include environmental toxicology, environmental microbiology, environmental biotechnology, and environmental nanotechnology. She earned her BSc in Microbiology and PhD in Environmental Science from Bharathidasan University, Tiruchirappalli, Tamil Nadu, India. She holds an MSc in Environmental Science from Bharathiar University, Coimbatore, Tamil Nadu, India. She is the recipient of a Tamil Nadu Government Scholarship and a Rajiv Gandhi National Fellowship of the University Grants Commission, Government of India, for her doctoral studies. She served as a Dr. D.S. Kothari Postdoctoral Fellow and UGC Postdoctoral Fellow at Karnatak University, Dharwad, South India, during 2012–2016 with funding from the University Grants Commission, Government of India, New Delhi, India.

Natália Cruz-Martins, PhD

Invited Professor, Faculty of Medicine, Researcher at the Laboratory of Neuropsychophysiology, Faculty of Psychology and Education Sciences (FPCEUP); and Institute for Research and Innovation in Health (i3S), University of Porto, Portugal

Natália Cruz-Martins, PhD, is currently an Invited Professor of the Faculty of Medicine, Researcher at the Laboratory of Neuropsychophysiology, Faculty of Psychology and Education Sciences (FPCEUP), and at the Institute for Research and Innovation in Health (i3S), University of Porto, Portugal. She has worked as a university professor since 2017. She was the advisor of eight MSc and PhD theses and is a member of the evaluation panel of the College of Nutritionists (Porto, Portugal). She has participated in six research projects funded by the Portuguese Foundation for Science and Technology (FCT–Portugal) and NORTE2020–Northern Regional Operational Program, and has published more than 120 articles in peer-reviewed, highly reputed international journals (h-index: 21) and eight book chapters and has presented more than 40 communications at national and international conferences. She obtained her BSc in Dietetics and Nutrition and her PhD in Chemical and Biological Engineering under the scope of applied chemistry and biochemistry, food science and nutrition, and clinical microbiology from the University of Minho, Braga, Portugal. She also held several specializations in evidence-based medicine, applied and clinical nutrition, and integrative medicine.

CONTENTS

CONTRIBUTORS

Noura H. Abdellah
Department of Pharmaceutics, Faculty of Pharmacy, Assiut University, Assiut–71526, Egypt

Charles Oluwaseun Adetunji
Applied Microbiology, Biotechnology, and Nanotechnology Laboratory, Department of Microbiology, Edo University Iyamho, PMB 04, Auchi, Edo State, Nigeria

Juliana Bunmi Adetunji
Nutrition and Toxicological Research Laboratory, Department of Biochemistry Sciences, Osun State University, Osogbo, Nigeria

Osikemekha Anthony Anani
Laboratory of Ecotoxicology and Forensic Biology, Department of Biological Science, Faculty of Science, Edo University, Iyamho, Edo State, Nigeria

B. Arredondo-Tamayo
Instituto Politécnico Nacional, Escuela Nacional de Ciencias Biológicas, Departamento de Ingeniería Bioquímica, Av. Wilfrido Massieu Esq, Cda, Manuel L. Stampa S/N, C.P. 07738, Mexico City, Mexico

Pedro Barata
Biomedical Research Center (CEBIMED)/Research Center of the Fernando Pessoa Energy, Environment, and Health Research Unit (FP-ENAS), Faculty of Health Sciences, University of Fernando Pessoa, Porto, Portugal; i3S-Instituto de Investigação e Inovação da Universidade do Porto, Portugal, Rua Alfredo Allen, 208, 4200-135 Porto, Portugal

Eduardo Fernandes Barbosa
Universidade Federal do Oeste da Bahia, Centro das Ciências Biológicas e da Saúde, Campus Reitor Edgard Santos, Rua Bertioga, 892, Morada Nobre I, Barreiras–47810-059, BA, Brazil

Gaber E. Batiha
Department of Pharmacology and Therapeutics, Faculty of Veterinary Medicine, Damanhour University, Damanhour–22511, AlBeheira, Egypt

Amany M. Beshbishy
National Research Center for Protozoan Diseases, Obihiro University of Agriculture and Veterinary Medicine, Nishi 2-13, Inada-Cho, Obihiro–080-8555, Hokkaido, Japan

Sneha Bhandari
Department of Biotechnology, Kumaun University, Nainital, Bhimtal Campus, Bhimtal–263136, Uttarakhand, India

Cínthia Caetano Bonatto
Laboratório de Nanobiotecnologia (LNANO), Embrapa Recursos Genéticos e Biotecnologia, Pq. Est. Biol. Final W5 Norte, Asa Norte, Brasília–70770-917, DF, Brazil; TecSinapse, Pesquisa Aplicada, São Paulo–04583-110, SP, Brazil

P. I. Cauich-Sánchez
Instituto Politécnico Nacional, Escuela Nacional de Ciencias Biológicas, Departamento de Microbiología, Plan de Ayala y Carpio S/N, C. P. 11340, Mexico City, Mexico

J. J. Chanona-Pérez
Instituto Politécnico Nacional, Escuela Nacional de Ciencias Biológicas, Departamento de Ingeniería Bioquímica, Av. Wilfrido Massieu Esq, Cda, Manuel L. Stampa S/N, C.P. 07738, Mexico City, Mexico

L. Dorantes-Álvarez
Instituto Politécnico Nacional, Escuela Nacional de Ciencias Biológicas, Departamento de Ingeniería Bioquímica, Av. Wilfrido Massieu Esq, Cda, Manuel L. Stampa S/N, C.P. 07738, Mexico City, Mexico

Leo Bey Fen
Faculty of Medicine, University of Malaya, Kuala Lumpur–50603, Malaysia; Nanotechnology and Catalysis Research Center, Institute for Advanced Studies, University of Malaya, Kuala Lumpur–50603, Malaysia

Maria Gabriela O. Fernandes
Pulmonology Department, Centro Hospitalar e Universitário de São João, Porto, Portugal; Faculty of Medicine, University of Porto, Porto, Portugal

Zahra Hajiahmadi
Department of Agricultural Biotechnology, Faculty of Agricultural Sciences, University of Guilan, Rasht–4199613776, Iran

M. A. Motalib Hossain
Nanotechnology and Catalysis Research Center, Institute for Advanced Studies, University of Malaya, Kuala Lumpur–50603, Malaysia

Saher Islam
Institute of Biochemistry and Biotechnology, Faculty of Biosciences, University of Veterinary and Animal Sciences, Lahore–54000, Pakistan

Maria Jacob
Pulmonology Department, Centro Hospitalar e Universitário de São João, Porto, Portugal

Parva Jani
Faculty of Pharmacy, The Maharaja Sayajirao University of Vadodara, Baroda, Gujarat, India

Ana Elena Jiménez-Carmona
Instituto Politécnico Nacional, Escuela Nacional de Ciencias Biológicas, Departamento de Ingeniería Bioquímica, Av. Wilfrido Massieu Esq, Cda, Manuel L. Stampa S/N, C.P. 07738, Mexico City, Mexico

Mohd. Rafie Johan
Nanotechnology and Catalysis Research Center, Institute for Advanced Studies, University of Malaya, Kuala Lumpur–50603, Malaysia

Oseni Kadiri
Department of Biochemistry, Faculty of Basic Medical Sciences, Edo University Iyamho, Nigeria

Victoria Baggi Mendonça Lauria
Laboratório de Nanobiotecnologia (LNANO), Embrapa Recursos Genéticos e Biotecnologia, Pq. Est. Biol. Final W5 Norte, Asa Norte, Brasília–70770-917, DF, Brazil; Programa de Pós-graduação em Nanociência e Nanobiotecnologia, Universidade de Brasília, Instituto de Ciências Biológicas, Asa Norte, Brasília–70910-900, DF, Brazil

Samuel Makinde
Informatics and Cyber-Physical Systems Laboratory, Department of Computer Science, Edo University Iyamho, PMB 04, Auchi, Edo State, Nigeria

Natália Cruz-Martins
Faculty of Medicine, University of Porto, Porto, Portugal; Institute for Research and Innovation in Health (i3S), University of Porto, Porto, Portugal; Laboratory of Neuropsychophysiology, Faculty of Psychology and Education Sciences, University of Porto, Portugal

Kelliane Almeida Medeiros
Hospital das Forças Armadas, Diretoria Técnica de Ensino e Pesquisa, Divisão de Pesquisa, Setor HFA, Sudoeste, Brasília–70673-900, DF, Brazil

J. V. Méndez-Méndez
Instituto Politécnico Nacional, Centro de Nanociencias y Micro y Nanotecnologías, Luis Enrique Erro S/N, Zacatenco, C.P. 07738, Gustavo A. Madero, Mexico City, Mexico

Olugbenga Samuel Micheal
Cardiometabolic Research Unit, Department of Physiology, College of Health Sciences, Bowen University, Iwo, Osun State, Nigeria

Ana Catarina Moreira
Hospital Garcia de Orta, E.P.E Almada, Lisboa, Portugal

Tapan K. Nailwal
Department of Biotechnology, Kumaun University, Nainital, Bhimtal Campus, Bhimtal–263136, Uttarakhand, India

Luísa Nascimento
Pulmonology Department, Centro Hospitalar de Trás-os-Montes e Alto Douro, Vila Real, Portugal

Nurul Izza Nordin
Industrial Biotechnology Research Center (IBRC), SIRIM Berhad, Shah Alam, Malaysia

Wilson Nwankwo
Informatics and Cyber-Physical Systems Laboratory, Department of Computer Science, Edo University Iyamho, PMB 04, Auchi, Edo State, Nigeria

Rita Oliveira
Biomedical Research Center (CEBIMED)/Research Center of the Fernando Pessoa Energy, Environment, and Health Research Unit (FP-ENAS), Faculty of Health Sciences, University of Fernando Pessoa, Porto, Portugal

Tatiane Melo Pereira
Laboratório de Nanobiotecnologia (LNANO), Embrapa Recursos Genéticos e Biotecnologia, Pq. Est. Biol. Final W5 Norte, Asa Norte, Brasília–70770-917, DF, Brazil; Programa de Pós-graduação em Ciências Biológicas (Biologia Molecular), Universidade de Brasília, Instituto de Ciências Biológicas, Asa Norte, Brasília–70910-900, DF, Brazil

Ana Cláudia Pimenta
Pulmonology Department, Centro Hospitalar de Trás-os-Montes e Alto Douro, Vila Real, Portugal

Sofia Pinheiro
Centro Hospitalar Vila Nova de Gaia, Researcher Unit of Biomedicine, Faculty of Medicine, University of Porto, Portugal

Vera Lúcia Perussi Polez
Laboratório de Prospecção de Compostos Bioativos (LPCB), Embrapa Recursos Genéticos e Biotecnologia, Pq. Est. Biol. Final W5 Norte, Asa Norte, Brasília–70770-917, DF, Brazil

Pradeep Rajole
Cytxon Biosolution Laboratory, Hubballi–580031, Karnataka, India

Ahmad Hazri Abd. Rashid
Industrial Biotechnology Research Center (IBRC), SIRIM Berhad, Shah Alam, Malaysia

Lalaji Rathod
Faculty of Pharmacy, The Maharaja Sayajirao University of Vadodara, Baroda, Gujarat, India

Jeyabalan Sangeetha
Department of Environmental Science, Central University of Kerala, Kasaragod–671316, Kerala, India

Thaís Ribeiro Santiago
Programa de Pós-Graduação em Fitopatologia, Universidade de Brasília, Instituto de Ciências Biológicas, Asa Norte, Brasília–70910-900, DF, Brazil

Krutika Sawant
Faculty of Pharmacy, The Maharaja Sayajirao University of Vadodara, Baroda, Gujarat, India

Reza Sayyad
School of Metallurgy and Materials Engineering, College of Engineering, University of Tehran, Tehran, Iran

Francine Schütz
Faculty of Medicine, Department of Biomedicine, University of Porto, Alameda Prof. Hernâni Monteiro–4200-319, Porto, Portugal

Arun Kashivishwanath Shettar
Department Applied Genetics, Karnatak University, Dharwad–580003, Karnataka, India

Reza Shirzadian-Khorramabad
Department of Agricultural Biotechnology, Faculty of Agricultural Sciences, University of Guilan, Rasht–4199613776, Iran

Luciano Paulino Silva
Laboratório de Nanobiotecnologia (LNANO), Embrapa Recursos Genéticos e Biotecnologia, Pq. Est. Biol. Final W5 Norte, Asa Norte, Brasília–70770-917, DF, Brazil;
Programa de Pós-graduação em Nanociência e Nanobiotecnologia, Universidade de Brasília, Instituto de Ciências Biológicas, Asa Norte, Brasília–70910-900, DF, Brazil;
Programa de Pós-graduação em Ciências Biológicas (Biologia Molecular), Universidade de Brasília, Instituto de Ciências Biológicas, Asa Norte, Brasília–70910-900, DF, Brazil

Ariane Pandolfo Silveira
Laboratório de Nanobiotecnologia (LNANO), Embrapa Recursos Genéticos e Biotecnologia, Pq. Est. Biol. Final W5 Norte, Asa Norte, Brasília–70770-917, DF, Brazil;
Programa de Pós-graduação em Nanociência e Nanobiotecnologia, Universidade de Brasília, Instituto de Ciências Biológicas, Asa Norte, Brasília–70910-900, DF, Brazil

Swati Sinha
Department of Biotechnology, Kumaun University, Nainital, Bhimtal Campus, Bhimtal–263136, Uttarakhand, India

Catarina Sousa
Pulmonology Department, Centro Hospitalar e Universitário de São João, Porto, Portugal

Devarajan Thangadurai
Department of Botany, Karnatak University, Dharwad–580003, Karnataka, India

H. H. Torres-Ventura
Instituto Politécnico Nacional, Escuela Nacional de Ciencias Biológicas,
Departamento de Ingeniería Bioquímica, Av. Wilfrido Massieu Esq, Cda,
Manuel L. Stampa S/N, C.P. 07738, Mexico City, Mexico

Anand Ishwar Torvi
Department of Chemistry, Karnatak University, Dharwad–580003, Karnataka, India

Syed Muhammad Kamal Uddin
Nanotechnology and Catalysis Research Center, Institute for Advanced Studies, University of Malaya, Kuala Lumpur–50603, Malaysia

Lívia Cristina De Souza Viol
Laboratório de Prospecção de Compostos Bioativos (LPCB), Embrapa Recursos Genéticos e Biotecnologia, Pq. Est. Biol. Final W5 Norte, Asa Norte, Brasília–70770-917, DF, Brazil

ABBREVIATIONS

3D	three dimensional
ABTS	2,2′-Azino-bis (3-ethylbenzothiazoline-6-sulfonic acid)
AC-ZnO	ZnO-loaded porous activated carbon
AEC2s	type II alveolar epithelial cells
AFM	atomic force microscopy
Ag NPs	silver nanoparticles
Ag	silver
Ag@CS/An	chitosan-AgNPs-antracol
$AgNO_3$	silver nitrate
AgNPs-PSAC	palm shell agro-waste derived carbon
AIEAS	active ingredients of *Eupatorium adenophorum* Spreng
Al	aluminum
Al_2O_3	aluminum oxide
ALK	anaplastic lymphoma kinase
AMF	arbuscular mycorrhizal fungi
APCs	antigen-presenting cells
Apt	aptamer
ASTM	American Society for Testing and Materials
ATDs	antitubercular drugs
ATP	adenosine triphosphate
Au NPs	gold nanoparticles
Au	gold
B	boron
Ba	barium
BBB	blood-brain barrier
BC	bacterial cellulose
BDP	beclomethasone dipropionate
BET	Brunauer Emmet teller
BMF	biomagnification factor
BNC PS/ZnO	bio-based nanocomposite
BNC	bacterial nanocellulose
B-SWCNTs	bombolitin II-single-walled carbon nanotubes
$C_{42}H_{70}O_{35}$	β-cyclodextrin
Ca	calcium

CAGR	compound annual growth rate
CalTech	California Institute of Technology
CaP	calcium phosphate
CB	carbon black
CBNs	carbon-based nanomaterials
C-dots	carbon quantum dots
CdSe	cadmium-selenium
CeO_2	cerium oxide
CIN	chitosan interferon-gamma-pDNA nanoparticles
CKD	chronic kidney disease
CLNPs	carrier lipid nanoparticles
CLSM	confocal laser scanning microscope
CNC	cellulose nanocrystals
CN-Cu-NCs	chitosan-copper nanocomposite
CNFs	cellulose nanofibers
CNS	central nervous system
CNT	carbon nanotube
CNWs	cellulose nanowhiskers
Co	cobalt
CO_2	carbon dioxide
Conc.	concentration
COPD	chronic obstructive pulmonary disease
Cu NPs	copper nanoparticles
COS	chitooligosaccharides
CpG	cytosine-phosphate-guanine
CS/PAA/Cu-HNCs	chitosan/polyacrylic acid/copper hydrogel nanocomposites
CsA	cyclosporin A
Cu	copper
CuO NPs	copper oxide nanoparticles
CuO	copper oxide
$CuSO_4$	copper sulfate
CV	cardiovascular
CVD	chemical vapor deposition
DMA	differential mobility analyzer
DMM	dimethomorph
DNA	deoxyribonucleic acid
DPPH	1,1-diphenyl-2-picrylhydrazyl
DSC	differential scanning calorimetry

DWNTs	double-walled nanotubes
ECEIA	electrochemical enzyme-linked immunoassay
EFSA	European Food Safety Authority
EGFR	epidermal growth factor receptor
EML4-ALK	echinoderm microtubule-associated protein-like 4-anaplastic lymphoma kinase
ENMs	engineered nanomaterials
ENPs	engineering nanoparticles
EPR	enhanced permeability and retention effect
FAO	Food and Agriculture Organization
FDA	Food and Drug Administration
Fe	iron
Fe^{2+}	ferrous
Fe_2O_3	iron oxide
Fe^{3+}	ferric
Fe_3O_4	iron oxide
FEA	finite element analysis
FeO	iron (II) oxide
Gd_2O_3	gadolinium (III) oxide
GDs	guideline documents
GFP	green fluorescent protein
GHR	growth hormone receptor
GI	gastrointestinal
GLP-2	glucagon-like peptide 2
GM	gut microbiome
GO	graphene oxide
GPD	gross domestic production
GSP	groundnut shells
GUS	β-glucuronidase
HNT	halloysite nanotube
IDEs	interdigitated electrodes
IL	interleukin
In_2O_3	indium oxide
INF	interferon
INH	isoniazid
IPN	Instituto Politécnico Nacional
ISO	International Organization for Standardization
IUCLID	International Uniform Chemical Information Database
K	potassium

L/D	length/diameter
La_2O_3	lanthanum (III) oxide
LC	lung cancer
LCA	life cycle assessment
LCWPS	lignocellulosic waste peanut shells
LDH NPs	layered double hydroxide nanoparticles
LNPs	lipid nanoparticles
LOD	loss on drying
LSPR	localized surface plasmon resonance
mAbs	monoclonal antibodies
MCP	monocyte chemotactic peptide
MgO	magnesium oxide
MHC	major histocompatibility complex
MNPs	metallic nanoparticles
MOA	mode of action
MRI	magnetic resonance imaging
MS	Murashige and Skoog
MSCs	mesenchymal stem cells
MSN	mesoporous silica nanoparticle
MTB	mycobacterium tuberculosis
MW	molecular weight
MWCNTs	multi-walled carbon nanotubes
MWNT	multi-walled nanotubes
N	nitrogen
nanoRA	risk prioritization assessment
NCC	nanocrystalline cellulose
NCs	nanocarriers
ND	nanodiamond
Nd:YAG	neodymium-doped yttrium aluminum garnet
NFC	nano-fibrillated cellulose
NGS	next-generation sequencing
Ni	nickel
NIST	National Institute of Standards and Technology
NM	nanomedicine
NMR	nuclear magnetic resonance
NNMs	nanoparticulate nanomedicines
NP-CNSs	nanoporous carbon nanosheets
NPs	nanoparticles
NS	nano-silica

NSCLC	non-small cell lung cancer
OVA	ovalbumin
P	phosphorus
PAA-b-PBA	poly(acrylic acid)-b-poly(butyl acrylate)
PAMAM	PEGylated poly(amidoamine)
Pb_2	lead
PDL-1	programmed cell death protein 1-ligand
PDMS	polydimethylsiloxane
pDNA	plasmid DNA
PECs	predicting the environmental concentrations
PEG	polyethylene glycol
PEG-PLA	polyethylene glycol-block-poly(D, L-lactic acid)
PEI	polyethylene imide
PEO	polyethylene oxide
PFT	PeakForce tapping
PHSNs	porous hollow silica NPs
PLA	polylactic acid
PLA-TPGS	polylactide tocopheryl polyethylene glycol 1000 succinate
PLGA	poly(lactic-co-glycolic acid)
PLGAPEG	poly(ethylene glycol) and PLGA
PLG-NPs	poly(lactide-co-glycolide) nanoparticles
PMMA	polymerizing methacrylic acid
PNECs	predicted no-effect concentrations
PNPs	polymeric nanoparticles
PPX	paclitaxel poliglumex
P-SWCNTs	polyvinyl alcohol-single-walled carbon nanotubes
PVA	polyvinyl alcohol
PVA/ST	poly(vinyl alcohol)/starch
PVAc-CA	poly(vinyl acetate-co-crotonic acid)
PVP	polyvinylpyrrolidone
PYY	peptide YY
PZA	pyrazinamide
QNM	quantitative nanomechanics
RD&I	research, development, and innovation
REACH	registration, evaluation, authorization, and restriction of chemicals
RIF	rifampicin
RMS	reference material standards

ROS	reactive oxygen species
SA	ScanAsyst
SAS	synthetic amorphous silica
SbD	safe-by-design
$scCO_2$	supercritical carbon dioxide
SCLC	small-cell lung cancer
SDR	spinning disc reactor
SDS	sodium dodecyl sulfate
SEDDS	solid self-emulsifying drug delivery system
SEM	scanning electron microscope
SERS	surface-enhanced Raman scattering
sgRNA	single guide RNA
Si	silicon
Si_3N_4	silicon nitride
SiO_2	silicon dioxide
siRNA	small interfering RNA
SLNPs	solid-lipid nanoparticles
SMPS	scanning mobility particle sizer
SOD	superoxide dismutase
SPM	scanning probe microscope
SWCNTs	single-walled carbon nanotubes
SWNT	single-walled nanotubes
TB	tuberculosis
TEM	transmission electron microscopy
TEMPO	2,2,6,6,-tertramethylpiperadine-1-oxyl
TGA	thermogravimetric analysis
TGs	test guidelines
Ti	titanium
Ti:Sapphire	titanium-doped sapphire
TiO_2 NPs	titanium oxide nanoparticles
TiO_2	titanium dioxide
TKI	tyrosine kinase inhibitor
TNF-α	tumor necrosis factor-α
TNM	tumor, nodes, metastasis
UP	unsaturated polyester
USD	United States dollar
UTS	ultimate tensile strength
VEGF	vascular endothelial growth factor
VEGFR2	vascular endothelial growth factor receptor 2

VSSA	volume-specific surface area
WHO	World Health Organization
WPMN	working party on manufactured nanomaterials
WS	worksheets
XPS	X-ray photoelectron spectroscopy
XRD	X-ray diffraction
YAG	yttrium aluminum garnet
Yb_2O_3	ytterbium oxide
Zn	zinc
ZnO NPs	zinc oxide nanoparticles
ZnO	zinc oxide
ZnS	zinc sulfide

SYMBOLS

τ	effective time constant
μg	microgram
μg/L	microgram/liter
μg/ml	microgram/milliliter
μL/L	microliter/liter
μm	micrometer
μM	micromolar
Å	angstrom
C_{60}	fullerenes
D	dimension
E	elastic modulus
F	force of the tip-sample
g	gram
g/L	gram/liter
h	hours
k	cantilever beam stiffness
k_1	cell spring constant
kDa	kilodalton
kHz	kilohertz
kPa	kilopascal
L	liter
mg/L	milligram/liter
mL	milliliter
mM	millimolar
N/m	newton per meter
ng	nanogram
nm	nanometer
nM	nanomolar
pH	power of hydrogen
ppm	parts per million
sp^2	hybrid orbitals
UV	ultraviolet
w/v	weight per volume
z	amount of deflection

PREFACE

Nanotechnology is believed as the next great revolution in biology, medicine, and agriculture. For the last two decades, expert groups have been engaged in the advancement of modern applications of novel nanoparticles for numerous biological applications. Nanoparticles are becoming progressively vital in medical treatments, with new diagnostic tools and drugs based on nanobiotechnology. Every year, hundreds of new ideas using nanomaterials are applied in the development of biosensors. An increasing number of new enterprises are also searching for market opportunities using these technologies. The prospect to exploit processes and structures of biomaterials for innovative functional bioelectronics, biosensors, and clinical applications has designed the rapidly emerging arena of nanobiotechnology. At present-day, nanobiotechnology concerns the exploitation of biological approaches optimized through progressions such as cellular components, cells, proteins, and nucleic acids to formulate functional nanostructures comprised of inorganic and organic materials. Nanobiotechnology also concerns the refinement and application of instruments, originally proposed to produce, and functionalize the nanostructured objects to applied research of vital biological processes.

Nanobiotechnology holds the potential of offering revolutionary awareness into different aspects of biological science ranging from basic queries of receptor jobs to personal medicine and drug discovery. Novel materials are assisting the bio-imaging of various cellular activities for longer intervals, directing to high-throughput cellular-based monitors for drug discovery and targeted delivery, and several diagnostic applications. As functionalized biomaterials, they signify one of the major empowering nanoscale elements for these novel technologies. Accordingly, many protocols highlight various strategies to manufacture and manipulate these nanoprobes for biotechnological applications. As with several areas fraught with growing hyperbole, it is crucial that the fundamental approaches be centered on solid and reproducible technologies.

In recent years, nanotechnology applications have spread to multiple areas, from engineering, technology to medicine. In fact, this growing interest in the use of nanotechnology is mainly due to the greater and better understanding of the structural properties and applications of nanoparticles and nanomaterials, where although there is not too much evidence on their safety, efficacy, and toxicological profiles, the reality is that the functional applications of nanotechnology seem promising. In view of this intense growth in the number of studies developed in this area, there is an urgent need to compile the data obtained so far, with a view not only to foster new studies, but also to facilitate the data interpretation through their summarization both for professionals, students, clinicians, and other target audiences.

Thus, this book intends to provide in a global way the most updated information possible regarding the use and applications of nanobiotechnology, starting from a careful characterization and introduction to the use of nanoparticles and nanomaterials, their nanomechanical properties in bacteria and biomedical applications (Chapters 1–4), progressively passing to its applications in targeted therapy to multiple pathologies, such as cancer, obstructive pulmonary diseases, chronic infectious diseases, until its impact on the modulation of the intestinal microbiota (Chapters 5–8). A special emphasis is also given to its potential in terms of promoting sustainability, such as the ability to improve plant systems in terms of tissue culture, moving to its added value in the transfer of macromolecules to plants, and finally in triggering the sustainable exploitation of agricultural, forestry, and food residues, ultimately promoting the green nanotechnology (Chapters 9–12).

—Devarajan Thangadurai, PhD
Saher Islam, PhD
Jeyabalan Sangeetha, PhD
Natália Cruz-Martins, PhD

CHAPTER 1

NANOPARTICLES AND NANOMATERIALS: AN UPDATE

ANAND ISHWAR TORVI,[1] JEYABALAN SANGEETHA,[2] ARUN KASHIVISHWANATH SHETTAR,[3] DEVARAJAN THANGADURAI,[4] and PRADEEP RAJOLE[5]

[1]*Department of Chemistry, Karnatak University, Dharwad–580003, Karnataka, India*

[2]*Department of Environmental Science, Central University of Kerala, Kasaragod–671316, Kerala, India*

[3]*Department of Applied Genetics, Karnatak University, Dharwad–580003, Karnataka, India*

[4]*Department of Botany, Karnatak University, Dharwad–580003, Karnataka, India*

[5]*Cytxon Biosolution Laboratory, Hubballi–580031, Karnataka, India*

1.1 INTRODUCTION

Nanotechnology is a known field of research since the last century. "Nanotechnology" was presented by Nobel Laureate Richard P. Feynman during his famous 1959 lecture "There's Plenty of Room at the Bottom," since then, various remarkable developments have been progressed in the nanotechnology field. The term nanoparticles (NPs) were coined from the Greek word 'nano' that means 'dwarf or small' and these NPs have one dimension less than 100 nm at least (Laurent et al., 2010). Based on the shape of the materials, NPs are classified as 0D, 1D, 2D, or 3D (Figure 1.1). These NPs are not simple molecules, but instead made of three different layers: (a) The upper layer is called surface layer, which further functionalized by different small molecules, surfactants, metals ions and polymers, (b) The shell layer, which is different from the core, and (c) The core, which is central portion of the NP and usually refers the NP itself.

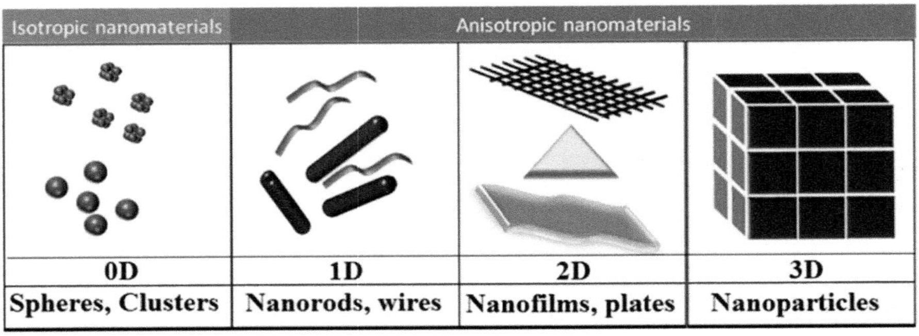

FIGURE 1.1 Different dimensions of nanomaterials.

Source: Reproduced with permission from: Sadhasivam et al. (2017); Copyright © Elsevier, 2017.

These NPs can be found in different size, structure, and shape. These NPs are observed in spherical, tubular, ribbon, ring, hallow, conical, flat, and spiral. The surface of these NPs can be uniform or non-uniform with variations in the surface morphology. Few nanoparticles are amorphous or crystalline with either single or multi-crystals enclosed either agglomerated or loosely. Nanoparticles have both solute and separate particle-phase properties. The surface-to-volume ratio of nanoparticles is 35–45% times higher as compared to large particles or atoms. The unique property "specific surface area" of nanoparticle is a major determining factor for its high intrinsic properties which is size-dependent (Auffan et al., 2009). Furthermore, the extraordinary features of these NPs are the main reason for its multifunctional properties and creating interest for its wide range of potential applications in medicines, cancer, energy, and nutrition (Chandra et al., 2006).

In order to explore novel physical properties, phenomena, and also to realize potential applications of nanoparticles, the synthetic method is very important. There exist a number of methods to synthesize the nanomaterials, which are categorized in two techniques "top-down and bottom-up." Solid-state route, ball milling comes in the category of top-down approach, while wet chemical routes like sol-gel and co-precipitation come in the category of bottom-up approach. Secondly, characterization of nanomaterials is necessary to analyze their various properties. Therefore, this chapter describes the classification, synthetic method, and characterization of nanomaterials.

1.2 SYNTHESIS OF NANOPARTICLES

The synthesis of NPs with definite size, shape, functional group, and structure is very important in the field of nanotechnology. Therefore, the synthetic methods gain a special attention. Traditionally nanoparticles were synthesized only from chemical and physical methods, including sol-gel, solvothermal process and ion sputtering technique. Generally, two main approaches for nanoparticles synthesis are bottom-up and top-down approaches. The bottom-up method is a synthesis of NPs through the condensation of atoms or molecular entities in a solution or a gas phase to form the NPs with the nanometer range. The second method, the top-down method involves the partition of solid mass into very smaller portions and form particles with nanometer size. The top-down method involves milling or attrition. The first method is more popular as compared to top-down approach because of its many benefits (Figure 1.2).

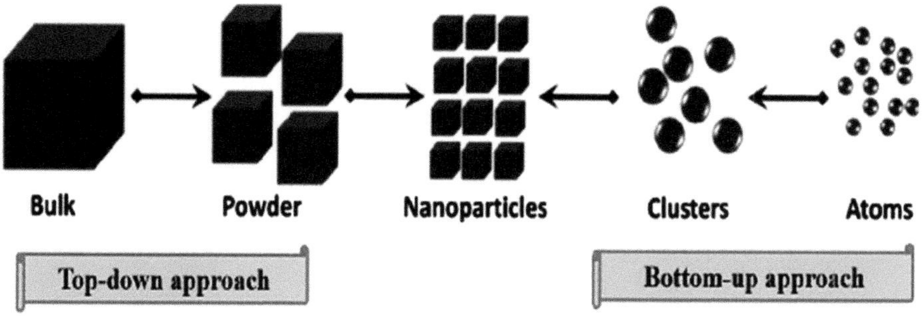

FIGURE 1.2 Schematic representation of bottom-up and top-down approaches for nanoparticle and nanomaterial synthesis.

1.2.1 BOTTOM-UP METHOD

Spinning, sol-gel, pyrolysis, biosynthesis, and chemical vapor deposition (CVD) are most generally employed bottom-up protocols for nanomaterials development.

1.2.1.1 SPINNING

The development of NPs is processed by spinning process using spinning disc reactor (SDR). The SDR contains rotating disc within the chamber where physical parameters like heat can be regulated. Reactor is commonly filled either with inert gases or nitrogen to remove the oxygen inside thereby to avoid any chemical reaction (Tai et al., 2007). The rotating disc within the chamber is get rotated at various speed levels where precursor is forced in. The spinning process makes atoms to fuse and precipitated then collected and finally dried (Mohammadi et al., 2014). The influencing operating parameters including flow rate of liquid, rotating speed of disc, feed location, liquid/ precursor proportion and disc surface determines the features of nanoparticles synthesized from SDR.

1.2.1.2 SOL-GEL

The sol means colloidal solutions of solids suspended in the liquid phase and gel means solid macromolecules submerged in the solvent. Sol-gel is the most widely used technique because of its ease of synthesis. Sol-gel technique involves a wet-chemical process which contains chemical solution that acts as a precursor. Metal chlorides and oxides are commonly used precursors for the sol-gel method (Ramesh et al., 2013). These precursors are then dispersed into host liquid by either stirring, sonication or shaking, and then the resultant structure contains solid and liquid phase. Phase separation assists to recover NPs by several methods, including filtration, centrifugation, and sedimentation, and moisture can be removed by further drying (Mann et al., 1997).

1.2.1.3 PYROLYSIS

It is most commonly employed industry process used for the development of nanomaterials in large scale. It involves the burning of precursor under flame. Precursor can be vapor or liquid that is processed with extreme pressure into the furnace via a small hole (Kammler et al., 2001). The combustion substances are further air classified for recovering nanoparticles. Few furnaces use plasma and laser instead of flame to provide extreme temperature for easy evaporation (D'Amato et al., 2013). The advantages of pyrolysis are simple, efficient, cost-effective, and continuous process with high yield.

1.2.1.4 BIOSYNTHESIS

Biosynthesis is eco-friendly and green approach for the development of nanoparticles which are biodegradable and nontoxic (Kuppusamy et al., 2014).

The biosynthesis process makes use of bacteria, fungi, and plant extracts along with precursors to synthesize nanomaterials instead of conventional chemicals for capping and bioreduction purposes (Figure 1.3). The developed biosynthesized nanomaterials contain unique as well as enhanced properties which are having many biomedical applications (Hasan et al., 2015).

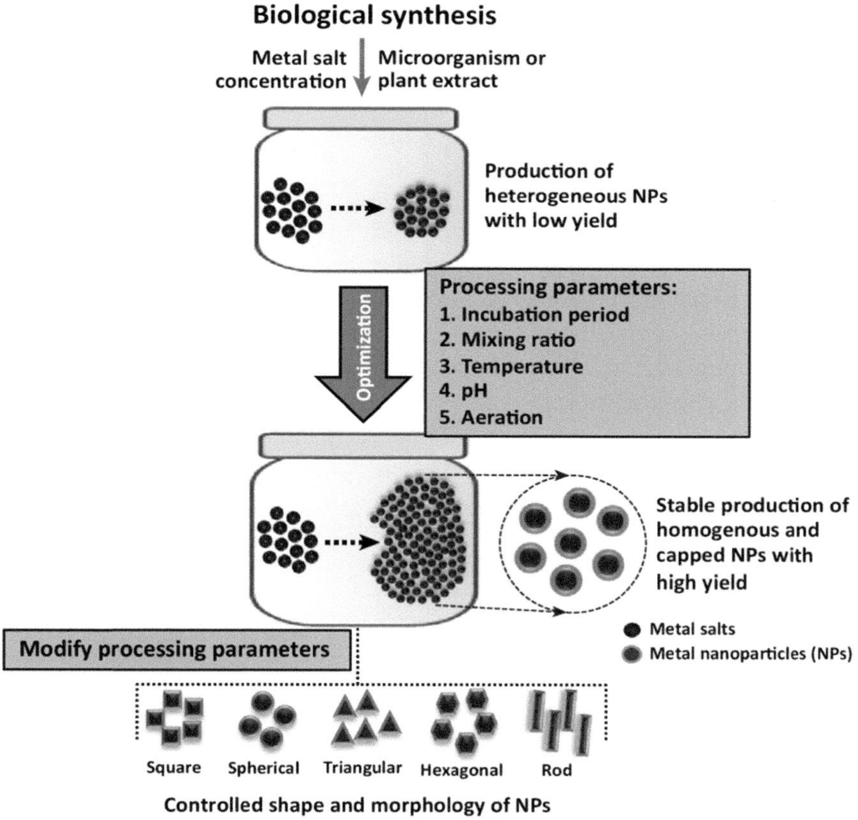

FIGURE 1.3 Optimization and production of homogenous nanoparticles from plants and microbes.

Source: Reproduced with permission from: Singh et al. (2016); Copyright © Elsevier, 2016.

1.2.1.5 CHEMICAL VAPOR DEPOSITION (CVD)

In this process, gaseous reactants are deposited onto a substrate and forms the thin film. Deposition is generally processed in certain reaction chamber to combine gas molecules at ambient temperature. A chemical reaction takes place when heated substrate molecules come into contact with the combined gas molecules (Bhaviripudi et al., 2007). This reaction gives thin film as a product on the surface of the substrate that can be recovered and then used. Substrate temperature is an important factor that influence in CVD process. The main advantage of the CVD process is that we can achieve highly uniform, pure, strong, and hard nanoparticles and the disadvantages include the requirement of distinctive equipment and gaseous by-products generated during the deposition are extremely toxic (Motoaki et al., 2003).

1.2.2 TOP-DOWN METHOD

In this method, bulk material is converted into small nano-sized particles. Preparation of nanoparticles is through size reduction of starting material by various physical and chemical treatments (Meyers et al., 2006). It includes methods such as mechanical milling, laser ablation, and thermal decomposition. Although top-down methods are easy to perform, is not a suitable method for preparing informal shaped and very small size particles. The major problem associated with this method is that change in surface chemistry and physicochemical properties of nanoparticles (Nadagouda et al., 2011).

1.2.2.1 MECHANICAL MILLING

Among the various top-down methods, mechanical milling approach is the most widely used technique to produce several nanoparticles. This technique is employed for milling as well as post-annealing of nanomaterials during the synthesis where various elements are get milled in an inert atmosphere (Yadav et al., 2012). The manipulating factor in this technique is a deformation of plastic that carries fracture in particle shape and tends to decrease the size of particles and cold-welding makes the increase in particle size.

1.2.2.2 LASER ABLATION

In the laser ablation method, laser irradiation is used to reduce the particle size to nano level. The solid target material is placed under a thin layer and then exposed to pulsed laser irradiation. Mainly Nd:YAG (neodymium-doped yttrium aluminum garnet) laser at 106 μm output and its harmonic,

Ti:Sapphire (titanium-doped sapphire) laser and copper vapor lasers are used (Simakin et al., 2004). The irradiation of substance to laser result in fragmentation in the form of nanostructures. The energy and laser pulse duration determines the relative quantity of ablated particles formed (Kruis et al., 1998). Several parameters such as wavelength ablation time, time duration of laser pulse, laser fluency and effective surrounding liquid medium with or without surfactant influences ablation efficiency and characteristic of metal particle formed (El-Nour et al., 2010).

1.2.2.3 THERMAL DECOMPOSITION

Thermal decomposition is an endothermic chemical decomposition produced by heat that breaks the chemical bonds in the compound (Salavati-Niasari et al., 2008). The specific temperature at which an element chemically decomposes is the decomposition temperature. Nanoparticles are synthesized by decomposing metals at a particular temperature undergoing chemical reaction that produce secondary products.

1.2.2.4 SPUTTERING

Sputtering is the deposition of nanomaterials on the surface by ejecting particles from it by frequent colliding with the ions (Shah et al., 2006). Sputtering is usually a deposition of a thin layer of nanoparticles followed by annealing. The thickness of the layer, temperature, and duration of annealing and substrate type determines size and shape of nanoparticles (Lugscheider et al., 1998).

1.2.2.5 PHYSICAL PROPERTIES

The physical properties include optical such as the color of the nanoparticle, its light penetration, absorption, reflection capabilities, UV absorption and reflection abilities in a solution or when coated onto a surface. It also includes the mechanical properties such as elastic, ductile, tensile strengths and flexibility that play a significant factor in their application. Other properties such as hydrophobicity, hydrophilicity, diffusion, suspension, and settling features have found ways in several modern things. Electric and magnetic properties like conductivity, semiconductivity, and resistivity have made nanomaterials to get utilized in renewable energy purposes.

1.2.2.6 CHEMICAL PROPERTIES

The chemical properties such as the reactivity of the nanoparticles with the target and stability and sensitivity to factors such as moisture, atmosphere, heat, and light determine its applications. The antibacterial, antifungal, disinfection, and toxicity properties of the nanoparticles are ideal for biomedical and environmental applications. Corrosive, anti-corrosive, oxidation, reduction, and flammability characteristics of the nanoparticles determine their respective usage.

1.3 CHARACTERIZATION

The characterization techniques play a very important role in determining the properties and applications of the NPs. The NPs characterization can be demonstrated from various techniques. The parameters of these NPs like size, shape, surface area, wettability can be analyzed through the characterization techniques and below we have explained in detail.

1.3.1 SIZE

The particle size is one of the most basic and important measurement for nanoparticle characterization. It determines the actual size of NPs and decides whether it comes under micro or nanoscale. The size of the NPs can be commonly determined using electron microscopy. The images of scanning electron microscope (SEM) and transmission electron microscope (TEM) are analyzed to determine the size of particles and clusters whereas laser diffraction methods are employed for analyzing the bulk samples in solid phase (Marsalek et al., 2014). The particles which are in the liquid phase are measured using photon correlation spectroscopy and centrifugation. The particles which are in the gaseous phase are difficult and irreverent to use the imaging techniques and hence a scanning mobility particle sizer (SMPS) is used, which provides a fast and accurate measurements compared to other methods.

1.3.2 SURFACE AREA

The surface area is also a very important factor in nanoparticle characterization. Surface area to volume ratio of a nanoparticle has a tremendous impact on its evaluating performance and properties. Generally, the surface area is determined by employing the Brunauer-Emmett-Teller (BET) technique. For the measurement of surface area of particles in the liquid phase, a conventional titration process is sufficient, but for accurate and reliable measurement of the particles, nuclear magnetic resonance (NMR) spectroscopy is used. To determine the surface area of the gaseous phase particles, a modified SMPS and differential mobility analyzer (DMA) is used.

1.3.3 COMPOSITION

The chemical composition determines the purity as well as performance of nanoparticles. The presence of undesired elements in nanoparticles may halve their efficiency and results in contamination and secondary reaction in process. The composition analysis is commonly carried out through X-ray photoelectron spectroscopy (XPS) (Sharma et al., 2014). Some techniques involve chemical digestion of the particles followed by wet chemical analysis such as mass spectrometry, atomic emission spectroscopy, and ion chromatography. The particles in the gaseous phase are collected either by filtration or electrostatically and spectrometric or wet chemical techniques are used for the analysis (Bzdek et al., 2011).

1.3.4 SURFACE MORPHOLOGY

The nanoparticles exhibit different shapes and structures which plays an important function in exploiting its properties. Generally, the particles exhibit spherical, ribbon, tube, cylindrical, flat, sheet, tubular like shapes. The surface is commonly analyzed with electron microscopy imaging techniques like TEM and SEM (Hodoroaba et al., 2014). The particles in the liquid phase are deposited on a surface and analyzed, whereas the particles in the gaseous phase are captured electrostatically or by filtration for imaging using electron microscopy.

1.3.5 SURFACE CHARGE

The surface charge or the charge of a nanoparticle determines its possible interactions with the target. Generally, a zeta potentiometer is used for the measurement of surface charges and its dispersion stability in a solution. The charge determination of particles in the gaseous phase is analyzed through DMA.

1.3.6 CRYSTALLOGRAPHY

Crystallography is the study of atoms and molecules arrangement in crystal solids. The crystallography of nanoparticles is carried out by a powder X-ray, electron, or neutron diffraction to determine the structural arrangement (Yano et al., 1996).

1.4 CLASSIFICATION OF NPs

NPs are broadly classified into various categories based on their surface morphology, structure, size, functional group, and chemical properties (Figure 1.4). The main classes of these NPs are explained below.

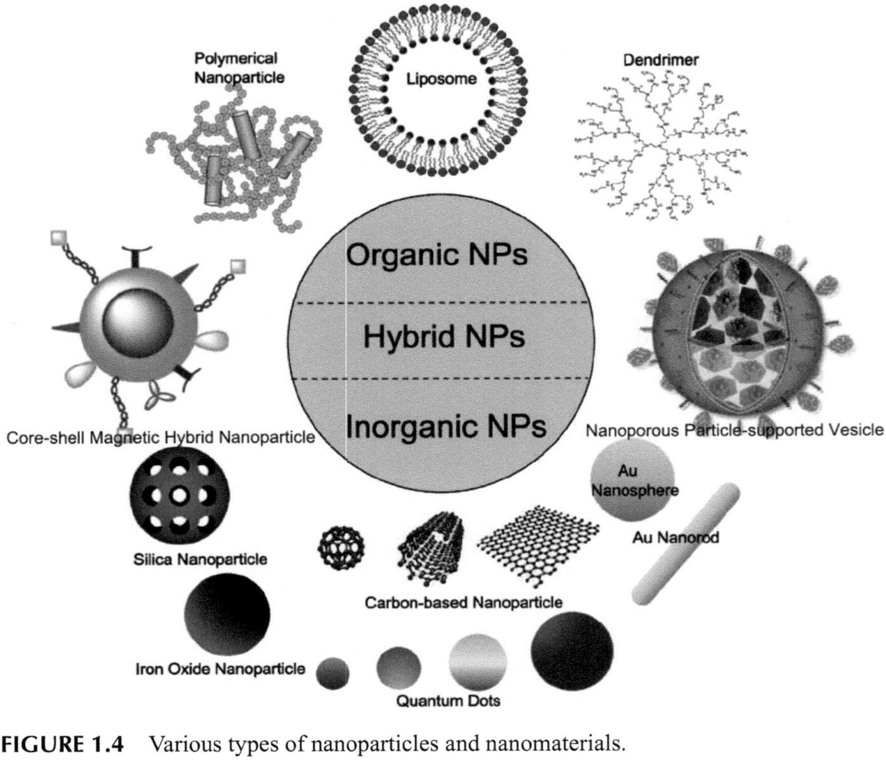

FIGURE 1.4 Various types of nanoparticles and nanomaterials.

Source: Reproduced with permission from: Ding and Ma (2017); Copyright © Royal Society of Chemistry (2017).

1.4.1 CARBON-BASED NANOPARTICLES (CBNs)

Carbon-based nanomaterials (CBNs) are important class, since these CBNs covers wide range of potential applications including energy and biomedicine. The physicochemical properties of these CBNs such as thermal, optical, electrical, and mechanical properties have attracted many researchers across the world to explore its wide range of applications. Therefore, extensive research attempts are being taken to use such materials for many commercial applications such as electronics and biomedical engineering.

1.4.1.1 CARBON NANOTUBES (CNTs)

Ever since their discovery, carbon nanotubes (CNTs) are the highly used CBNs (Iijima et al., 1993; Saito et al., 1998). These CNTs are commonly prepared from arc discharge or CVD of graphite. These CNTs have structural resemblance with graphite sheets rolling upon each other. These rolled sheets are either single, double or many walls, and therefore these CNTs are sub-classified as single-walled (SWNTs), double-walled (DWNTs) or multiwalled carbon nanotubes (MWCNTs), respectively. These CNTs possess cylindrical carbon structure and exhibits optical, electrical, and other applications. The tunable physical properties of these CNTs (diameter, size) are having many advantages for its applications. For example, CNTs are well known for their superb mechanical strength, rigidity, and flexibility are greater than that of some commercially available high-strength materials (e.g., high tensile steel, carbon fibers, and Kevlar®). Thus, they have been utilized as reinforcing elements for composite materials such as plastics and metal alloys, which have already led to several commercialized products (Yamamoto et al., 2007).

1.4.1.2 GRAPHENE

Graphene has already grabbed the attention of many material scientists working in various research areas. The ground-breaking work by Geim and Novoselov provided a simple method for extracting graphene from graphite via exfoliation and explored its unique electrical properties (Novoselov et al., 2004; Geim et al., 2007). Similar to CNTs, graphene also exhibits electrical, optical, and thermal properties. Graphene is structurally robust yet highly flexible, which makes it attractive for engineering thin, flexible materials (Eda et al., 2008; Kim et al., 2009).

1.4.1.3 OTHER CARBON-BASED NANOMATERIALS (CBNs)

Fullerenes (C_{60}) are carbon-based molecules that have spherical shape and composed of joint carbon atoms with sp^2 hybridization. Around 28–1500 carbon particles form spherical assembly with 8.2 nm diameter for single-layered and 4 nm–36 nm diameter for multi-layered fullerenes. The popularity of C_{60} has somewhat diminished in recent years with the rise of more scalable and practical CBNs such as CNTs and graphene. However, its uniform size and shape, as well as availability for chemical modification, led many scientists to develop C_{60} derivatives for therapeutic purposes (Jensen et al., 1996).

Nanodiamond (ND) has also grabbed the interest in the field of biomedical engineering in recent years (Mochalin et al., 2012). NDs are synthesized by high energy treatment of graphite, most commonly via detonation, and are smaller than 10 nm. They have similar physical properties as bulk diamond, such as fluorescence and photoluminescence, as well as biocompatibility. Unlike other CBNs, NDs are made up mostly of tetrahedral clusters of sp^3-carbon. The surface of NDs, however, is functionalized with various functional groups or sp^2-carbon for colloidal stability, which enables chemical modification for targeted drug and gene delivery and tissue labeling.

1.4.2 METAL NANOPARTICLES

Metallic nanoparticles (MNPs) or metal nanoparticles, a new terminology has been originated in the nanotechnology field in recent few years. The noble metal like silver, gold, and platinum having beneficial effects on health are utilized for synthesis of nanomaterials and designated as MNPs (Bhattacharya et al., 2008). Currently researchers are more focusing on MNPs and nanostructures synthesis due to their conspicuous characteristics which are useful in catalysis (Narayanan et al., 2004), polymer preparations (Moura et al., 2017), disease diagnosis (Banerjee et al., 2017), sensor technology (Gomez et al., 2001; Shaikh et al., 2016) and labeling of optoelectronic recorded media (Gracias et al., 2000).

1.4.3 POLYMER NANOPARTICLES (PNPs)

The PNPs are widely used as biomaterials due to their potential characteristics for simple elaboration, design, and good biocompatibility, extensive variety of functional groups and noticeable bio-imitative characteristics. These PNPs have potential applications in smart delivery of drugs and had a major role as they are capable of conducting therapeutic substances right into purposed position in the human body.

Characteristics as well as advantages of PNPs are several (Kayser et al., 2005). For bioavailability and efficacy, they show a remarkable enhancement over oral and intravenous administration routes. Moreover, they transport the active components to targeted tissue with specified concentration. PNPs are considered as suitable candidates for vaccines and targeted antibiotics delivery and cancer therapy compliant with the choice of polymer and adjustment of drug release.

The required necessities for PNPs delivery system design are to efficiently control their superficial feature and particulate size to manage infiltration, enhance elasticity, adjust solubility, and manage remedies release pattern from the PNPs to acquire selected specific action and the designated target site at a desired time (Bennet and Kim, 2014).

1.4.4 LIPID-BASED NPs

Lipid-based NPs involve lipid moieties and these NPs have many biomedical applications. Generally, lipid NPs are spherical in shape and have 10 nm to 1,000 nm diameters. Similar to polymeric NPs, lipid NPs contain a solid core, which is made up of lipids, and the matrix comprises soluble lipophilic particles. The external core of these lipid NPS will be surfactants and emulsifiers.

1.5 CONCLUSION AND FUTURE PERSPECTIVES

Nowadays, there is increasing scope for nanotechnology, which makes use of its potential applications in various fields such as chemistry, physics, biomedicine, and material science. The synthesis of nanostructures is a very dynamic and complex process. Several approaches, including CVD, hydrothermal method, chemical reduction, solvothermal method, sol-gel methods, spray pyrolysis, laser ablation, sputtering, and green synthesis method, have been developed to synthesize nanoparticles of correct morphology. Each approach has its particular advantages and limitations. Greener synthesis of nanoparticles is eco-friendly, sustainable, free from contaminants, and inexpensive. Though many methods have been successfully developed for the synthesis of nanoparticles, still more precise synthetic methods are needed to overcome the limitations of present approaches, and proper care should be taken to implement these at the industrial level.

KEYWORDS

- **carbon nanotube**
- **chemical vapor deposition**
- **crystallography**
- **fullerenes**
- **graphene**
- **laser ablation**
- **nanomaterial**
- **nanoparticles**
- **polymer nanoparticles**
- **pyrolysis**
- **sol-gel**
- **sputtering**

REFERENCES

Adachi, M., Tsukui, S., & Okuyama, K., (2003). Nanoparticle synthesis by ionizing source gas in chemical vapor deposition. *Japanese Journal of Applied Physics, 42*(1A), L77.

Auffan, M., Rose, J., Wiesner, M. R., & Bottero, J. Y., (2009). Chemical stability of metallic nanoparticles: A parameter controlling their potential cellular toxicity *in vitro*. *Environmental Pollution, 157*(4), 1127–1133.

Banerjee, K., Das, S., Choudhury, P., Ghosh, S., Baral, R., & Choudhuri, S. K., (2017). A novel approach of synthesizing and evaluating the anticancer potential of silver oxide nanoparticles *in vitro*. *Chemotherapy, 62*(5), 279–289.

Bennet, D., & Kim, S., (2014). Polymer nanoparticles for smart drug delivery. In: Sezer, A. D., (ed.), *Application of Nanotechnology in Drug Delivery* (pp. 257–310). InTech Open, London.

Bhattacharya, R., & Mukherjee, P., (2008). Biological properties of "naked" metal nanoparticles. *Advanced Drug Delivery Reviews, 60*(11), 1289–1306.

Bhaviripudi, S., Mile, E., Steiner, S. A., Zare, A. T., Dresselhaus, M. S., Belcher, A. M., & Kong, J., (2007). CVD synthesis of single-walled carbon nanotubes from gold nanoparticle catalysts. *Journal of the American Chemical Society, 129*(6), 1516–1517.

Bzdek, B. R., Zordan, C. A., Iii, G. W. L., Murray, V., & Johnston, M. V., (2011). Nanoparticle chemical composition during new particle formation. *Aerosol Sci. Technol., 45*, 1041–1048.

Chandra, R., Chawla, A. K., & Ayyub, P., (2006). Optical and structural properties of sputter-deposited nanocrystalline Cu_2O films: Effect of sputtering gas. *Journal of Nanoscience and Nanotechnology, 6*(4), 1119–1123.

D'Amato, R., Falconieri, M., Gagliardi, S., Popovici, E., Serra, E., Terranova, G., & Borsella, E., (2013). Synthesis of ceramic nanoparticles by laser pyrolysis: From research to applications. *Journal of Analytical and Applied Pyrolysis, 104*, 461–469.

Ding, H. M., & Ma, Y. Q., (2017). Computational approaches to cell-nanomaterial interactions: Keeping balance between therapeutic efficiency and cytotoxicity. *Nanoscale Horizons, 3*(1), 6–27.

Dresselhaus, G., Dresselhaus, M. S., & Saito, R., (1998). *Physical Properties of Carbon Nanotubes*. World Scientific, Singapore.

Eda, G., Fanchini, G., & Chhowalla, M., (2008). Large-area ultrathin films of reduced graphene oxide as a transparent and flexible electronic material. *Nature Nanotechnology, 3*(5), 270–274.

El-Nour, K. M. A., Eftaiha, A. A., Al-Warthan, A., & Ammar, R. A., (2010). Synthesis and applications of silver nanoparticles. *Arabian Journal of Chemistry, 3*(3), 135–140.

Geim, A. K., & Novoselov, K. S., (2010). The rise of graphene. In: Rodgers, P., (ed.), *Nanoscience and Technology* (pp. 11–19). https://doi.org/10.1142/9789814287005_0002.

Gomez-Romero, P., (2001). Hybrid organic-inorganic materials in search of synergic activity. *Advanced Materials, 13*(3), 163–174.

Gracias, D. H., Tien, J., Breen, T. L., Hsu, C., & Whitesides, G. M., (2000). Forming electrical networks in three dimensions by self-assembly. *Science, 289*(5482), 1170–1172.

Hasan, S., (2015). A review on nanoparticles: Their synthesis and types. *Res. J. Recent Sci., 2277*, 2502.

Hodoroaba, V. D., Rades, S., & Unger, W. E., (2014). Inspection of morphology and elemental imaging of single nanoparticles by high-resolution SEM/EDX in transmission mode. *Surface and Interface Analysis, 46*(10, 11), 945–948.

Iijima, S., & Ichihashi, T., (1993). Single-shell carbon nanotubes of 1-nm diameter. *Nat. Nanotechnol., 363*, 603–605.

Jensen, A. W., Wilson, S. R., & Schuster, D. I., (1996). Biological applications of fullerenes. *Bioorganic and Medicinal Chemistry, 4*(6), 767–779.

Kammler, B. H. K., Mädler, L., & Pratsinis, S. E., (2001). Flame synthesis of nanoparticles. *Chem. Eng. Technol., 24*(6), 583–596.

Kayser, O., Lemke, A., & Hernandez-Trejo, N., (2005). The impact of nanobiotechnology on the development of new drug delivery systems. *Current Pharmaceutical Biotechnology, 6*(1), 3–5.

Kim, K. S., Zhao, Y., Jang, H., Lee, S. Y., Kim, J. M., Kim, K. S., et al., (2009). Large-scale pattern growth of graphene films for stretchable transparent electrodes. *Nature, 457* (7230), 706–710.

Kruis, F. E., Fissan, H., & Peled, A., (1998). Synthesis of nanoparticles in the gas phase for electronic, optical and magnetic applications: A review. *Journal of Aerosol Science, 29*(5, 6), 511–535.

Kuppusamy, P., Yusoff, M. M., Maniam, G. P., & Govindan, N., (2016). Biosynthesis of metallic nanoparticles using plant derivatives and their new avenues in pharmacological applications: An updated report. *Saudi Pharmaceutical Journal, 24*(4), 473–484.

Laurent, S., Forge, D., Port, M., Roch, A., Robic, C., Vander, E. L., & Muller, R. N., (2008). Magnetic iron oxide nanoparticles: Synthesis, stabilization, vectorization, physicochemical characterizations, and biological applications. *Chemical Reviews, 108*(6), 2064–2110.

Lugscheider, E., Bärwulf, S., Barimani, C., Riester, M., & Hilgers, H., (1998). Magnetron-sputtered hard material coatings on thermoplastic polymers for clean room applications. *Surface and Coatings Technology, 108*, 398–402.

Mann, S., Burkett, S. L., Davis, S. A., Fowler, C. E., Mendelson, N. H., Sims, S. D., et al., (1997). Sol-gel synthesis of organized matter. *Chemistry of Materials, 9*(11), 2300–2310.

Marsalek, R., (2014). Particle size and zeta potential of ZnO. *APCBEE Procedia, 9*, 13–17.

Meyers, M. A., Mishra, A., & Benson, D. J., (2006). Mechanical properties of nanocrystalline materials. *Progress in Materials Science, 51*(4), 427–556.

Mochalin, V. N., Shenderova, O., Ho, D., & Gogotsi, Y., (2012). The properties and applications of nanodiamonds. *Nature Nanotechnology, 7*(1), 11–23.

Mohammadi, S., Harvey, A., & Boodhoo, K. V., (2014). Synthesis of TiO_2 nanoparticles in a spinning disc reactor. *Chemical Engineering Journal, 258*, 171–184.

Moura, D., Souza, M. T., Liverani, L., Rella, G., Luz, G. M., Mano, J. F., & Boccaccini, A. R., (2017). Development of a bioactive glass-polymer composite for wound healing applications. *Materials Science and Engineering: C, 76*, 224–232.

Nadagouda, M. N., Speth, T. F., & Varma, R. S., (2011). Microwave-assisted green synthesis of silver nanostructures. *Accounts of Chemical Research, 44*(7), 469–478.

Narayanan, R., & El-Sayed, M. A., (2004). Shape-dependent catalytic activity of platinum nanoparticles in colloidal solution. *Nano Letters, 4*(7), 1343–1348.

Novoselov, K. S., Geim, A. K., Morozov, S. V., Jiang, D., Zhang, Y., Dubonos, S. V., et al., (2004). Electric field effect in atomically thin carbon films. *Science, 306*(5696), 666–669.

Ramesh, S., (2013). Sol-gel synthesis and characterization of $Ag_{3(2+x)}Al_xTi_{4-x}O_{11+\delta}$ ($0.0 \le x \le 1.0$) nanoparticles. *Journal of Nanosience, 2013*, 929321. http://dx.doi.org/10.1155/2013/929321.

Sadhasivam, T., Kim, H. T., Jung, S., Roh, S. H., Park, J. H., & Jung, H. Y., (2017). Dimensional effects of nanostructured Mg/MgH_2 for hydrogen storage applications: A review. *Renewable and Sustainable Energy Reviews, 72*, 523–534.

Salavati-Niasari, M., Davar, F., & Mir, N., (2008). Synthesis and characterization of metallic copper nanoparticles via thermal decomposition. *Polyhedron, 27*(17), 3514–3518.

Shah, P., & Gavrin, A., (2006). Synthesis of nanoparticles using high-pressure sputtering for magnetic domain imaging. *Journal of Magnetism and Magnetic Materials, 301*(1), 118–123.

Shaikh, S. F., Mane, R. S., Min, B. K., Hwang, Y. J., & Joo, O. S., (2016). D-sorbitol-induced phase control of TiO_2 nanoparticles and its application for dye-sensitized solar cells. *Scientific Reports, 6*, 20103.

Sharma, V., & Rao, L. J. M., (2014). An overview on chemical composition, bioactivity and processing of leaves of *Cinnamomum tamala*. *Critical Reviews in Food Science and Nutrition, 54*(4), 433–448.

Simakin, A. V., Voronov, V. V., Kirichenko, N. A., & Shafeev, G. A., (2004). Nanoparticles produced by laser ablation of solids in liquid environment. *Applied Physics A, 79*(4–6), 1127–1132.

Singh, P., Kim, Y. J., Zhang, D., & Yang, D. C., (2016). Biological synthesis of nanoparticles from plants and microorganisms. *Trends in Biotechnology, 34*(7), 588–599.

Tai, C. Y., Tai, C. T., Chang, M. H., & Liu, H. S., (2007). Synthesis of magnesium hydroxide and oxide nanoparticles using a spinning disk reactor. *Industrial and Engineering Chemistry Research, 46*(17), 5536–5541.

Yadav, T. P., Yadav, R. M., & Singh, D. P., (2012). Mechanical milling: A top down approach for the synthesis of nanomaterials and nanocomposites. *Nanoscience and Nanotechnology, 2*(3), 22–48.

Yamamoto, T., Watanabe, K., & Hernández, E. R., (2007). Mechanical properties, thermal stability and heat transport in carbon nanotubes. In: Jorio, A., Dresselhaus, G., & Dresselhaus, M. S., (eds.), *Carbon Nanotubes* (pp. 165–195). Springer, Berlin, Heidelberg. https://doi.org/10.1007/978-3-540-72865-8_5.

Yano, F., Hiraoka, A., Itoga, T., Kojima, H., Kanehori, K., & Mitsui, Y., (1996). Influence of ion-implantation on native oxidation of Si in a clean-room atmosphere. *Applied Surface Science, 100*, 138–142.

CHAPTER 2

NANOTECHNOLOGY AND PROBIOTICS

FRANCINE SCHÜTZ,[1] SOFIA PINHEIRO,[2] RITA OLIVEIRA,[3] and PEDRO BARATA[3,4]

[1]*Faculty of Medicine, Department of Biomedicine, University of Porto, Alameda Prof. Hernâni Monteiro–4200-319, Porto, Portugal*

[2]*Centro Hospitalar Vila Nova de Gaia, Researcher Unit of Biomedicine, Faculty of Medicine, University of Porto, Portugal*

[3]*Biomedical Research Center (CEBIMED)/Research Center of the Fernando Pessoa Energy, Environment, and Health Research Unit (FP-ENAS), Faculty of Health Sciences, University of Fernando Pessoa, Porto, Portugal*

[4]*i3S-Instituto de Investigação e Inovação da Universidade do Porto, Portugal, Rua Alfredo Allen, 208, 4200-135 Porto, Portugal*

2.1 INTRODUCTION

The human body is colonized by 10 to 100 trillion microorganisms, including bacteria, viruses, archaea, and eukaryotic microbes. The human microbiota holds a crucial role in human health maintenance, with dysbiosis being linked to the pathophysiology of many disorders (Wang et al., 2017).

Actually, the gut microbiome (GM) is perhaps the most studied, displaying a key function in the digestion of fibrous foods, synthesis of vitamins, lipid storage and metabolism, energy harvest and supply, pathogenic bacteria suppression, and maintenance of intestinal barrier integrity. Most importantly, it holds a central duty in both gut and systemic immune systems modulation (Lazar et al., 2018). Indeed, human GM includes approximately

150 times more genes than the whole human genome, having even been labeled an essential organ as it interacts with several organs and systems, such as lungs, brain, bone, liver, and cardiovascular (CV) system (Feng et al., 2018). Furthermore, it is dynamic and changes with age, bearing modulation by a multitude of factors. Although it was previously believed that the human embryo is sterile, current research suggests otherwise, with microbes being found in semen, placenta, umbilical cord blood, amniotic fluid, and meconium (Lakshminarayanan et al., 2014; Kundu et al., 2017; Nagpal et al., 2018).

Shifts in gut microbiota profile are influenced by a variety of perinatal factors, like gestational age and maturity. For example, the gut microbiota profile of a full-term delivery is dominated by *Bacteroides*, *Parabacteroides*, and *Christensenellaceae*, while that of a preterm delivery is composed of *Lactobacillus*, *Streptococcus*, and *Carnobacterium* during the 1st, 2nd, and 4th year of life, respectively (Fouhy et al., 2019).

The chosen feeding method (formula milk or breastfeeding) also influences the colonization process in the newborn. Recent evidence suggests that the breast milk microbiota directly colonizes and modulates the infant gut microbiota, with dose-dependent effects (Pannaraj et al., 2017). Exclusive breastfeeding for longer periods is correlated with a decrease in diarrhea-inducing gut dysbiosis (Ho et al., 2018). The introduction to solid food into the infant's diet then produces an additional level of complexity to the GM (Palmer et al., 2007).

In young adults, there is a peak in gut microbiota richness and complexity, resulting in the formation of a robust GM mostly consisting of bacteria from *Firmicutes, Bacteroidetes, Actinobacteria, Verrucomicrobia, Proteobacteria,* and *Fusobacteria* phyla (Kundu et al., 2017). However, with age, there is a loss in microbiota diversity, with a shift in dominant species and a reduction in commensal bacteria (Salazar et al., 2014). Moreover, a plethora of environmental factors, including diet, early antibiotics' exposure, alcohol consumption, practice of physical activity and geographical location can consequently result in gut microbiota dysbiosis and disease (D'Argenio and Salvatore, 2015; Quigley, 2017). Indeed, gut dysbiosis have been associated with various diseases, including gastrointestinal (GI) (Richard et al., 2019), cancer (Helmink et al., 2019), atopic dermatitis (Weidinger et al., 2018), liver (Tripathi et al., 2018), food allergies (Bunyavanich, 2019), diabetes (Abdellatif, 2019), obesity (Fayfman et al., 2019) and autism (Liu et al., 2019).

It is therefore important to establish the healthy GM properties, providing the pathway to probiotic intervention in disease prevention and adjunctive treatment (Bäckhed et al., 2012; Lloyd-Price et al., 2016; Cani, 2018; George, 2018). According to the FAO/WHO, probiotics can be defined as live microorganisms conferring health benefits to hosts when given at proper dosage (FAO, 2006). Probiotics can modulate GM and maintain human health through several mechanisms, including the blockage of pathogenic bacteria adhesion to the intestinal epithelium; production of bacterial inhibitory agents (Reid, 2016); enhancement of GI immune response (Borchers et al., 2009); intestinal permeability repair (George et al., 2018); promotes the maintenance of appropriate levels of short-chain fatty acids (LeBlanc et al., 2017); suppression of opportunistic bacteria growth (Sikorska, 2013) and modulation of the local and systemic immune system, with suppression of intestinal pro-inflammatory cytokines production (Yan and Polk, 2011).

On the other side, probiotic intervention has proven useful in the adjunctive treatment and prevention of various GI diseases, i.e., irritable bowel syndrome (Didari et al., 2015), necrotizing enterocolitis (Thomas et al., 2017), inflammatory bowel disease (Derwa et al., 2017), prophylaxis of *Helicobacter pylori* infection (Chakravarty and Gaur, 2019), constipation (Martínez-Martínez et al., 2017), antibiotic-associated diarrhea (Mantegazza et al., 2018), radiotherapy-induced diarrhea (Qiu et al., 2019), *Clostridium difficile*-associated diarrhea (Hudson et al., 2019) and travelers' diarrhea (McFarland and Goh, 2019). Additionally, probiotic intervention has proven beneficial for preventing and treating allergic disorders, such as eczema (Akelma and Topçu, 2016), at the same time that improves symptoms and the quality-of-life of allergic rhinitis patients (Zajac et al., 2015). Moreover, the administration of probiotics during the prenatal period and/or breastfeeding mothers and their infants during the postnatal period has been associated with a decreased incidence of eczema in children (Sohn and Underwood, 2017). In addition, some evidence has also shown that children with cow's milk allergy demonstrate symptom relief following probiotics supplementation (Tan-Lim and Esteban-Ipac, 2017).

To what concerns to metabolic disorders, probiotic intervention has also shown useful in obesity, with gut microbiota modulation resulting in lower insulin resistance and increased satiety, and as a consequence in a decrease in fat mass and body weight (Ejtaheda et al., 2019; Mazloom et al., 2019). In the colon, probiotics can also increase glucagon like peptide 2 (GLP-2) production with a higher expression of zonula occludens-1, recovery of the tight junction between epithelial cells and reduced gut permeability (Cani et al., 2009). Furthermore, the supplementation with some probiotics can increase the secretion of GLP-1 and peptide YY (PYY) by the gut, thus affecting insulin resistance and beta-cell function (Molinaro et al., 2012).

Probiotic intervention has even shown useful in neurological disorders, with the microbiota-gut-brain axis being described as an interactive, bi-directional exchange of regulatory signals between the GI tract and central nervous system (CNS) (Dinan and Cryan, 2017). Psychobiotic are bacteria which act on a gut-brain axis through the production and delivery of neuroactive substances, thus modulating mood and cognition, sleep, and appetite (Dinan et al., 2013; Kali, 2016).

Currently, the consumption of probiotics has increased worldwide due to their abilities to promote health and well-being (Suez et al., 2019). However, some key factors, like cell viability and enough number of bacteria in intestinal tract influence in the health claims attributed to probiotics. Indeed, the oral administration of some probiotic products is restricted by a wide loss in cell viability after the passage through the stomach, due to high bile salt and acid concentrations, which markedly affects their effectiveness (Cook et al., 2012). In this way, the development of effective strategies which allow the preservation of probiotic viability during all stages of manufacturing, long-term storage, and passage over the GI tract is therefore of key importance.

2.2 NANOTECHNOLOGY AND PROBIOTICS

Microencapsulation is a promising process which allows for higher survival rates and controlled release of probiotic microorganisms throughout the passage over the GI tract. The microencapsulation is a technology in which bacteria can be immobilized into a polymer coating, and thus the structure is retained in the stomach before degrading in the gut (Solanki et al., 2013; Ipailien and Petraityt, 2018).

2.2.1 METHODS OF MICROENCAPSULATION OF PROBIOTICS

2.2.1.1 FREEZE DRYING

Freeze-drying, or lyophilization, is a process capable of enhancing the probiotic products shelf-life. There are three phases involved in this technology: freezing, primary drying, and secondary drying via sublimation under vacuum (Martín et al., 2015). Differently to spray drying, the freeze-drying is less damaging to microbial cells, as low temperatures are used (Dianawati et al., 2016). Moreover, the processing conditions during the freeze-drying process are milder and therefore provide a raised survival rate of probiotic bacteria when compared to other technologies, like spray drying (Liu et al., 2019). However, this technique does bear some disadvantages. For instance, the high osmolarity may damage cell membrane, so that to mitigate this issue, excipients, such as whey protein, trehalose, maltodextrin, skim milk powder, and glucose, have been used due to their capacity to accumulate inside the cells, thus reducing cell membrane damages (Capela et al., 2006). During this process, ice crystal formation is another problem that may also damage the cell membrane. To minimize this question, a high freezing rate has been used to boost smaller ice crystals formation and to reduce cell damage. Moreover, to ensure probiotic viability during dehydration, cryo-protectants have also been utilized in the process (Broeckx et al., 2016).

2.2.1.2 SPRAY DRYING

Current studies also suggest that the microencapsulation of probiotic cells in a matrix can protect them from damage and ameliorate their overall survival and viability (Islam et al., 2010; Coghetto et al., 2016; Dianawati et al., 2016; Sarao and Arora, 2017). In the food industry, the most often used microencapsulation method is spray drying due to a variety of factors, which include lower energy consumption (6 to 10 times lower in comparison to freeze-drying), continuous operation line, lower operational costs, and versatility (Peanparkdee et al., 2016). In this method, material is dispersed, resulting in the formation of an emulsion; then the liquid is homogenized, the mixture is atomized into the drying chamber, and finally, the solvent evaporated. In order to guarantee cell viability, this method requires control of several parameters, such as drying temperature, duration of heat exposure, product feed, gas flow, water activity, carrier, and storage conditions (Martín et al., 2015). The probiotic bacterial strain utilized and its tolerance to stress conditions must also be taken into consideration (Anekella and Orsat, 2013). The monitorization and adjustment of inlet and outlet temperatures are also important to improve the probiotic encapsulation' viability with ideal particle size distribution (Martín et al., 2015).

2.2.1.2.1 Spray Chilling

This method of microencapsulation consists in the atomization of a mixture of bioactive ingredient and molten carrier inside a chamber maintained at temperatures lower than the melting point of the carrier. Atomization produces droplets that solidify quickly in contact with cold air, forming microparticles. This encapsulation technique has some advantages, namely, it is comparatively cheap, can be applied at an industrial scale, and can generate small-sized beads. However, this method presents some disadvantages, such as low capacity of encapsulation and this can result in the loss of the core material during storage due to the characteristic arrangement of the lipid ingredients during the solidification and crystallization processes (Gouin, 2004; Martín et al., 2015).

2.2.1.2.2 Spray Freeze Drying

This technique of microencapsulation combines both the freeze-drying and spray drying processes, producing a fine powder preparation without heat-induced damage. As main advantages, this technique allows a larger specific surface area of the powder and higher fine particle fraction in comparison to the spray-dried capsules. Furthermore, it prevents the agglomerates drying. However, like all techniques, it also possesses disadvantages, which include a high energy use, longer time to process and moreover, it is a pricey technology (Her et al., 2015).

2.2.1.2.3 Ultrasonic Vacuum Spray Dryer

The ultrasonic vacuum spray dryer microencapsulation method produces a dry powder containing highly viable probiotic cells. This technique uses low temperatures, vacuum atmosphere, and an ultrasonic nozzle in the chamber during drying. The drying process can be divided into two steps: (1) the vacuum spray drying of the solution, and (2) the fluidized-bed drying of the powder. This division allows for the minimization of oxidative and thermal stresses in the drying process (Semyonov et al., 2011). The high probiotic cell viability and high survival rates during storage that this technique boasts can also be attributed to the formation of a glassy matrix produced from a combination of maltodextrin and trehalose, which provides optimal protection for probiotic cells during dehydration.

2.2.1.3 EXTRUSION METHOD

The extrusion method is the most popular probiotic cell microencapsulating technique, due to its simplicity, low operational costs, high probiotic viability, lack of deleterious solvents use and versatility (it can be done under aerobic and anaerobic conditions) (Krasaekoopt et al., 2003; Sarao and Arora, 2017). This method involves the addition of probiotic microorganisms into a hydrocolloid preparation. Then, cell suspension passes through a syringe needle forming droplets that are dripped into the hardening solution with cations, such as calcium (Heidebach et al., 2012). This technique is also known as prilling because of its controlled manner to droplet formation, unlike spray drying. A variety of methods can be used to extract the hydrocolloid-cell droplet from needle-via coaxial flow, nozzle vibration, jet pulsation, or electrostatic field (Martín et al., 2015). The size and shape of the capsules formed will be influenced by the diameter of the needle orifice, distance between the needle and hardening solution as well as the viscosity of the hydrocolloid-cell mixture (Heidebach et al., 2012). However, this method presents a significant disadvantage - the slow formation of the microbeads makes it complex to scale up the production (Sarao and Arora, 2017).

2.2.1.4 EMULSION METHOD

In the emulsification method, beads are formed from two stages: aqueous phase dispersion, containing bacterial cells, and a polymeric suspension, within an organic phase, such as oil, leading to a water-in-oil emulsion (Picone et al., 2017). The beads are then collected by filtration (Mandal and Hati, 2017). Emulsification generally results in small diameter beads, and the size is controlled by the speed of agitation (Heidebach et al., 2012).

2.2.1.5 ELECTROSPINNING

Electrospinning is a highly versatile method which combines two techniques-electrospray and spinning (Agarwal et al., 2008; López-Rubio et al., 2012). The technique comprises the application of electrostatic forces in the solution through the application of high voltage, where an electrode connected to a positive or negative high voltage source is coupled to the needle tip that will pass the polymeric solution, thus these charges are induced inside of the solution. At first, the solution is kept in the form of a drop on the needle tip, due to its surface tension. With the gradual increase in electrical voltage, the droplet amplifies, forming a cone. When the electrostatic forces overcome the surface tension, a jet from the tip of the cone is ejected towards the region of lower potential, being most of the time a grounded metallic collector. The moment the jet is transported, the solvent evaporates, and the polymer solidifies, producing a membrane composed of nanocapsules in powder form that are deposited in the collection (Agarwal et al., 2008). The advantage of this method is that it is not necessary to use heat, which is important for structure preservation and to obtain high encapsulation efficacy of the active ingredients during storage. Furthermore, this process is capable of producing fibers and/or capsules with submicron-scale diameters (Chen et al., 2017).

2.2.1.6 IMPINGING AEROSOL TECHNOLOGY

The impinging aerosol technique is similar to extrusion and emulsion methods, and consists in the cross-linking of sodium alginate solution using calcium chloride, resulting in the formation of microbeads with a diameter of less than 40 μm (Liu et al., 2019). This technique uses two separate aerosols: in one, the probiotic microbes are alginate solution' suspended, in the other, the probiotics are suspended in a calcium chloride. Generally, the alginate mixture is injected from the top of the cylinder, whereas the calcium chloride mixture is injected from the base of the cylinder (Martín et al., 2015). This method is suitable for heat-labile and solvent sensitive materials encapsulation (Sohail et al., 2012). However, this method does have some disadvantages, which include a wide range of particle size because of microcapsules aggregation and loss of materials due to adhesion to the walls of the spray chamber (Sohail et al., 2011).

2.2.1.7 INTERPOLYMER COMPLEX IN SUPERCRITICAL CARBON DIOXIDE (CO_2)

Most microencapsulation methods of probiotics imply the use of water and/or solvents at extreme temperatures, potentially compromising the survival of the encapsulated cells. Supercritical fluids are characterized by their temperature and pressure being higher than the corresponding critical values. Above the critical point, there is no longer surface tension and separation between the liquid and gas phases in equilibrium, forming only one supercritical phase, whose properties, such as density and diffusion coefficient alter with change in temperature or pressure (Moolman et al., 2006; Liu et al., 2019). The interpolymer complex in the supercritical

CO_2 method is based on interpolymer complex formation between poly-(vinylpyrrolidone) (PVP) and poly(vinyl acetate-co-crotonic acid) (PVAc-CA) in supercritical carbon dioxide ($scCO_2$). The interpolymer complex presents some advantages comparing with the two-individual polymer, such as lower solubility, has better barrier characteristic and is biodegradable (Vidhyalakshmi et al., 2009; Mamvura et al., 2011). In the supercritical technique, the process of interpolymer complex immobilizes the probiotic cells and then $scCO_2$ is gasified through depressurizing, forming the probiotic microcapsule (Thantsha et al., 2009). The supercritical technology presents various advantages, namely, it improves the heat stability of the probiotic product, and bacterial survival in simulated GI fluids at the same time that permits large-scale manufacturing. However, the high-pressure operation entails high investment costs (Thantsha et al., 2011).

2.2.1.8 HYBRIDIZATION SYSTEM

The hybridization system consists of a high-speed rotor, where the powder mixture is subjected to a high airstream generated by the blades. During the process, the particles mix by incorporating or aggregating on the surface of host particles (Ann et al., 2007). Comparing to other microencapsulation techniques, including spray drying, the hybridization system leads to high microcapsule yields, in addition to minimizing bacterial damage, since the hybridization has a cooling system that keeps the temperature below 30 °C (Park et al., 2002).

2.2.2 FLUID BED

This method is performed by suspended solid particles in the drying air through a heated gas wight passed through a bed of solid particles, with controlled velocity (Broeckx et al., 2016). This method has some advantages, namely low operational costs, easy scale-up and production of a multilayer coating with distinct functional properties (Liu et al., 2019). The disadvantages of this technique are that the methodology is difficult to control and the reasonably long time to process can trigger significant damages to probiotic bacteria. However, viability loss can be mitigated by adding protectants and controlling parameters, like temperature and atomizing air pressure (Martín et al., 2015).

2.3 CONCLUSION

The probiotics administration has proved to be extremely useful in the treatment and prevention of several pathologies; however, its administration has not always proven to be easy. The development of effective strategies which allow the preservation of probiotic viability during all stages of product manufacturing, long-term storage, and passage over the GI tract is therefore of vital importance. In this sense, with the aim of improving the quality of supplements and the effectiveness of probiotics consumption, a variety of technologies have been evaluated and used with substantial success. Anyway, there is still a long evolution that needs and will be done, in order to further increase the quality and the effectiveness of probiotics delivery, but in the near future, more and more interesting and less expensive solutions will appear both in the market and in scientific publications.

ACKNOWLEDGMENTS

N.M. acknowledges the Portuguese Foundation for Science and Technology under the Horizon 2020 Program (PTDC/PSI-GER/28076/2017).

KEYWORDS

- **electrospinning**
- **freeze-drying**
- **gut microbiome**
- **microencapsulation**
- **nanomaterials**
- **probiotics**
- **spray chilling**
- **spray drying**
- **spray freeze drying**
- **ultrasonic vacuum spray dryer**

REFERENCES

Abdellatif, A. M., & Sarvetnick, N. E., (2019). Current understanding of the role of gut dysbiosis in type 1 diabetes. *Journal of Diabetes, 11*(8), 632–644.

Agarwal, S., Wendorff, J. H., & Greiner, A., (2008). Use of electrospinning technique for biomedical applications. *Polymer, 49*, 5603–5621. https://doi.org/10.1016/j.polymer.2008.09.014d.

Akelma, A. Z., & Biten, A. A., (2016). Probiotics and infantile atopic eczema. *Pediatric Health, Medicine and Therapeutics, 6*(1), 75–82.

Anekella, K., & Orsat, V., (2013). Optimization of microencapsulation of probiotics in raspberry juice by spray drying. *Food Science and Technology, 50*, 17–24. https://doi.org/10.1016/j.lwt.2012.08.003.

Ann, E. Y., Kim, Y., Oh, S., Imm, J. Y., Park, D. J., Han, S. H., & Kim, S. H., (2007). Microencapsulation of *Lactobacillus acidophilus* ATCC 43121 with prebiotic substrates using a hybridization system. *International Journal of Food Science and Technology, 42*(4), 411–419. doi: https://doi.org/10.1111/j.1365-2621.2007.01236.x.

Assadpour, E., & Jafari, S. M., (2019). Advances in spray-drying encapsulation of food bioactive ingredients: From microcapsules to nanocapsules. *Annual Review of Food Science and Technology, 10*, 103–131.

Bäckhed, F., Fraser, C. M., Ringel, Y., Sanders, M. E., Sartor, R. B., Sherman, P. M., Versalovic, J., et al., (2012). Defining a healthy human gut microbiome: Current concepts, future directions, and clinical applications. *Cell Host and Microbe, 12*(5), 611–622.

Borchers, A. T., Selmi, C., Meyers, F. J., Keen, C. L., & Gershwin, M. E., (2009). Probiotics and immunity. *Journal of Gastroenterology, 44*(1), 26–46.

Broeckx, G., Vandenheuvel, D., Claes, I. J., Lebeer, S., & Kiekens, F., (2016). Drying techniques of probiotic bacteria as an important step towards the development of novel pharmabiotics. *International Journal of Pharmaceutics, 505*(1, 2), 303–318.

Bunyavanich, S., (2019). Food allergy: Could the gut microbiota hold the key? *Nature Reviews Gastroenterology and Hepatology, 16*(4), 201–202.

Cani, P. D., Possemiers, S., Van De, W. T., Guiot, Y., Everard, A., Rottier, O., Geurts, L., et al., (2009). Changes in gut microbiota control inflammation in obese mice through a mechanism involving GLP-2-driven improvement of gut permeability. *Gut, 58*(8), 1091–1103.

Cani. P. D., (2018). Human gut microbiome: Hopes, threats and promises. *Gut, 67*(9), 1716–1725.

Capela, P., Hay, T. K., & Shah, N. P., (2006). Effect of cryoprotectants, prebiotics and microencapsulation on survival of probiotic organisms in yoghurt and freeze-dried yoghurt. *Food Research International, 39*(2), 203–211. doi: https://doi.org/10.1016/j.foodres. 2005.07.007.

Chakravarty, K., & Gaur, S., (2019). Role of probiotics in prophylaxis of *Helicobacter pylori* infection. *Current Pharmaceutical Biotechnology, 20*(2), 137–145.

Chen, J., Wang, Q., Liu, C. M., & Gong, J., (2017). Issues deserve attention in encapsulating probiotics: Critical review of existing literature. *Critical Reviews in Food Science and Nutrition, 57*(6), 1228–1238.

Coghetto, C. C., Brinques, G. B., & Ayub, M. A., (2016). Probiotics production and alternative encapsulation methodologies to improve their viabilities under adverse environmental conditions. *International Journal of Food Sciences and Nutrition, 67*(8), 929–943.

Cook, M. T., Tzortzis, G., Charalampopoulos, D., & Khutoryanskiy, V. V., (2012). Microencapsulation of probiotics for gastrointestinal delivery. *Journal of Controlled Release: Official Journal of the Controlled Release Society, 162*(1), 56–67.
D'Argenio, V., & Salvatore, F., (2015). The role of the gut microbiome in the healthy adult status. *Clinica Chimica Acta-International Journal of Clinical Chemistry, 451*(Pt.: A), 97–102.
Derwa, Y., Gracie, D. J., Hamlin, P. J., & Ford, A. C., (2017). Systematic review with meta-analysis: The efficacy of probiotics in inflammatory bowel disease. *Alimentary Pharmacology and Therapeutics, 46*(4), 389–400.
Dianawati, D., Mishra, V., & Shah, N. P., (2016). Survival of microencapsulated probiotic bacteria after processing and during storage: A review. *Critical Reviews in Food Science and Nutrition, 56*(10), 1685–1716.
Didari, T., Mozaffari, S., Nikfar, S., & Abdollahi, M., (2015). Effectiveness of probiotics in irritable bowel syndrome: Updated systematic review with meta-analysis. *World Journal of Gastroenterology, 21*(10), 3072–3084.
Dinan, T. G., & Cryan, J. F., (2017). The microbiome-gut-brain axis in health and disease. *Gastroenterology Clinics of North America, 46*(1), 77–89.
Dinan, T. G., Stanton, C., & Cryan, J. F., (2013). Psychobiotics: A novel class of psychotropic. *Biological Psychiatry, 74*(10), 720–726.
Ejtaheda, H. S., Angoorania, P., Sorousha, A. R., Atlasi, R., Hasani-Ranjbar, S., Mortazavian, A. M., & Larijani, B., (2019). Probiotics supplementation for the obesity management: A systematic review of animal studies and clinical trials. *Journal of Functional Foods, 52*, 228–242. doi: Https://doi.org/10.1016/j.jff.2018.10.039.
Fayfman, M., Flint, K., & Srinivasan, S., (2019). Obesity, motility, diet, and intestinal microbiota-connecting the dots. *Current Gastroenterology Reports, 21*(4), 15.
Feng, Q., Chen, W. D., & Wang, Y. D., (2018). Gut microbiota: An integral moderator in health and disease. *Frontiers in Microbiology, 9*, 151.
Fouhy, F., Watkins, C., Hill, C. J., O'Shea, C. A., Nagle, B., Dempsey, E. M., O'Toole, P. W., et al., (2019). Perinatal factors affect the gut microbiota up to four years after birth. *Nature Communications, 10*(1), 1517.
George, K. R., Patra, J. K., Gouda, S., Park, Y., Shin, H. S., & Das, G., (2018). Benefaction of probiotics for human health: A review. *Journal of Food and Drug Analysis, 26*(3), 927–939.
Gouin, S., (2004). Microencapsulation: industrial appraisal of existing technologies and trends. *Trends in Food Science and Technology, 15*, 330–334. doi: https://doi.org/10.1016/j.tifs.2003.10.005.
Gupta, V., & Garg, R., (2009). Probiotics. *Indian Journal of Medical Microbiology, 27*(3), 202–209.
Heidebach, T., Först, P., & Kulozik, U., (2012). Microencapsulation of probiotic cells for food applications. *Critical Reviews in Food Science and Nutrition, 52*(4), 291–311.
Helmink, B. A., Khan, M., Hermann, A., Gopalakrishnan, V., & Wargo, J. A., (2019). The microbiome, cancer, and cancer therapy. *Nature Medicine, 25*(3), 377–388.
Her, J. H., Kim, M. S., & Lee, K. G., (2015). Preparation of probiotic powder by the spray freeze-drying method. *Journal of Food Engineering, 150*, 70–74. doi: https://doi.org/10.1016/j.jfoodeng.2014.10.029.
Hill, C., Guarner, F., Reid, G., Gibson, G. R., Merenstein, D. J., Pot, B., Morelli, L., et al., (2014). Expert consensus document. The international scientific association for probiotics and prebiotics consensus statement on the scope and appropriate use of the term probiotic. *Nature Reviews Gastroenterology and Hepatology, 11*(8), 506–514.

Ho, N. T., Li, F., Lee-Sarwar, K. A., Tun, H. M., Brown, B. P., Pannaraj, P. S., Bender, J. M., et al., (2018). Meta-analysis of effects of exclusive breastfeeding on infant gut microbiota across populations. *Nature Communications, 9*(1), 4169.
Hudson, S. L., Arnoczy, G., Gibson, H., Thurber, C., Lee, J., & Kessell, A., (2019). Probiotic use as prophylaxis for *Clostridium difficile*-associated diarrhea in a community hospital. *American Journal of Infection Control, 47*(8), 1028–1029.
Islam, M. A., Yun, C. H., Choi, Y. J., & Cho, C. S., (2010). Microencapsulation of live probiotic bacteria. *Journal of Microbiology and Biotechnology, 20*(10), 1367–1377.
Islam, S. U., (2016). Clinical uses of probiotics. *Medicine, 95*(5), e2658.
Kali, A., (2016). Psychobiotics: An emerging probiotic in psychiatric practice. *Biomedical Journal, 39*(3), 223–224.
Krasaekoopt, W., Bhandari, B., & Deeth, H., (2003). Evaluation of encapsulation techniques of probiotics for yoghurt. *International Dairy Journal, 13*(1), 3–13. doi: https://doi.org/10.1016/S0958-6946(02)00155-3.
Kundu, P., Blacher, E., Elinav, E., & Pettersson, S., (2017). Our gut microbiome: The evolving inner self. *Cell, 171*(7), 1481–1493.
Lakshminarayanan, B., Stanton, C., O'Toole, P. W., & Ross, R. P., (2014). Compositional dynamics of the human intestinal microbiota with aging: Implications for health. *The Journal of Nutrition, Health and Aging, 18*(9), 773–786.
LeBlanc, J. G., Chain, F., Martín, R., Bermúdez-Humarán, L. G., Courau, S., & Langella, P., (2017). Beneficial effects on host energy metabolism of short-chain fatty acids and vitamins produced by commensal and probiotic bacteria. *Microbial Cell Factories, 16*(1), 79.
Liu, F., Li, J., Wu, F., Zheng, H., Peng, Q., & Zhou, H., (2019). Altered composition and function of intestinal microbiota in autism spectrum disorders: A systematic review. *Translational Psychiatry, 9*(1), 43.
Liu, H., Cui, S. W., Chen, M., Li, Y., Liang, R., Xu, F., & Zhong, F., (2019). Protective approaches and mechanisms of microencapsulation to the survival of probiotic bacteria during processing, storage and gastrointestinal digestion: A review. *Critical Reviews in Food Science and Nutrition, 59*(17), 2863–2878.
Lloyd-Price, J., Abu-Ali, G., & Huttenhower, C., (2016). The healthy human microbiome. *Genome Medicine, 8*(1), 51.
López-Rubio, A., Sanchez, E., Wilkanowicz, S., Sanz, Y., & Lagaron, J. M., (2012). Electrospinning as a useful technique for the encapsulation of living Bifidobacteria in food hydrocolloids. *Food Hydrocolloids, 28*, 159. doi: https://doi.org/10.1016/j.foodhyd.2011.12.008.
Mamvura, C. I., Moolman, F. S., Kalombo, L., Hall, A. N., & Thantsha, M. S., (2011). Characterization of the poly-(vinylpyrrolidone)-poly-(vinylacetate-Co-crotonic acid) (PVP:PVAc-CA) interpolymer complex matrix microparticles encapsulating a *Bifidobacterium lactis* Bb12 probiotic strain. *Probiotics and Antimicrobial Proteins, 3*(2), 97–102.
Mandal, S., & Hati, S., (2017). Microencapsulation of bacterial cells by emulsion technique for probiotic application. *Methods in Molecular Biology, 1479*, 273–279.
Mantegazza, C., Molinari, P., D'Auria, E., Sonnino, M., Morelli, L., & Zuccotti, G. V., (2018). Probiotics and antibiotic-associated diarrhea in children: A review and new evidence on *Lactobacillus rhamnosus* GG during and after antibiotic treatment. *Pharmacological Research, 128*, 63–72.
Martín, M. J., Villoslada, F. L., Ruiz, M. A., & Morales, M. E., (2015). Microencapsulation of bacteria: A review of different technologies and their impact on the probiotic effects.

Innovative Food Science and Emerging Technologies, 27, 15–25. doi: https://doi.org/10.1016/j.ifset.2014.09.010.
Martínez-Martínez, M. I., Calabuig-Tolsá, R., & Cauli, O., (2017). The effect of probiotics as a treatment for constipation in elderly people: A systematic review. *Archives of Gerontology and Geriatrics, 71*, 142–149.
Mazloom, K., Siddiqi, I., & Covasa, M., (2019). Probiotics: How effective are they in the fight against obesity? *Nutrients, 11*(2), 258.
McFarland, L. V., & Goh, S., (2019). Are probiotics and prebiotics effective in the prevention of travelers' diarrhea: A systematic review and meta-analysis. *Travel Medicine and Infectious Disease, 27*, 11–19.
Molinaro, F., Paschetta, E., Cassader, M., Gambino, R., & Musso, G., (2012). Probiotics, prebiotics, energy balance, and obesity: Mechanistic insights and therapeutic implications. *Gastroenterology Clinics of North America, 41*(4), 843–854.
Moolman, F. S., Labuschagne, P. W., Thantsha, M. S., Van, D. M. T. L., Rolfes, H., & Cloete, T. E., (2006). Encapsulating probiotics with an interpolymer complex in supercritical carbon dioxide. *South African Journal of Science, 102*, 349–354.
Nagpal, R., Mainali, R., Ahmadi, S., Wang, S., Singh, R., Kavanagh, K., Kitzman, D. W., et al., (2018). Gut microbiome and aging: Physiological and mechanistic insights. *Nutrition and Healthy Aging, 4*(4), 267–285.
Palmer, C., Bik, E. M., DiGiulio, D. B., Relman, D. A., & Brown, P. O., (2007). Development of the human infant intestinal microbiota. *PLoS Biology, 5*(7), e177.
Pannaraj, P. S., Li, F., Cerini, C., Bender, J. M., Yang, S., Rollie, A., Adisetiyo, H., et al., (2017). Association between breast milk bacterial communities and establishment and development of the infant gut microbiome. *JAMA Pediatrics, 171*(7), 647–654.
Park, D. J., An, E. Y., Kim, J. S., Imm, J. Y., Han, K. S., Kim, S. H., & Oh, S. J., (2002). Dry enteric coating process of lactic acid bacteria by hybridization system. *Korean Journal of Food Science and Technology, 34*, 856–861.
Picone, C. S. F., Bueno, A. C., Michelon, M., & Cunha, R. L., (2017). Development of a probiotic delivery system based on gelation of water-in-oil emulsions. *Food Science and Technology, 86*, 62–68. doi: https://doi.org/10.1016/j.lwt.2017.07.045.
Qiu, G., Yu, Y., Wang, Y., & Wang, X., (2019). The significance of probiotics in preventing radiotherapy-induced diarrhea in patients with cervical cancer: A systematic review and meta-analysis. *International Journal of Surgery, 65*, 61–69.
Quigley, E. M. M., (2017). Gut microbiome as a clinical tool in gastrointestinal disease management: Are we there yet?. *Nature Reviews Gastroenterology and Hepatology, 14*(5), 315–320.
Reid, G., (2016). Probiotics: Definition, scope and mechanisms of action. *Best Practice and Research Clinical Gastroenterology, 30*(1), 17–25.
Richard, M. L., & Sokol, H., (2019). The gut mycobiota: Insights into analysis, environmental interactions and role in gastrointestinal diseases. *Nature Reviews Gastroenterology and Hepatology, 16*(6), 331–345.
Rosenfeld, C. S., (2017). Gut dysbiosis in animals due to environmental chemical exposures. *Frontiers in Cellular and Infection Microbiology, 7*, 396.
Salazar, N., Arboleya, S., Valdés, L., Stanton, C., Ross, P., Ruiz, L., Gueimonde, M., & De Reyes-Gavilán, C. G. L., (2014). The human intestinal microbiome at extreme ages of life. Dietary intervention as a way to counteract alterations. *Frontiers in Genetics, 5*, 406.

Sarao, L. K., & Arora, M., (2017). Probiotics, prebiotics, and microencapsulation: A review. *Critical Reviews in Food Science and Nutrition, 57*(2), 344–371.

Semyonov, D., Ramon, O., & Shimon, E., (2011). Using ultrasonic vacuum spray dryer to produce highly viable dry probiotics. *Food Science and Technology, 44*, 1844–1852. https://doi.org/10.1016/j.lwt.2011.03.021.

Sikorska, H., & Smoragiewicz, W., (2013). Role of probiotics in the prevention and treatment of meticillin-resistant *Staphylococcus aureus* infections. *International Journal of Antimicrobial Agents, 42*(6), 475–481.

Singh, V. P., Sharma, J., Babu, S., Rizwanulla, & Singla, A., (2013). Role of probiotics in health and disease: A review. *The Journal of the Pakistan Medical Association, 63*(2), 253–257.

Šipailienė, A., & Petraitytė, S., (2018). Encapsulation of probiotics: Proper selection of the probiotic strain and the influence of encapsulation technology and materials on the viability of encapsulated microorganisms. *Probiotics and Antimicrobial Proteins, 10*(1), 1–10.

Sohail, A., Turner, M. S., Coombes, A., Bostrom, T., & Bhandari, B., (2011). Survivability of probiotics encapsulated in alginate gel microbeads using a novel impinging aerosols method. *International Journal of Food Microbiology, 145*(1), 162–168.

Sohail, A., Turner, M. S., Prabawati, E. K., Coombes, A. G., & Bhandari, B., (2012). Evaluation of *Lactobacillus rhamnosus* GG and *Lactobacillus acidophilus* NCFM encapsulated using a novel impinging aerosol method in fruit food products. *International Journal of Food Microbiology, 157*(2), 162–166.

Sohn, K., & Underwood, M. A., (2017). Prenatal and postnatal administration of prebiotics and probiotics. *Seminars in Fetal and Neonatal Medicine, 22*(5), 284–289.

Solanki, H. K., Pawar, D. D., Shah, D. A., Prajapati, V. D., Jani, G. K., Mulla, A. M., & Thakar, P. M., (2013). Development of microencapsulation delivery system for long-term preservation of probiotics as biotherapeutics agent. *BioMed Research International, 2013*, 620719.

Suez, J., Zmora, N., Segal, E., & Elinav, E., (2019). The pros, cons, and many unknowns of probiotics. *Nature Medicine, 25*(5), 716–729.

Tan-Lim, C., & Esteban-Ipac, N., (2018). Probiotics as treatment for food allergies among pediatric patients: A meta-analysis. *The World Allergy Organization Journal, 11*(1), 25.

Thantsha, M. S., Cloete, T. E., Moolman, F. S., & Labuschagne, P. W., (2009). Supercritical carbon dioxide interpolymer complexes improve survival of *B. longum* Bb-46 in simulated gastrointestinal fluids. *International Journal of Food Microbiology, 129*(1), 88–92.

Thantsha, M. S., Labuschagne, P. W., & Mamvura, C. I., (2014). Supercritical CO_2 interpolymer complex encapsulation improves heat stability of probiotic bifidobacteria. *World Journal of Microbiology and Biotechnology, 30*(2), 479–486.

Thomas, J. P., Raine, T., Reddy, S., & Belteki, G., (2017). Probiotics for the prevention of necrotizing enterocolitis in very-low-birth-weight infants: A meta-analysis and systematic review. *Acta Paediatrica, 106*(11), 1729–1741.

Tripathi, A., Debelius, J., Brenner, D. A., Karin, M., Loomba, R., Schnabl, B., & Knight, R., (2018). The gut-liver axis and the intersection with the microbiome. *Nature Reviews Gastroenterology and Hepatology, 15*(7), 397–411.

Vidhyalakshmi, R., Bhakyaraj, R., & Subhasree, R. S., (2009). Encapsulation "the future of probiotics": A review. *Advances in Biological Research, 3*, 96–103.

Wang, B., Yao, M., Lv, L., Ling, Z., & Lil, L., (2017). The human microbiota in health and disease. *Engineering, 1*(3), 71–82. doi.org/10.1016/J.ENG.2017.01.008.

Weidinger, S., Beck, L. A., Bieber, T., Kabashima, K., & Irvine, A. D., (2018). Atopic dermatitis. *Nature Reviews Disease Primers, 4*(1), 1.
Yan, F., & Polk, D. B., (2011). Probiotics and immune health. *Current Opinion in Gastroenterology, 27*(6), 496–501.
Zajac, A. E., Adams, A. S., & Turner, J. H., (2015). A systematic review and meta-analysis of probiotics for the treatment of allergic rhinitis. *International Forum of Allergy and Rhinology, 5*(6), 524–532.
Zampini, A., Nguyen, A. H., Rose, E., Monga, M., & Miller, A. W., (2019). Defining dysbiosis in patients with urolithiasis. *Scientific Reports, 9*(1), 5425.

CHAPTER 3

NANOCELLULOSE: A VERSATILE BIOPOLYMER

LALAJI RATHOD, PARVA JANI, and KRUTIKA SAWANT

Faculty of Pharmacy, The Maharaja Sayajirao University of Vadodara, Baroda, Gujarat, India

3.1 INTRODUCTION

Today, due to pollution and other environmental issues related to the non-biodegradability of materials, the researcher's main concern with any substance is that at the end of its utility, it should be non-toxic and biodegradable or reusable and recyclable. Cellulose is known to mankind for more than 100 years. It is a widely available organic polymer, representing about 1.5×10^{12} tons of the annual biomass production (Ball, 2005). Cellulose is the most significant molecular skeleton constituent in plants which is a polymeric material with good properties and structure (Kim et al., 2015). It is formed by the recurring D-glucose units, attached to each other by β-1-4-linkages. Formation of hydrogen bond between the three hydroxyl groups of D-glucopyranose determines the crystalline arrangement and physical properties of cellulose (John and Thomas, 2008; Lavoine et al., 2012); the extremely functionalized homopolymer with linear rigid-chain is recognized by its chirality, biodegradability, hydrophilicity, wide chemical modifying capacity, and formation of versatile semi-crystalline fiber morphologies (Khlemm et al., 2009). Nanocellulose has been a topic of increasing interest in recent years because of its unique characteristics as compared to ordinary cellulose polymer like high mechanical strength, hydrophilicity, the formation of versatile semi-crystalline fiber morphologies, high modification capability, large surface area, etc. Nanocellulose is available in nano range dimension as an elementary building block of natural cellulose (Kim et al., 2015). It is naturally present and can be obtained from cellulose of wood, various plants, bacteria, and algae. As per the Web of Science database, more than 5000 scientific articles were published in between 1994 and 2016, and also more than 500 patent databases were reported as per Google patents database on nanocellulose (Abbati et al., 2017).

3.2 STRUCTURE AND TYPES OF NANOCELLULOSE

3.2.1 STRUCTURE

Plant cell walls are composed of cellulose microfibers, hemicelluloses, and lignin, a cellulose fiber composite. The macrofibers consist of microfibrils which are strings of cellulose crystal linked with the microfibril axis by disordered amorphous domains, and at nano level, those microfibrils are made up of nanofibrils of cellulose (Azizi et al., 2005). A cellulose fiber usually has a diameter around 25–30 μm and is made up of a bundle of cellulose microfibers having 0.1 to 1 μm diameter. Nano-cellulose fibrils with a diameter of 10–70 nm and lengths of thousands of nanometers are constituting microfibers (Sandolo et al., 2007; Vazquez et al., 2013). Hierarchical structure of wood cellulose is shown in Figure 3.1. Nano-cellulose can be also obtained from bacteria and algae.

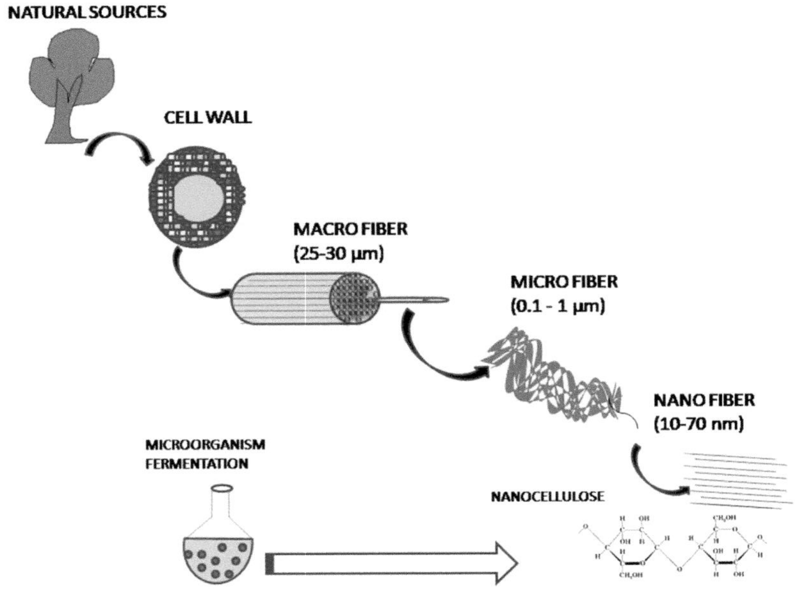

FIGURE 3.1 Hierarchical structure of nanocellulose.

3.2.2 TYPES OF NANOCELLULOSE

Cellulose nanofibrils have crystalline as well as amorphous components in a row. Mainly four types of cellulose crystalline polymorphs are observed: Cellulose I, II, III, and IV. Cellulose I or Natural cellulose is naturally produced by different organisms and is thermodynamically metastable and mostly converted into cellulose II or III. Cellulose II which is manufactured by regeneration and mercerization, is the most stable polymorph (Kim et al., 2015). The different types of nanocellulose can be classified into cellulose microcrystal, cellulose microfibril, cellulose nanocrystal, cellulose nanofibril and bacterial nanocellulose (BNC) as per their dimension, shape, and method of preparation (Kargarzadeh et al., 2017).

3.2.2.1 CELLULOSE NANOCRYSTALS (CNCS)

It is also known as cellulose nanowhiskers (CNWs) and nanocrystalline cellulose (NCC), which is produced by selective hydrolysis of amorphous cellulose regions by heat-controlled techniques or acid hydrolysis (mainly sulfuric acid) giving highly crystalline structured particles with source dependent dimensions (Sandolo et al., 2007). It can be 5–50 nm in diameter and 100–500 nm in length (Kargarzadeh et al., 2017).

3.2.2.2 CELLULOSE NANOFIBRILS (CNFS)

Also known as nano-fibrillated cellulose (NFC), it contains amorphous and crystalline cellulose domains having a width of 20–50 nm and a length of 500–2000 nm (Kargarzadeh et al., 2017). It can be extracted from cellulosic fibers by mechanical treatment (like homogenization, grinding, milling), chemical treatment like 2,2,6,6,-tetramethylpiperidine-1-oxyl (TEMPO) oxidation or combination of both (Ioelovich, 2014; Siro et al., 2014). Its aqueous solution of low concentration (1.0%) has high viscosity (Habibi et al., 2006).

3.2.2.3 BACTERIAL NANOCELLULOSE (BNC)

This is produced by microbial fermentation. A main advantage of BC over plant source cellulose is that it is in pure cellulose form without lignin and hemicelluloses and hence pretreatment is not required to remove lignin and hemicelluloses. Average diameter of BC ranges between 20 nm and 100 nm with micrometer length.

3.3 PREPARATION OF NANOCELLULOSE: SOURCE AND METHOD

3.3.1 DIFFERENT SOURCES OF NANOCELLULOSE

Plant including wood, hemp, rice husk, kenaf, flax, and coconut husk; agricultural crops and by-products; Tunicates and algae are different sources of nanocellulose. Bio nanocellulose fibers are secreted as extracellular structures by some bacterial organisms from genera Acetobacter, Rhizobium, Alcaligenes, Agrobacterium, Pseudomonas, or Sarcina (Kargarzadeh et al., 2017). *Acetobacter xylinum* (or *Gluconacetobacter xylinus*), which is a Gram-negative strain of acetic acid-producing bacteria is the most effective produce of bacterial cellulose (BC) (Bajpai, 2013).

3.3.2 METHOD OF PREPARATION OF NANOCELLULOSE

Nanocellulose is produced from the lignocellulosic biomass using a single or combination of more than one method. There is presence of cellulose, hemicelluloses, and lignin in wood fibers. Different properties of nanocellulose such as shape, size, surface morphology, yield, strength, etc., depend on the composition of lignocellulosic biomass, pretreatment, and preparation process parameters. Different pretreatment like pulping process, bleaching, steam explosion, alkaline/acid-alkaline pretreatment, enzymatic pretreatment, ionic liquid, and oxidation are given to raw cellulose. Pretreatment is followed by chemical method and/or mechanical method for the preparation of nanocellulose (Sandolo et al., 2007).

3.3.2.1 CHEMICAL METHOD

Removal of lignin from the amorphous part of cellulose fibers is the first step for preparation of nanocellulose. It makes extraction process of microfibrillated cellulose easier. This is followed by the dissolution of microfibrilated cellulose or microcrystalline cellulose which separates into nanofibers. This process is done by various methods such as the kraft pulping process

(Iwamoto et al., 2005), acid hydrolysis, hydrolysis with solid acids, hydrolysis with gaseous acid, hydrolysis with metal salt catalyst and cellulose enzymes treatment (Sandolo et al., 2007).

3.3.2.2 MECHANICAL METHOD

Shearing force is used for delaminating inter-fibrillar hydrogen bonding of cellulose microfibers which releases the CNF. Here also, certain pretreatments like TEMPO oxidation (Saito et al., 2009), mild alkali treatment, enzymatic pretreatment (Paakko et al., 2007) are used to reduce energy consumption. Different mechanical methods like high pressure homogenization (Nakagaito and Yano, 2005), microfluidization, grinding, cryocrushing (Alemdar and Sain, 2008), high-intensity ultrasonication (Li et al., 2011), electrospinning, aqueous counter collision, grinding, bead milling, pearl milling, ball milling and twin-screw extrusion can be used. TEMPO oxidation method is also used for surface modification like introduction of carboxylate and aldehyde functional groups into the native cellulose chain using mild or strong acid hydrolysis (Henriksson et al., 2007).

3.3.2.3 BACTERIAL CELLULOSE (BC)

Nowadays, nanocellulose is produced using a very popular microbial fermentation method in which cellulose is secreted as an extracellular product. Generally, microorganisms used for fermentation are bacteria of the genera *Acetobacter*, *Escherichia*, *Azotobacter*, *Salmonella*, *Agrobacterium*, *Achromobacter*, *Sarcina*, *Aerobacter*, and *Rhizobium* (Corujo et al., 2016). Cellulose produced by this method consists of nanometric width ribbons which are free of hemicelluloses, lignin, and pectin present in vegetal cellulose. BNC produced by Static culture is obtained as an entangled nanoribbon network grown in the air-liquid interface of the fermentation vial, whereas by agitated fermentation method, final product is obtained in the form of pellets (Corujo et al., 2016). Different methods of nanocellulose preparation are summarized in Table 3.1. In all the above methods, nanocellulose is normally produced as water dispersed materials so that different drying techniques such as oven drying, supercritical drying, drying at room temperature, spray drying, freeze-drying can be used to get dry powder.

TABLE 3.1 Different Methods of Nanocellulose Preparation

Method of Preparation	References
One-pot oxidative hydrolysis	Chen and Lee (2018)
Chemical-free technology	Jongaroontaprangsee et al. (2018)
Steam explosion technique	Deepa et al. (2011)
Enzymatic hydrolysis process	Patel and Joshi (2020)
TEMPO-mediated oxidation	Saito et al. (2007)
Heterogenous liquefaction	Li et al. (2015)
Acid hydrolysis	Wulandari et al. (2016)
Mechanical milling process	Nagendra et al. (2014)
High pressure homogenization	Wang et al. (2015)
Combination of mechanical and chemical method	Hettrich et al. (2014)
Catalytic synthesis	Abd Hamid et al. (2014)
Mechanical blending	Jiang et al. (2013)
Microfluidization	Lavoine et al. (2013)
Cryocrushing	Alemdar et al. (2007)
High-intensity ultrasonication	Chen et al. (2013)

3.4 PROPERTIES OF NANOCELLULOSE

Nanocellulose's properties depend on its raw material and method used for preparation. It has some useful properties such as high surface area to volume ratio, high aspect ratio (fiber length to width ratio), high elastic modulus, chemical reactivity, optical transparency, high strength, biodegradation, low thermal expansion coefficient, and good absorptivity (Barbash et al., 2017). There are different techniques to evaluate these properties (Table 3.2).

3.5 APPLICATIONS IN DIFFERENT FIELDS

Nanocellulose has great potential to be replaced or modified with existing materials in different fields such as nonwovens, forest products, packaging, consumer goods, architecture, construction, cosmetics, automotive, sporting goods, medical, pharmaceuticals, paint, coatings, adhesives, electronics, aerospace, oil, gas, textiles, apparel, photonics, etc., due to its physicochemical properties like renewable nature, biocompatibility, surface modification, mechanical strength, interesting optical properties and anisotropic shape (Miller, 2014). There are many industries worldwide such as

TABLE 3.2 Different Techniques to Characterize Nanocellulose

Properties of Nanocellulose	Techniques for Characterization	References
Morphology	Transmission electron microscopy and scanning electron microscopy	Schyrr et al. (2014)
Topographical Characterization	Atomic force microscopy	Barbash et al. (2017)
Chemical composition	α-cellulose (by ASTM D1103-55T), hemicelluloses (by ASTM D1104-56) and lignin (by ASTM D1106-56)	Abraham et al. (2011)
Moisture content	Karl-Fisher Titration, LOD	Abraham et al. (2011)
Particle size and size distribution	Dynamic light scattering/laser diffraction	Hornsby et al. (2012)
Thermal characterization	Differential scanning calorimetry (DSC) and thermogravimetric analysis (TGA)	Chirayil et al. (2014)
Mechanical strength	As per ISO 527-1	Chirayil et al. (2014)
Degree of crystallinity	X-ray diffraction (XRD)	Hornsby et al. (2012)
Water absorptivity	Percent swelling index	Liu et al. (2016)
Dielectric properties	Dielectric spectroscopy	Le Bras et al. (2015)

Borregaard, CelluForce INC, UPM-Kymmene Oyj, Daicel Finechem Ltd, American Process Inc., Nippon Paper Industries, Co. Ltd., Axcelon Biopolymers Corporation, Innventia AB, Melodea Ltd, Oji Holdings Corporation, Sappi Global and Stora Enso which are working on the cellulose-based products (Miller, 2014). The recent advanced research on novel applications of nanocellulose has been enumerated in Table 3.3. Some advanced applications of nanocellulose are:

1. Energy and Electronic Applications: Includes cellulose for energy storage, cellulose for display devices, cellulose for energy harvesting, paper, and transistor (Jose et al., 2014).
2. Nanocellulose Photonics: Different optical functionalities like chirality (Usov et al., 2015), fluorescence (Abitbol et al., 2013; Schyrr et al., 2014), surface plasmons (Padalkar et al., 2010; Lam et al., 2012), low refractive index (Buskens et al., 2015) and UV-blocking (Nevo et al., 2015) of CNC-based systems can be utilized in anti-counterfeiting technologies, particle tracking and sensing, greenhouse plastics, optoelectronics, coating, and packaging (Saito et al., 2011).

TABLE 3.3 Research on Advanced Applications of Nanocellulose

Research Work on Nanocellulose	References
Super-hydrophobic and super oleophilic nanocellulose sponge for oil/water separation	Phanthong et al. (2018)
Cellulose nanofibers composite films based on thermoplastic starch	Nasri-Nasrabadi et al. (2014)
Biosensors based on porous cellulose nanocrystal-PVA scaffolds for fluorescence-based sensing schemes	Schyrr et al. (2014)
Transparent conductive nanocellulose paper for electronic devices	Li and Lee (2017)
PVA reinforced with crystalline nanocellulose for 3D printing application	Cataldi et al. (2018)
Nanocellulose prepared from wood and algae as an electrical insulator	Le Bras et al. (2015)
Magnetic nanocellulose for oil adsorption	Gu et al. (2020)
Nanocellulose for removal of contaminants from wastewater	Tshikovhi et al. (2020)
Wound healing properties of nanocellulose composite	Taheri et al. (2020)

3. Films of Nanocellulose: Nanocellulose has some unique properties such as chiral nematic organization, water sorption, gas barrier, mechanical strength, thermal conductivity, optical, alignment in the presence of external forces like magnetic, electric, and shear. CNC's long axes arrange themselves parallel to the electric field (Bordel et al., 2006) while perpendicular to the magnetic field (Sugiyama et al., 1992) because of its negative magnetic susceptibility. CNCs also arrange themselves in presence of external shear forces (Hoeger et al., 2011). Due to these versatile properties, CNCs films and foams can be used to prepare flexible electronic devices, food packaging materials, and super-capacitor electrodes.
4. Nanocellulose Functionalization: Hydroxyl groups and large surface area permits surface modification of nanocellulose using covalent bonding by oxidation, esterification, silylation as well by non-covalent bonding (Kargarzadeh et al., 2018).
5. Polymer Nanocomposites: These materials are prepared by reinforcing nanomaterials into polymer matrix (Mondal, 2018). Physicochemical properties of matrix such as mechanical properties, dimensional stability and gas-solvent barrier properties can be improved by incorporating nanocellulose due to its high aspect ratio, high surface area,

plenty of hydroxyl groups, economical deposition into polymer matrix by different methods like electrospinning, melt mixing, solution processing and *in situ* polymerization (Lee et al., 2005; Dufresne, 2012, 2013; Bhanvase and Sonawane, 2014; Mondal, 2018).

3.6 COMMERCIAL APPLICATIONS

It was estimated that global production of cellulosic nanomaterials will reach up to 2600 tons per day by 2020 from 340 tons per year in 2013. Some of the industries which commercially produce nanocellulose in tons are: Innventia (Sweden), FiberLean Technologies (UK), Borregaard (Norway), Nippon paper (Japan), UPM (Finland), Stora Enso (Finland), Daicel (Japan), Oji paper, VTTT, CelluForce, Blue Goose Biorefineries, Kruger (Canada), Melodia, USDA Forest Products Lab and Alberta Innovates-Technology Futures. It was predicted that nanocellulose market will range from 18 to 56 million tons (dry equivalent) in the future. Commercially, nanocellulose is used in the following industries: paints and coatings, cement manufacturing, automotive parts, paper, and paper packaging (Abbati et al., 2017). As per the report published by transparency market research with title "nanocellulose technology market-global industry analysis, size, share, growth, trends, and forecast, 2017–2025," the market of nanocellulose will give stable CAGR (compound annual growth rate) of over 30% during 2017 to 2025 with respect to its market valued around 50 million USD in 2016 (Miller, 2014). Some examples of commercialized nanocellulose products are shown in Table 3.4.

TABLE 3.4 Marketed Products Containing Nanocellulose

Marketed Product Name	Description	Manufacturer
Hada Care Acty	Functional cellulose nanofibers (TEMPO CNF) in deodorant sheet of adult diapers	Nippon Paper Crecia Co., Ltd. Japan
Rheocrysta™	Ballpoint pen ink with nanocellulose as thickening agent	Mitsubishi Pencil Co and DKS
New Natura™	Liquid packaging board enhanced with micro-fibrillated cellulose (MFC) for extra strength and lower weight	Stora Enso., Finland
Nanocellulose face masque	Hydrating face mask as cosmetic	DeLeon™ Cosmetics, USA

3.7 FUTURE PERSPECTIVES

Apart from medical applications, due to its extraordinary properties, such as very good strength to weight ratio, which is 8 times higher than stainless steel, nanocellulose can also be used in body armor. It will be not surprising if we would see super-flexible screen and bendable batteries which are made from nanocellulose due its transparency and lightweight. Nanocellulose can also be utilized to make filters for the purification of almost all kinds of liquids and gases. Nowadays researchers are also conducting trials for obtaining bio-fuel from nanocellulose. Super absorbent aerogels with a swelling capacity of 10,000 times can be made using nanocellulose.

3.8 CONCLUSION

The development of nanocellulose from cellulose which was discovered more than 100 years ago, is a landmark in the field of pharmaceuticals, chemical, paper, automobile, electrical, oil, and ink industries. Nanocellulose is an ideal excipient as it is cheap, biodegradable, biocompatible, and easy to synthesize and scale-up. It is a versatile excipient which can be exploited in a variety of applications in medicine and nonmedical applications. Nanofibers can be used in drug delivery systems, advanced wound management, medical devices, etc. Many pharmaceutical formulations based on nanocellulose are already in the market, and numerous formulations are under development and clinical trials. We can conclude that nanocellulose is the excipient of the future owing to its unique physicochemical properties and versatility.

KEYWORDS

- **cellulose**
- **crystalline nanocellulose**
- **functional cellulose nanofibers**
- **nanocellulose**
- **nanocrystalline cellulose**
- **physicochemical properties**

REFERENCES

Abbati, C., Houtman, C., Philips, R., Bilek, E., Rojas, O., Pal, L., Peresin, M., et al., (2017). Conversion economics of forest biomaterials: Risk and financial analysis of CNC manufacturing. *Biofuels, Bioproducts and Biorefining, 11*(4), 682–700.

Abd, H. S., Al Amin, M., & Ali, M., (2014). Zeolite supported ionic liquid catalyst for the synthesis of nanocellulose from palm tree biomass. *Micro/Nano Science and Engineering, 925*, 52–56.

Abitbol, T., Palermo, A., Moran-Mirabal, J., & Cranston, E., (2013). Fluorescent labeling and characterization of cellulose nanocrystals with varying charge contents. *Biomacromolecules, 14*(9), 3278–3284.

Abraham, E., Deepa, B., Pothan, L., Thomas, S., John, M., Anandjiwala, R., & Cvelbar, U., (2011). Extraction of nanocellulose fibrils from lignocellulosic fibres: A novel approach. *Carbohydrate Polymer, 86*(4), 1468–1475.

Alemdar, A., & Sain, M., (2008). Biocomposites from wheat straw nanofibers: morphology, thermal and mechanical properties. *Composites Science and Technology, 68*(2), 557–565.

Azizi, S. A., Alloin, F., & Dufresne, A., (2005). Review of recent research into cellulosic whiskers, their properties and their application in nanocomposite field. *Biomacromolecules, 6*(2), 612–626.

Bajpai, P., (2013). Biorefinery opportunities in the pulp and paper industry. In: Bajpai, P., (ed.), *Biorefinery in the Pulp and Paper Industry* (pp. 11–15). Academic Press, Boston.

Ball, P., (2005). Material witness: In praise of wood. *Nature Materials, 4*, 515.

Barbash, V., Yaschenko, O., & Shniruk, O., (2017). Preparation and properties of nanocellulose from organosolv straw pulp. *Nanoscale Research Letters, 12*(1), 241.

Bhanvase, B., & Sonawane, S., (2014). Ultrasound-assisted *in situ* emulsion polymerization for polymer nanocomposite: A review. *Chemical Engineering and Processing: Process Intensification, 85*, 86–107.

Bordel, D., Putaux, J. L., & Laurent, H., (2006). Orientation of native cellulose in an electric field. *Langmuir, 22*(11), 4899–4901.

Buskens, P., Maurice, M., Meulendijks, N., Ee, R., Burghoorn, M., Verheijen, M., & Veldhoven, E., (2015). Highly porous, ultra-low refractive index coatings produced through random packing of silicated cellulose nanocrystals. *Colloids and Surfaces A: Physicochemical and Engineering Aspects, 487*, 1–8.

Cataldi, A., Rigotti, D., Nguyen, V., & Pegoretti, A., (2018). Polyvinyl alcohol reinforced with crystalline nanocellulose for 3D printing application. *Materials Today Communications, 15*, 236–244.

Chen, P., Yu, H., Liu, Y., Chen, W., Wang, X., & Ouyang, M., (2013) Concentration effects on the isolation and dynamic rheological behavior of cellulose nanofibers via ultrasonic processing. *Cellulose, 20*(1), 149–157.

Chen, Y. W., & Lee, H. V., (2018). Revalorization of selected municipal solid wastes as new precursors of "green" nanocellulose via a novel one-pot isolation system: A source perspective. *International Journal of Biological Macromolecules, 107*, 78–92.

Chirayil, C., Mathew, L., & Thomas, S., (2014). Review of recent research in nano cellulose preparation from different lignocellulosic fibers. *Reviews on Advanced Materials Science, 37*, 20–28.

Corujo, V. F., Cerrutti, P., Foresti, M. L., & Vázquez, A., (2016). Production of bacterial nanocellulose from non-conventional fermentation media. In: Puglia, D., Fortunati, E.,

& Kenny, J., (eds.), *Multifunctional Polymeric Nanocomposites Based on Cellulosic Reinforcements* (pp. 39–59). William Andrew Publishing, USA.

Cowie, J., Bilek, E. M., Wegner, T., & Shatkin, J. A., (2014). Market projections of cellulose nanomaterial-enabled products: Part 2: Volume estimates. *TAPPI Journal, 13*(6), 57–69.

Deepa, B., Abraham, E., Cherian, B., Bismarck, A., Blaker, J., Pothan, L., Leao, A., et al., (2011). Structure, morphology and thermal characteristics of banana nano fibers obtained by steam explosion. *Bioresource Technology, 102*(2), 1988–1997.

Dufresne, A., (2012). Potential of nanocellulose as a reinforcing phase for polymers. *J-FOR, 2*, 6–16.

Dufresne, A., (2013). Nanocellulose: A new ageless bionanomaterial. *Materials Today, 16*(6), 220–227.

Gu, H., Zhou, X., Lyu, S., Pan, D., Dong, M., Wu, S., Ding, T., et al., (2020). Magnetic nanocellulose-magnetite aerogel for easy oil adsorption. *Journal of Colloid and Interface Science, 560*, 849–856.

Habibi, Y., Chanzy, H., & Vignon, M., (2006). TEMPO-mediated surface oxidation of cellulose whiskers. *Cellulose, 13*(6), 679–687.

Henriksson, M., Henriksson, G., Berglund, L., & Lindström, T., (2007). An environmentally friendly method for enzyme-assisted preparation of microfibrillated cellulose (MFC) nanofibers. *European Polymer Journal, 43*(8), 3434–3441.

Hettrich, K., Pinnow, M., Volkert, B., Passauer, L., & Fischer, S., (2014). Novel aspects of nanocellulose. *Cellulose, 21*, 2479–2488.

Hoeger, I., Rojas, O., Efimenko, K., Velev, O., & Kelley, S., (2011). Ultrathin film coatings of aligned cellulose nanocrystals from a convective-shear assembly system and their surface mechanical properties. *Soft Matter, 7*(5), 1957–1967.

Hornsby, P., Muhamad, M., McMullan, C., Sharma, S., & Carmichael, E., (2012). Nano-cellulose reinforced polymers derived from banana tree and other fibre sources. *Proceedings of the 15th European Conference on Composite Materials*. http://www.escm.eu.org/eccm15/data/assets/250.pdf (accessed on 8 July 2021).

Ioelovich, M., (2014). Peculiarities of cellulose nanoparticles. *TAPPI J., 13*(5), 45–52.

Iwamoto, S., Nakagaito, A., Yano, H., & Nogi, M., (2005). Optically transparent composites reinforced with plant fiber-based nanofibers. *Applied Physics A, 81*(6), 1109–1112.

Jiang, F., & Hsiech, Y., (2013). Chemically and mechanically isolated nanocellulose and their self-assembled structures. *Carbohydrate Polymers, 95*(1), 32–40.

John, M., & Thomas, S., (2008). Biofibres and biocomposites. *Carbohydrate Polymers, 71*(3), 343–364.

Jongaroontaprangsee, S., Chiewchan, N., & Devahastin, S., (2018). Production of nanocellulose from lime residues using chemical-free technology. *Materials Today: Proceedings, 5*(5), 11095–11100.

Kargarzadeh, H., Ahmad, I., Thomas, S., Ioelovich, M., & Dufresne, A., (2017). Methods for extraction of nanocellulose from various sources. In: Kargarzadeh, H., Ahmad, I., Thomas, S., & Dufresne, A., (eds.), *Handbook of Nanocellulose and Cellulose Nanocomposites* (pp. 1–51). Wiley-VCH, Germany.

Kargarzadeh, H., Mariano, M., Gopakumar, D., Ahmad, I., Thomas, S., Dufresne, A., Huang, J., & Lin, N., (2018). Advances in cellulose nanomaterials. *Cellulose, 25*(4), 2151–2189.

Kim, J. H., Shim, B. S., Kim, H. S., Lee, Y. J., Min, S. K., Jang, D., Abas, Z., & Kim, J., (2015). Review of nanocellulose for sustainable future materials. *International Journal of Precision Engineering and Manufacturing-Green Technology, 2*(2), 197–213.

Klemm, D., Schumann, D., Kramer, F., Heßler, N., Koth, D., & Sultanova, B., (2009). Nanocellulose materials-different cellulose, different functionality. *Macromolecular Symposia, 280*, 60–71.

Lam, E., Hrapovic, S., Majid, E., Chong, J., & Luong, J., (2012). Catalysis using gold nanoparticles decorated on nanocrystalline cellulose. *Nanoscale, 4*(3), 997–1002.

Lavoine, N., Desloges, I., Dufresne, A., & Bras, J., (2012). Micro fibrillated cellulose-its barrier properties and applications in cellulosic materials: A review. *Carbohydrate Polymers, 90*(2), 735–764.

Le Bras, D., Strømme, M., & Mihranyan, A., (2015). Characterization of dielectric properties of nanocellulose from wood and algae for electrical insulator applications. *The Journal of Physical Chemistry B, 119*(18), 5911–5917.

Lee, L., Zeng, C., Cao, X., Han, X., Shen, J., & Xu, G., (2005). Polymer nanocomposite foams. *Compos. Sci. Technol., 65*(15, 16), 2344–2363.

Li, S., & Lee, P. S., (2017). Development and applications of transparent conductive nanocellulose paper. *Science and Technology of Advanced Materials, 18*(1), 620–633.

Li, W., Wang, R., & Liu, S., (2011). Nanocrystalline cellulose prepared from softwood kraft pulp via ultrasonic-assisted acid hydrolysis. *Bioresources, 6*(4), 4271–4281.

Li, W., Zhang, Y., Li, J., Zhou, Y., Li, R., & Zhou, W., (2015). Characterization of cellulose from banana pseudo-stem by heterogeneous liquefaction. *Carbohydrate Polymers, 132*, 513–519.

Liu, J., Chinga-Carrasco, G., Cheng, F., Xu, W., Willför, S., Syverud, K., & Xu, C., (2016). Hemicellulose-reinforced nanocellulose hydrogels for wound healing application. *Cellulose, 23*(5), 3129–3143.

Miller, J., (2014). Nanocellulose: Technology applications, and markets. *RISI Latin American Pulp and Paper Outlook Conference*. http://www.mktintell.com/files/Miller_20Presentation.pdf (accessed on 8 July 2021).

Mondal, S., (2018). Review on nanocellulose polymer nanocomposites. *Polymer-Plastics Technology and Engineering, 57*(13), 1–15.

Nagendra, S., Satyarasad, V., Pusty, G., & Ramji, K., (2014). Synthesis of bio-degradable banana nanofibers. *International Journal of Innovative Technology and Research, 2*(1), 730–734.

Nakagaito, A., & Yano, H., (2005). Novel high-strength biocomposites based on microfibrillated cellulose having nano-order-unit web-like network structure. *Applied Physics A, 80*(1), 155–159.

Nasri-Nasrabadi, B., Behzad, T., & Bagheri, R., (2014). Preparation and characterization of cellulose nanofiber reinforced thermoplastic starch composites. *Fibers and Polymers, 15*(2), 347–354.

Nevo, Y., Peer, N., Yochelis, S., Igbaria, M., Meirovitch, S., Shoseyov, O., & Paltiel, Y., (2015). Nano bio optically tunable composite nanocrystalline cellulose films. *RSC Advances, 5*(10), 7713–7719.

Paakko, M., Ankerfors, M., Kosonen, H., Nykanen, A., Ahola, S., Osterberg, M., Ruokolainen, J., et al., (2007). Enzymatic hydrolysis combined with mechanical shearing and high-pressure homogenization for nanoscale cellulose fibrils and strong gels. *Biomacromolecules, 8*(6), 1934–1941.

Padalkar, S., Capadona, J. R., Rowan, S. J., Weder, C., Won, Y. H., Stanciu, L. A., & Moon, R. J., (2010). Natural biopolymers: Novel templates for the synthesis of nanostructures. *Langmuir, 26*(11), 8497–8502.

Patel, B. H., & Joshi, P. V., (2020). Banana nanocellulose fiber/PVOH composite film as soluble packaging material: Preparation and characterization. *Journal of Packaging Technology and Research, 4*(1), 95–101.

Phanthong, P., Reubroycharoen, P., Kongparakul, S., Samart, C., Wang, Z., Hao, X., Abudula, A., & Guan, G., (2018). Fabrication and evaluation of nanocellulose sponge for oil/water separation. *Carbohydrate Polymers, 190,* 184–189.

Saito, T., Hirota, M., Tamura, N., Kimura, S., Fukuzumi, H., Heux, L., & Isogai, A., (2009). Individualization of nano-sized plant cellulose fibrils by direct surface carboxylation using TEMPO catalyst under neutral conditions. *Biomacromolecules, 10*(7), 1992–1996.

Saito, T., Kimura, S., Nishiyama, Y., & Isogai, A., (2007). Cellulose nanofibers prepared by TEMPO-mediated oxidation of native cellulose. *Biomacromolecules, 8*(8), 2485–2491.

Saito, T., Uematsu, T., Kimura, S., Enomae, T., & Isogai, A., (2011). Self-aligned integration of native cellulose nanofibrils towards producing diverse bulk materials. *Soft Matter, 7*(19), 8804–8809.

Sandolo, C., Matricardi, P., Alhaique, F., & Coviello, T., (2007). Dynamo-mechanical and rheological characterization of guar gum hydrogels. *European Polymer Journal, 43*(8), 3355–3367.

Schyrr, B., Pasche, S.,Voirin, G.,Weder, C., Simon, Y. C., & Foster, E. J., (2014). Biosensors based on porous cellulose nanocrystal-poly(vinyl alcohol) scaffolds. *ACS Applied Materials and Interfaces, 6*(15), 12674–12683.

Shatkin, J. A., Wegner, T. H., Bilek, E. M., & Cowie, J., (2014). Market projections of cellulose nanomaterial-enabled products: Part 1: Applications. *TAPPI Journal, 13*(5), 9–16.

Siró, I., & Plackett, D., (2010). Micro fibrillated cellulose and new nanocomposite materials: A review. *Cellulose, 17*(3), 459–494.

Sugiyama, J., Chanzy, H., & Maret, G., (1992). Orientation of cellulose microcrystals by strong magnetic fields. *Macromolecules, 25*(16), 4232–4234.

Taheri, P., Jahanmardi, R., Koosha, M., & Abdi, S., (2020). Physical, mechanical, and wound healing properties of chitosan/gelatin blend films containing tannic acid and/or bacterial nanocellulose. *International Journal of Biological Macromolecules, 154,* 421–432.

Tshikovhi, A., Mishra, S. B., & Mishra, A. K., (2020). Nanocellulose-based composites for the removal of contaminants from wastewater. *International Journal of Biological Macromolecules, 152,* 616–632.

Usov, I., Nyström, G., Adamcik, J., Handschin, S., Schütz, C., Fall, A., Bergström, L., & Mezzenga, R., (2015). Understanding nanocellulose chirality and structure-properties relationship at the single fibril level. *Nature Communications, 6,* 7564.

Vazquez, A., Foresti, M., Cerrutti, P., & Galvagno, M., (2013). Bacterial cellulose from simple and low cost production media by *Gluconacetobacter xylinus*. *Journal of Polymers and the Environment, 21,* 545–554.

Wang, H., Zhang, X., Jiang, Z., Li, W., & Yu, Y., (2015). A comparison study on the preparation of nanocellulose fibrils from fibers and parenchymal cells in bamboo (*Phyllostachys pubescens*). *Industrial Crops and Products, 71,* 8088.

Wulandari, W. T., Rochliadi, A., & Arcana, I. M., (2016). Nanocellulose prepared by acid hydrolysis of isolated cellulose from sugarcane bagasse. *IOP Conference Series: Materials Science and Engineering, 107*(1), 012045.

CHAPTER 4

ATOMIC FORCE MICROSCOPY PRINCIPLES AND RECENT STUDIES OF IMAGING AND NANOMECHANICAL PROPERTIES IN BACTERIA

H. H. TORRES-VENTURA,[1] J. J. CHANONA-PÉREZ,[1] L. DORANTES-ÁLVAREZ,[1] J. V. MÉNDEZ-MÉNDEZ,[2] B. ARREDONDO-TAMAYO,[1] P. I. CAUICH-SÁNCHEZ,[3] and ANA ELENA JIMÉNEZ-CARMONA[1]

[1]*Instituto Politécnico Nacional, Escuela Nacional de Ciencias Biológicas, Departamento de Ingeniería Bioquímica, Av. Wilfrido Massieu Esq, Cda, Manuel L. Stampa S/N, C.P. 07738, Mexico City, Mexico*

[2]*Instituto Politécnico Nacional, Centro de Nanociencias y Micro y Nanotecnologías, Luis Enrique Erro S/N, Zacatenco, C.P. 07738, Gustavo A. Madero, Mexico City, Mexico*

[3]*Instituto Politécnico Nacional, Escuela Nacional de Ciencias Biológicas, Departamento de Microbiología, Plan de Ayala y Carpio S/N, C. P. 11340, Mexico City, Mexico*

4.1 INTRODUCTION

The microorganisms can be found practically within all the places, in several industries, medical processes, water treatments, domestic environments, and mainly in several foodstuffs, surfaces, and water that are in contact with food. Within industries of production, packing, storage, and delivery of dairy, meats, barnyard fowl, vegetables, and seafood products, the bacterial contamination could be prejudicial for the processes, due to they can be responsible for products deterioration, decreased efficiency of the process, corrosion or blockages pipe and equipment breakdown (Garrett et al., 2008). Thus, the ingestion of food contaminated with pathogen bacteria or their toxins could lead to morbidity and mortality, representing an important public health problem with high economic and social costs around the world, for these reasons, much research has focused on the study of bacterial cells behavior to control its growth. However, there are probiotics bacteria that have beneficial effects in industries process and for human organism, and its study is also important. Some bacterial cells distinct characteristics are their viability, motility, and mechanical properties under different medium. Understanding the bacteria mechanical properties will raise our knowledge about the physical and structural functions of cells to know their behavior. Nevertheless, the bacterial cells surface mechanical properties have been complicated to study due to their small size. So, the atomic force microscopy (AFM) turns out to be a great analysis technique to study the bacteria mechanical properties and morphology, since it provides important information of the surface structures of living cells at nanoscale resolution. In recent years, AFM has emerged as a more suitable technique to obtain higher-resolution imaging in the analysis and identification of live cells on microbiological systems than optical microscopy which the resolution is restricted by the wavelength of its light source, and the electronic microscopy that although could reach high-resolution cells images, its convoluted sample preparation process and vacuum conditions required may cause significant changes in the living specimen and it would not be possible to observe cells in their native environments. In addition, the AFM technique has the advantage that the sample preparation is easier than in electronic microscopy and its capability to obtain images and mechanical analysis of live bacteria. Therefore, they can be analyzed in their proper environmental conditions in both air and aqueous environment in real-time, or in the interaction with antimicrobials to observe their effects on the bacterial surface (Liu and Yang, 2019).

4.2 ATOMIC FORCE MICROSCOPY (AFM) PRINCIPLES

The AFM technique was developed in 1985 by Binnig, Quate, and Gerber, it is considered a high-resolution microscopy. It is part of the scanning probe microscopes (SPM) group, and its components are detailed in Figure 4.1(a). AFM employs a nanoscale sharp tip as a force sensor to analyze sample topography properties, as well as electric and magnetic force gradients and mechanical properties, among others, through the interaction of the cantilever tip with the sample surface.

The AFM is not a conventional optical microscope, it has the capable to measure the sample microscopic characteristics, particularly, in the study of bacterial cell topographic imaging and the evaluation of their mechanical properties, the AFM techniques provide valuable and detailed information about bacterial cell wall, for instance, it is possible to study the effect of the antimicrobial compounds and antibiotics on the morphology structure of bacteria (Figure 4.1(b)), since basically the tip scans the surface of the sample at a specific distance known as non-contact mode or it can be in direct contact which is known as contact mode, depending on the characteristics of the surface to be evaluated (James et al., 2017). AFM important feature is that images of sample surface are acquired by sensing it without light requirement (Figure 4.1(c)). It is possible to obtain images of surface topographic that reach a resolution of several tenths of a nanometer. In addition to surface topography with high resolution, and unlike other microscopic techniques, AFM is able to expose different features of the specimen, and it is possible to obtain the hardness, stiffness, adhesion, and strength of the interaction between the surface of the sample and the tip of the cantilever (Figure 4.1(d)). It is a technique that has become an important tool for researchers in many biological disciplines. There are two main modes in where it operates, imaging and force spectroscopy; the first one is divided into three major mode categories, which are contact, non-contact, and intermittent mode and can be adapted for air or liquid media. Other AFM imaging techniques have been implemented, which map properties such as conductance and friction over a surface, nevertheless, they are the contact and intermittent mode which have been used mainly for the study of cell surfaces (James et al., 2017; Liu and Yang, 2019). During measurement, the AFM records the attractive and repulsive forces at the nanoscale of the specimen by the deflection of the pointed tip of the cantilever, which is systematically scanned across the sample surface (Figure 4.2(a)).

These forces are measured by the displacement of a reflected laser in the upper of the cantilever gold-coated side toward a quadratic photosensitive photodiode (Figure 4.1(a)). The contact mode is the simplest and original AFM imaging technique, where the cantilever deflection is detected while the specimen is probed horizontally by a piezoelectric scanner, this ensures that the force applied to the sample is uniform and controlled to avoid damage to the sample, thus, image is generated by mapping the vertical distance during scanning, and axial and longitudinal deflections in the cantilever are

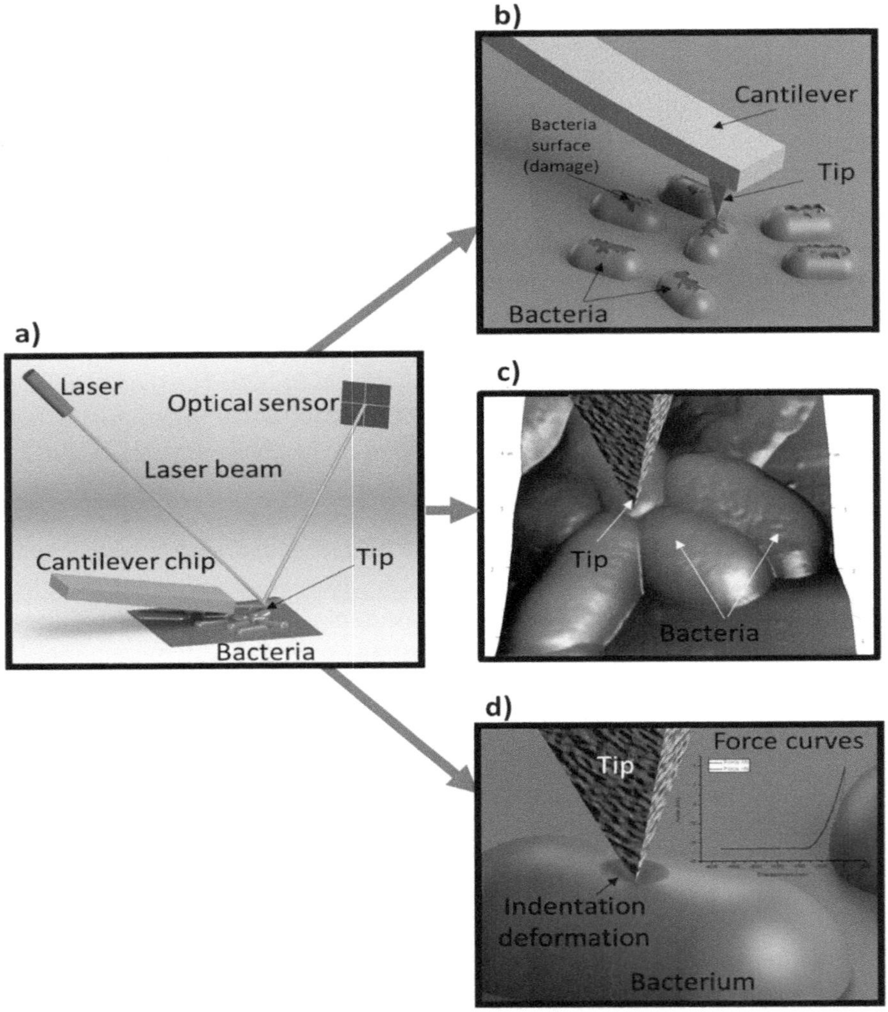

FIGURE 4.1 (a) Simplified diagram of atomic force microscope components for imaging acquisition; (b) zooming of cantilever-tip and interaction with bacterial cells surfaces with different local topography; (c) tip-bacteria interaction during the capture of height images; (d) diagrammatic representation of indentation deformation of a bacterium by the tip and force curves obtained.

registered across the photodiode (Figure 4.2(b)) (Aguayo et al., 2016). While non-contact mode is a technique in which the integrated cantilever vibrates very close to its resonance frequency and it is driven by a piezoelectric actuator. Topographic images are generated by mapping the vertical distance the scanner moves, as it maintains a constant amplitude of oscillation at each lateral data point (Rogers et al., 2004). Tapping mode considerably reduces the problems associated with electrostatic and lateral shear forces, adhesion, and friction present in conventional AFM techniques when the tip is lifted to avoid dragging it over the surface (Boussu et al., 2005). However, due to the mobility of some bacterial species and the usually weak forces that bind cells to the surface, it is almost always necessary to immobilize the cells, the bacterial cells must be strongly fixed to a surface of some substrate in order to support the frictional forces of the tip and thus avoid the destruction of the samples. Different bacterial immobilization strategies have been described in the literature (Kailas et al., 2009).

4.3 IMAGING AND NANOINDENTATION TECHNIQUES

Samples hardness, surface features, and origin of the materials are essential to select the adequate AFM operation mode to obtain quality images and good measurements of the properties of interest. The primary modes used in AFM for imaging and evaluation of mechanical properties are contact, tapping, and non-contact techniques. Contact mode is most appropriate for hard surface samples, however in this method, the tip can be contaminated when the sample surface is sticky, or the measurements are carried out in fluid media. The tapping method is more suitable for imaging of soft biological samples and with low surface adhesion such as DNA (deoxyribonucleic acid) chains, biopolymers, bacteria, yeast, cells, and tissues of animal and vegetable origin. The non-contact method is commonly used for imaging soft samples (Jembrek et al., 2015; Cardenas-Perez et al., 2019). As a function of the interaction tip-sample, a description of these modes is presented below.

4.3.1 CONTACT MODE

It is carried out at the moment in which the tip of the sample makes continuous contact or contact without intermittent vibration in the cantilever, the following modes of operation in AFM can be classified in this section: Contact mode, force curves, nanoindentation, force volume. In the contact method, there is a direct interaction of the cantilever tip with the sample, because the tip touches the surface of the sample literally. The process involves the following sequence, when the tip touches the sample, it experiences a repulsive force, so it can be said that the contact method works with repulsive forces and the force applied to scan a surface must be greater than the repulsive force experienced by the tip (Escamilla-Garcia et al., 2015). When the tip is transported towards the sample (cantilever is in undeflected stage), it starts to experience an attractive force, and the cantilever is deflected. If the movement continues towards the sample, the tip experiences a repulsive force (Figure 4.2(a)). The force at the tip is dependent on the distance amid the cantilever and the sample (Mendez, 2010). For this, the tip on the cantilever scans over the surface of the sample and can damage it, if the contact force is very high or if the sample is made of a very soft material or sticky like bacteria or biopolymers. However, when the contact force is adequate, it is possible to obtain the topography of its surface by deflecting the cantilever in the photodetector located in the atomic force microscope (Figure 4.2(b)).

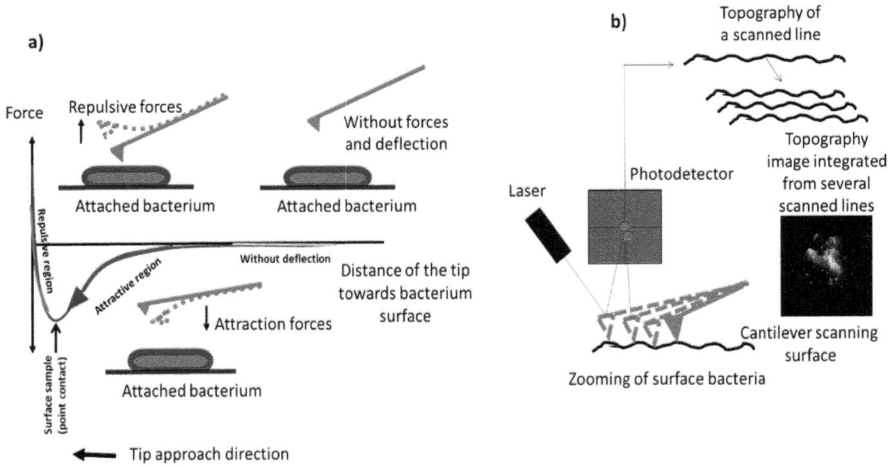

FIGURE 4.2 (a) Diagram of forces involves during tip approach in contact mode; and (b) scheme of generation topography image in contact mode in an atomic force microscope.

4.3.2 *WITHIN CONTACT MODE THERE ARE OTHER TWO VARIANTS OF OPERATION*

The constant height method, where attractive forces are implicated in the interaction tip-sample (Figure 4.3(a)). In this method, the distance amid sample-tip is constant. The tip experiences a force of attraction, and the topography of the sample is obtained from which the cantilever is deflected, a disadvantage is that for samples with large corrugated surfaces or large variations of the height, this operation mode could be limited. Another variant in the constant method is called constant force or constant deflection mode, in this method of operation, the interaction of the tip is naturally repulsive. Thus, as soon as the sample is close to the tip, a force of repulsion is also experienced, the force is maintaining constant moving the scanner up-down according with the topography or roughness of the sample (Figure 4.3(b)).

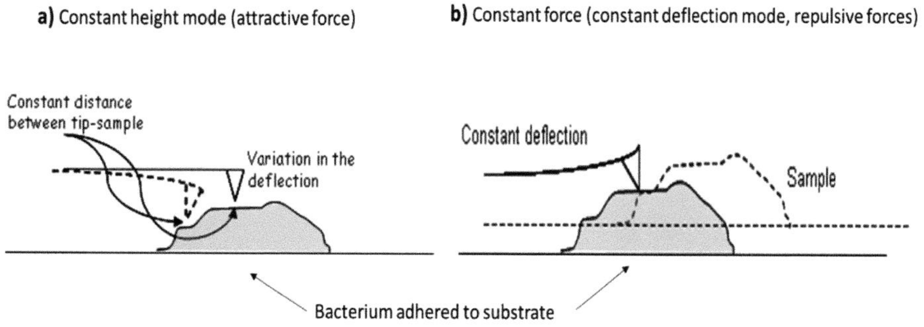

FIGURE 4.3 Diagrams of (a) constant height (attraction force); and (b) constant force (repulsion forces) included in an atomic force microscope.

The force curves are the key factor for measuring the mechanical properties of materials. These curves are considered as a contact technique, although they also can be obtaining by in tapping mode. A force curve is a representation of the force applied to the sample in terms of the separation tip-sample, and it is possible to measure the force detected by the tip attached to the cantilever. The force of the tip-sample (F) is measured with this relationship: $F = k\ z$, where k is the spring constant of the cantilever and z is how much it deflects. This approach is widely used in order to have an evaluation of different mechanical properties, mainly of soft materials, such as bacteria, plant, and animal cells (Raman et al., 2011; Chopinet et al., 2013; Cardenas-Perez et al., 2016, 2017; Chang and Liu, 2018). At the moment of interacting sample and tip, two curves are generated, an approach and retraction curves (Figure 4.4), from these curves theoretically several mechanical properties can be extracted (Figure 4.5) such as deformation, energy dissipation elasticity, maximum peak, and adhesion forces, Young's modulus (E) or stiffness and the plasticity index (Cardenas-Perez et al., 2019; Nicolas-Alvarez et al., 2019) (Figure 4.6).

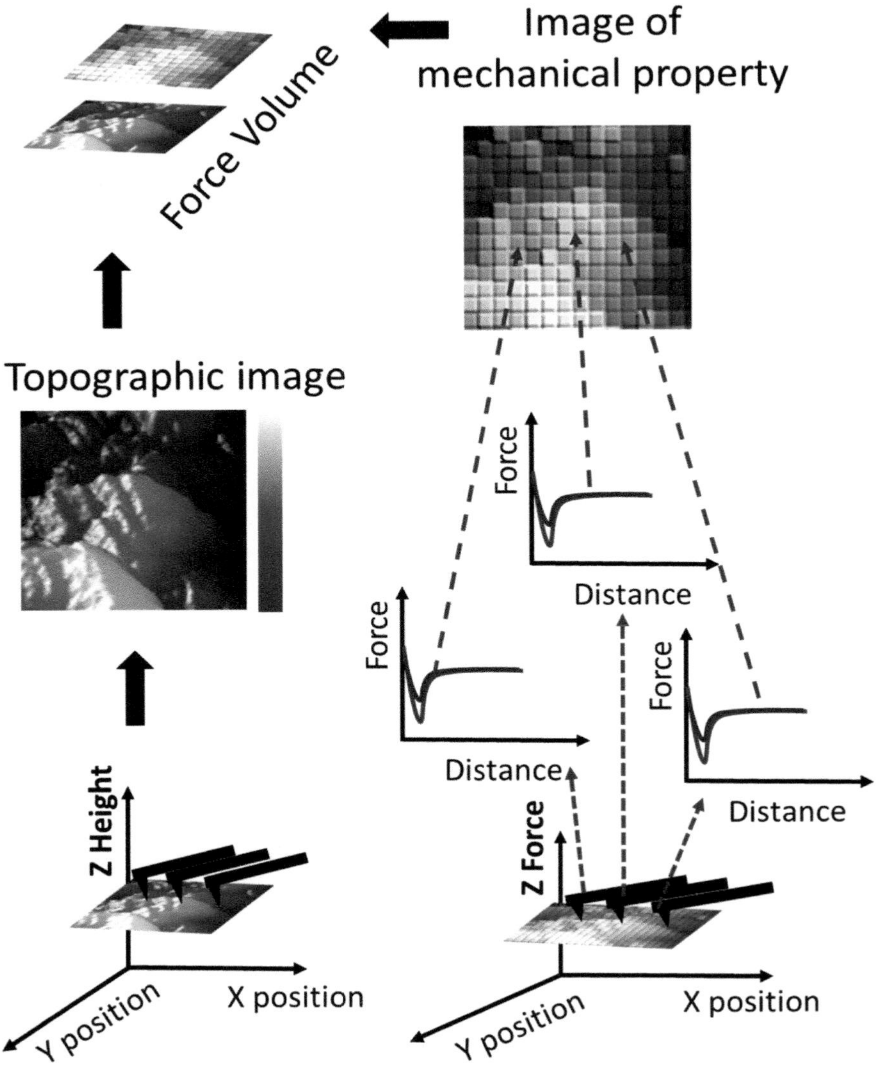

FIGURE 4.4 Overall diagram of generation of the force volume image in contact mode.

When force curves are obtained on biological materials such as bacteria, soft biopolymers, animal, and plant cells, the sample are indented gently by the tip of silicon and the nanomechanical properties are extracted from deflection of cantilever, sometimes a residual fingerprint is left on the material surface but as the soft materials are elastic the residual fingerprint disappear quickly. In the case of the true indentation mode the hard samples are penetrated by the diamond tip and a mark with the geometry of tip is left on the surface of the material (Figure 4.5), thus from this residual fingerprint the mechanical properties are determined, usually nanoindentation is preferred for hard materials such as metals, plastics of high density, woods, calcified materials (bones, marine, and eggshells) lignocellulose structures (spines and nuts shells) (Severa et al., 2010; Imbert et al., 2014; Athanasiadou et al., 2018; Marin-Bustamante et al., 2018; Nicolas-Bermudez et al., 2018).

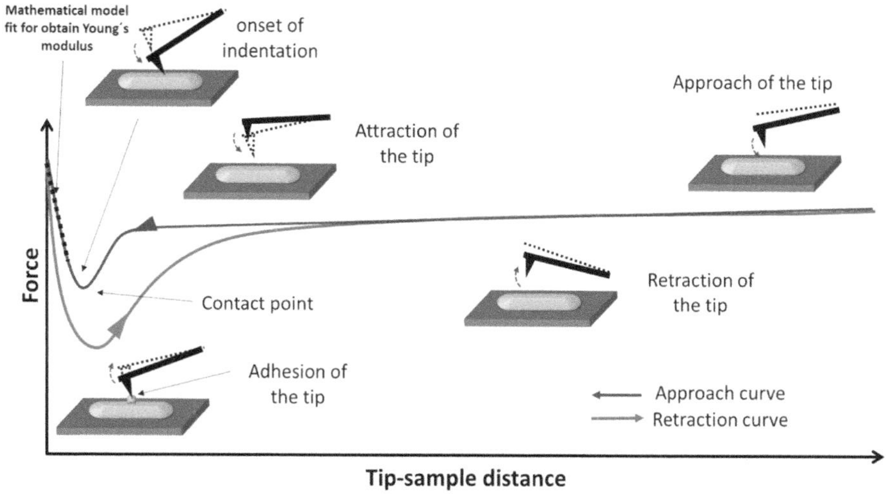

FIGURE 4.5 Representation of force curves in contact mode, main positions of cantilever-tip during the indentation of a bacterium and Young's module analysis.

A reliable technique commonly used in biomechanical studies is the force volume imaging that linked the force indentation and surface topography by simultaneously collects a set of information containing topographic data and force-distance curves. This technique is known as force volume mode, in this method each of the force curves is measured in a single position in the scanning zone (x, y) and the force and distance curves are an additional axis or another dimension (different mechanical properties) in order to create 3D-maps. Therefore, the force volume data set looks like a stack of horizontal cuts, a stack of horizontal cuts, each of which represents the force data matrix at a certain height. A single force volume image represents one of these cuts, showing the distribution (x, y) of the force data over the scanning zone at that height (Figure 4.4), force volume mode is similar to optimized mode called PeakForce quantitative nanomechanics (QNM) that works in tapping mode. Thereby, force volume images provide simultaneously useful information about topography and mechanical properties (Figure 4.4). Also, this technique can be used to know how the molecules or cells respond to mechanical stimuli or external forces, and in this way, topography images and individual force curves at any point could be obtained (Jembrek et al., 2015).

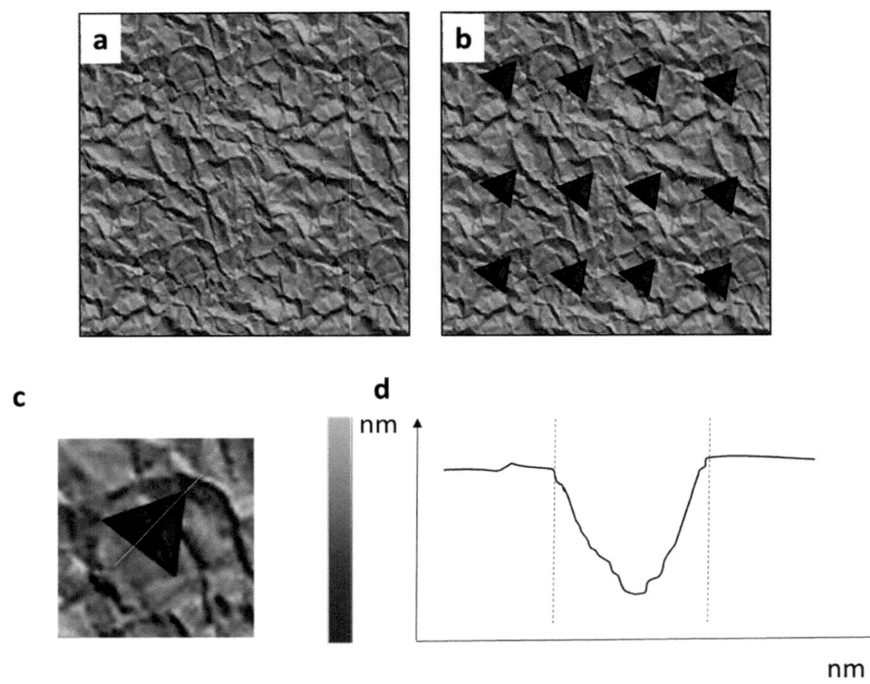

FIGURE 4.6 Example of nanoindentation technique to hard material; (a) before nanoindentation with diamond tip; (b) after, where a matrix of residual fingerprints can be observed; (c) zooming of residual fingerprint after indentation; and (d) height profile of residual fingerprint after indentation.

4.3.3 INTERMITTENT CONTACT MODES: TAPPING, PHASE CONTRAST, PEAKFORCE TAPPING (PFT) AND SCANASYST, PEAKFORCE QUANTITATIVE NANOMECHANICS (QNM)

The tapping mode allow to obtain topographic mapping by having contact with the surface of the sample by means of an oscillating tip, where the cantilever is excited with a piezoelectric controller and is known as resonance oscillation. The cantilever has oscillation amplitude, which changes according to the relief of the sample surface, and the topographical image is obtained by controlling this variation and closing the z feedback loop in order to reduce them. In Figure 4.7(a) it can see a diagram that illustrates the operation of the tapping mode for obtaining images, the variation of the cantilever in its amplitude of oscillation is the feedback signal, in addition the variation also measures a part of the force with that the tip has contact with the surface of the sample. The tapping mode reduces the lateral forces between tip and sample, where the oscillation of the cantilever is at a frequency close to that of its resonance. The essential principle should be that the tip on the cantilever has intermittent contact with the surface of the sample at each oscillation. The amplitude of oscillation, and the association of energy with oscillation, must be enough to overcome the forces of adhesion, and with amplitude control, the force that is applied to the sample increases or decreases, this being the feedback signal. This results in oscillations whose amplitudes are typically from 20 to 100 nm.

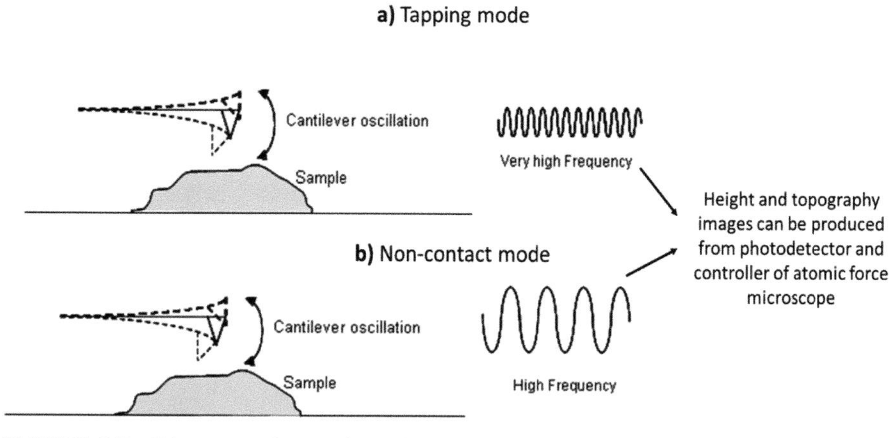

FIGURE 4.7 Diagrams of operation principles of: (a) tapping mode; and (b) non-contact mode of atomic force microscopy.

The tapping mode makes it possible to determine height, 3D (three dimensional) topographic, phase, and PeakForce error images in fragile samples (Figure 4.8(a–d)) that can be damaged by the lateral force which is typically applied in contact mode, also uses higher scanning speeds than contact mode. The tip-sample interactions induce variations of the amplitude as an effect of different forces, such as Van der Waals, electrostatic, chemical, and drag (viscous fluids). These forces decrease the oscillation amplitude and piezoelectric detects the variation to maintain a constant distance between tip-sample, approximately within the order of 1–10 nm, by controlling the loop. Tapping mode is used in cases where it is necessary to overcome the friction or adhesion forces that make it difficult to obtain the image, for instance, is recommended to obtain topography images of biological and soft materials including biopolymers, bacteria, microorganisms, animal, and plant cells in air and liquid media (Mendez, 2010; Escamilla-Garcia et al., 2015).

Phase-contrast image or phase only (currently obsolete term) is considered as a secondary mode of AFM, where the offset of the cantilever oscillation in relation to the signal sent to the piezo-controller is used as the basis for obtaining images. These phase images can be obtained due to changes in material properties like viscoelasticity, adhesion, and friction, that all may cause the phase offset. In addition, these images can be generated at the same time as the microscope is operating in any cantilever vibration mode, such as tapping mode. Variations in phase offset also show modifications in the surface of the sample, specifically in its mechanical properties. The feedback loop of the system operates as usual, i.e., it considers the variations in cantilever deflection or amplitude of vibration to measure the topography of the sample. The phase offset is monitored at the same time as the topographic image is captured (Figure 4.8(c)), so that topography images and material properties are obtained at the same time. This alternative and useful imaging technique has several applications, such as the determination of contaminants, differentiating regions of high or low adhesion and even surface hardness, the mapping of various elements in composite materials, as well as electrical and magnetic properties. Overall, phase imaging mode is useful to characterize the distribution and differences of composition on the surface in different materials. Despite the benefits that offer phase imaging, in the qualitative characterization of composition of materials, in biological material has been scarce applied, probably because they are composed by light elements that do not show significant differences in their molecular weights (MW).

Due to the fact that atomic force microscopes interact directly with the samples, a plethora of operating modes have been developed, in addition the simplicity of its operation allows to observe all kinds of samples in different media (vacuum, air, and liquids), in a lateral resolution range that includes, from 100 microns to 0.1 nm, and to obtain an enormous amount of structural, physical, compositional, mechanical (Figures 4.8(e,f) and 4.9), electrical, magnetic, rheological properties, etc. It is even possible to generate nanometric patterns (nanolithography) and manipulate the samples at the atomic level. This has led to AFM having multiple applications in a wide range of areas of knowledge, so it is currently used to characterize from hard to soft materials, of inorganic and biological nature.

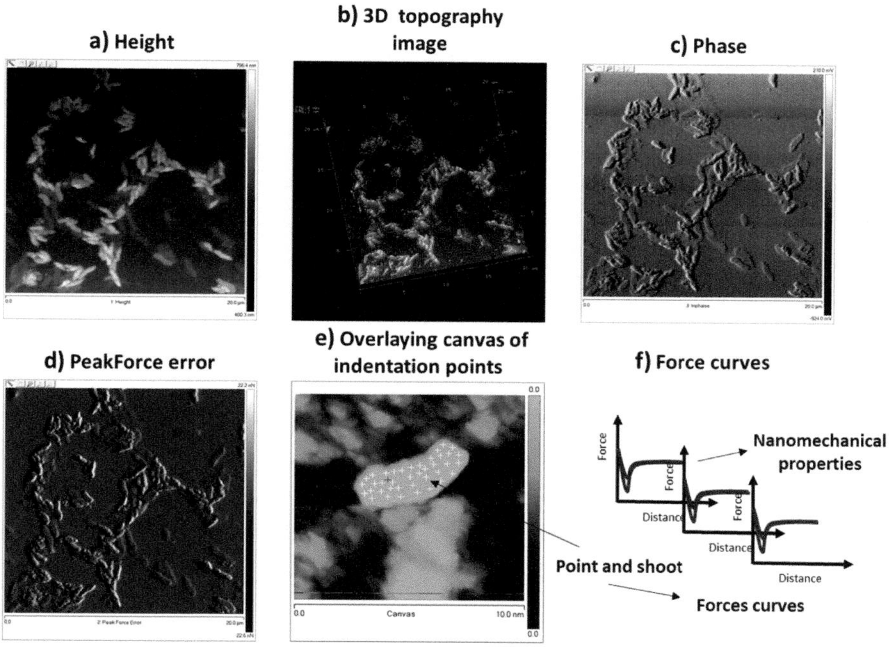

FIGURE 4.8 Image gallery of *Campylobacter jejuni* obtained with PeakForce tapping mode in an atomic force microscope (Bruker, Bioscope Catalyst ScanAsyst, USA): (a) height; (b) 3D topography; (c) phase; (d) PeakForce error; (e) overlaying canvas of indentation points obtained using point and shoot method; and (f) force curves from where several nanomechanical properties can be extracted.

All of this makes of the AFM an all-rounder of the characterization microscopy techniques, which has led several companies dedicated to the development of analytical instruments to generate different patented techniques, algorithms, and operation modes. There are more than 20,000 published articles related to the AFM; above 500 patents have been issued related to various types of microscopies, such as SPM and atomic force microscopes (AFM); many companies are involved in its development, such as Bruker, JPK, Park System, Asylum Research, WITec among others, with a worldwide turnover of $250–300 million per year and approximately 10,000 trading systems sold (Nanowerk, 2008). The wide range of AFM techniques and operation modes can generate confusion in novices' users, but as previously mentioned, three main operation modes are contact, non-contact, and tapping, others are variants of these modes, algorithms, and modes optimized according to the developments of each manufacturer, and specific techniques to measure several properties (magnetic, electrical, mechanical, electrochemical, among others) in all types of materials. Particularly, as for the optimized modes for imaging and measurement of mechanical properties (Figures 4.8 and 4.9), Bruker has introduced two AFM techniques: PeakForce Tapping™ (PFT) and ScanAsyst™ (SA).

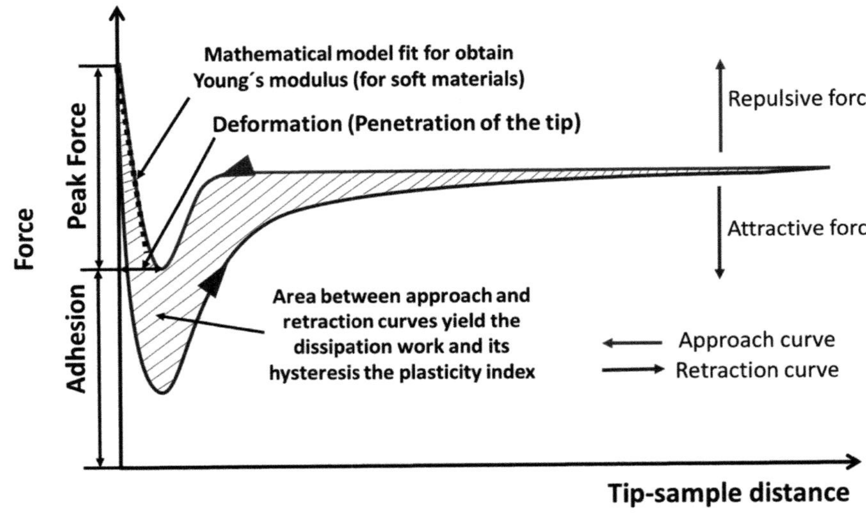

FIGURE 4.9 Diagram of a typical force curve and nanomechanical properties that theoretically can be obtained from different modes of operation of an atomic force microscope: contact, tapping, PeakForce tapping and PeakForce quantitative nanomechanics (QNM).

PeakForce tapping (PFT) is an enhanced AFM operating mode that uses force as a feedback variable to reduce peak-sample interaction forces, while maintaining the scanning speeds used in tapping mode. PFT and TappingMode has a similar operating principle, because both avoid lateral forces and intermittent contact with the sample. However, PFT works differently than TappingMode in that it works with a non-resonance mode. The PFT is an oscillating technique that joint the advantages of contact and TappingMode imaging modes, mainly in terms of direct force control and prevention of lateral force damage. The difference with the conventional force curve and the force volume imaging is that the z position is modulated by a sine wave, which avoids unwanted resonances at the tipping points.

It is possible to obtain continuous force curves at frequencies from 1 kHz to 10 kHz, which also allows image speeds in the TappingMode range. Therefore, the most important differences between the two modes are: (1) At the PFT operating frequency, the tip oscillates outside the resonance frequency (1 to 10 kHz), which is lower than the standard tapping mode frequency in fluids, and (2) the parameter that remains continuous in the feedback cycle in the PFT is the maximum peak load force, but not the amplitude of the cantilever oscillations. Image speed and resolution in both modes are similar. The amplitude of oscillation of the cantilever in PFT has variations from 0.1 nm to 3 μm peak to peak. The general operation principle is illustrated in Figure 4.10.

ScanAsyst is an optimized operating mode that automates AFM imaging, automatically adjusts scanning parameters such as set-points, feedback gains, and scanning speed. The algorithm continuously inspects the image quality status to make the appropriate parameter settings. This makes it easy to obtain images by simply selecting the region and scanning area in either air or fluid operation. ScanAsyst uses the same mechanism as PFT, to make an automatic adjustment of all critical parameters of the image. Because peak force feedback directly controls the interaction force, there is no difference for the peak force of the soft or hard areas of the sample. The direct interaction force control provides the option of delivering an evenly optimized feedback loop for all areas of the heterogeneous sample. A real-time feedback loop continuously inspects and adjusts the gain to keep the data within a predefined noise level (as opposed to manual gain adjustment, where a gain is commonly used for a full image). ScanAsyst improves the gain according to the current state of the sample in different sections. The ScanAsyst algorithm also improves the setpoint to the minimal force needed to scan the sample (surface), controls the scanning speed and if necessary, automatically reduces the z limit. As a result, it displays high quality images without the need to adjust external or user image parameters.

4.3.4 PEAKFORCE QUANTITATIVE NANOMECHANICS (QNM)

PeakForce quantitative nanomechanics (QNM) is a valuable and quantitative AFM technique to obtain nanomechanical properties of biological samples such biopolymers and bacteria, with an image acquisition speed similar to tapping mode, since QNM operate with both PFT and ScanAsyst mode. So, from PeakForce-QNM different imaging of nanomechanical properties can be obtained such as PeakForce, DMTModulus, LogDMTModulus, adhesion,

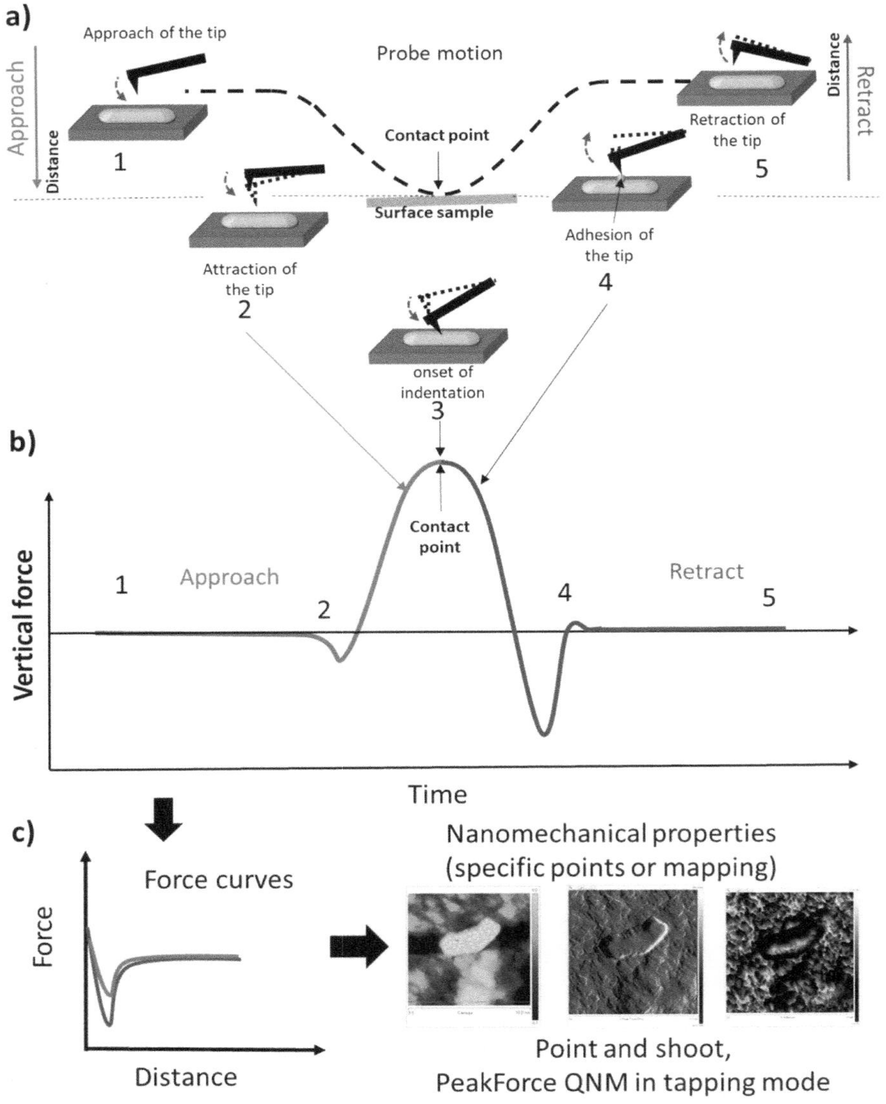

FIGURE 4.10 General operation principle of PeakForce tapping: (a) probe motion and different position of the tip during the interaction tip-sample; (b) plot of force and piezo Z position as a function of time; (c) force curves and different images where is possible obtain several nanomechanical properties by indentation of specific regions or mapping of whole image by means of point and shoot method or PeakForce quantitative nanomechanics (QNM) in PeakForce tapping.

deformation, and dissipation, the information that can be extracting from PeakForce-QNM overcomes other AFM modes. With PeakForce-QNM, it is possible to obtain quantitative nanomechanical data directly from organic or biological surfaces without damaging them. In PeakForce-QNM tapping, the probe oscillates at a lower frequency, meantime force curves are generated each time the AFM tip contacts the sample surface (Figure 4.10(c)). As the probe oscillates, a force curve is recorded for each pixel in the image. Several quantitative parameters can be obtained in real-time from each force curves (Figures 4.9 and 4.10). Nowadays, PeakForce-QNM is the most effective AFM technique for characterization of nanomechanical properties (Jembrek et al., 2015). It has huge potential for novel applications in life sciences, specifically in microbiology, for instance, it is possible to evaluate the morphology, topography, and nanomechanical properties of diverse microorganisms, including different bacterial species, in native state or when they are subjected to antimicrobials, in this way the effects of these compounds on the structure and its cell mechanics can be estimated quickly and efficiently. An example is provide in Figure 4.11 where is shown an image gallery of *Campylobacter jejuni* in native state was obtained with PFT mode in an atomic force microscope (Bruker, Bioscope Catalyst ScanAsyst, USA), using the PeakForce-QNM module, where images of height, 3D-topography, and mappings of PeakForce, DMTModulus, LogDMTModulus, adhesion, deformation, and dissipation are showed, in these images each point representing the mechanical properties and a force curve for the entire surface of a bacterium can be obtained directly from the microscope software.

4.3.5 NON-CONTACT MODE

The non-contact mode has a unique advantage over other contemporary scanning probe techniques, such as AFM contact mode and STM (Martin et al., 1987). The lack of repulsive forces (presented in contact mode) in non-contact samples leads to a use in soft image samples. This mode operates by the detection principle known as amplitude modulation. The corresponding detection scheme takes advantage of the change in the amplitude of the oscillation of a cantilever, due to the tip-sample interaction to create the image (Figure 4.7(b)). The cantilever oscillates on the surface of the sample at a distance within the attraction dependence of the intermolecular force curve. However, it uses variations in van der Waals forces that pull the tip by attraction during the approximation to the surface to determine the topography of the surface (Jembrek et al., 2015), and are detected by an oscillating probe, which is excited at its resonance frequency by a piezoelectric and is approximate to the sample. Commonly the probe oscillation amplitudes are within the range of 1 nm and even less, and furthermore, the resolution in this mode is defined by the tiny separation (in nanometers) between tip and sample. Operation requires more precise control of the tip force and no contamination of the surface at distances close to the sample. This technique is used mostly in high vacuum. Usually, the cantilever selected for this mode should be with a high spring constant in the range of 20–100 N/m, so that the surface of the sample does not adhere to low amplitudes. The cantilevers used commonly in this mode are faster and less noisy, but low frequency cantilevers are used with AFM systems that do not tolerate probes with short cantilever arms. So, the imaging in non-contact mode is done with silicon probes which are normally used for operation in tapping mode.

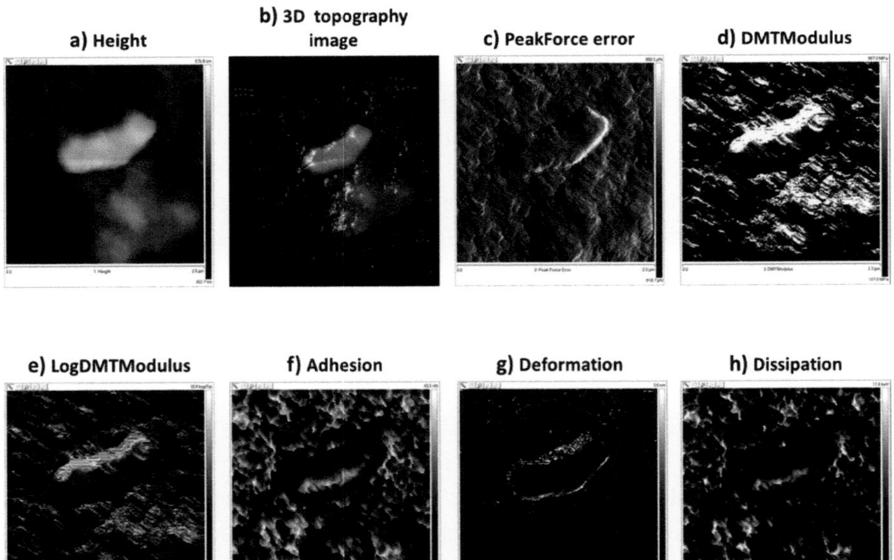

FIGURE 4.11 Image gallery of *Campylobacter jejuni* obtained with PeakForce tapping mode (Bruker, Bioscope Catalyst ScanAsyst, USA), using PeakForce quantitative nanomechanics (QNM) module, where: (a) and (b) correspond to imaging technique, while from (c) to (h) are mapping of several nanomechanical properties of an isolated bacterium.

4.4 STRATEGIES FOR BACTERIAL CELL FIXATION

The immobilization of the bacterial cells is an important step in the samples preparation for the AFM analysis, both to obtain the topographic imaging and the force curves of bacterial cells under physiological conditions, in order to withstand the friction by the vertical and lateral shear forces of the scanning tip in operation. In addition, it is required that the immobilization technique does not affect the chemical and structural integrity of the cells surface. There are several and different mechanisms that have been used for bacteria immobilization. The commonly used bacteria immobilization method by the AFM studies is the chemical fixing across electrostatic interactions employing poly-L-lysine or poly-ethylene imide (PEI), the glass surfaces are coated with these polymers creating positive charges that promote an irreversible adhesion of the bacterial cells to the substrate (Figure 4.12(a)), other strategy to increase the adherence of the bacteria to the glass substrate is the use of gelatin or glutaraldehyde (Figure 4.12(b)); it has also been used others coated substrates with these substances to attach cells, such as, gold, mica, and microstructured polydimethylsiloxane (PDMS). In addition, there is a mechanism to immobilize the bacteria through cross-linking the carboxyl groups of their surfaces with amine groups coupled to the glass slides treated with aminosilanes (Figure 4.12(c)) (Vadillo-Rodríguez et al., 2004; Meyer et al., 2010). For example, Chopinet et al. (2013) immobilized *Escherichia coli* cells on PEI coated glass slides under physiological conditions, while, Fantner et al. (2010) and Raman et al. (2011) applied glass coverslips coated with poly-L-lysine to increase the adherence of *E. coli* to the substrate; Vadillo-Rodriguez et al. (2004, 2008) also used poly-L-lysine-coated slides for *Klebsiella terrigena* and *Pseudomonas aeruginosa* immobilization, Van Der Hofstadt et al. (2015) immobilized *E. coli* in three types of substrates, glass, gold, and mica, all of them coated with gelatin, and, Yamashita et al. (2012) inoculated *Magnetospirillum magneticum* on mica coated by poly-L-lysine and glutaraldehyde to attach the cells to the substrate.

Another immobilization technique of the bacterial cells is by their fixation to the AFM cantilever tip, for this method is necessary the functionalization of a cantilever with a confluent layer of bacteria for probing them against an interest surface. This methodology is not simple and, hence, previous treatment of both the bacteria cells and the cantilevers is required, the bacteria are treated with glutaraldehyde solution, and for the preparation of cantilevers tip is necessary to coat it with poly-L-lysine, PEI, or adhesive polymer poly-dopamine solutions for the firm attachment of bacterial cells (Vadillo-Rodríguez et al., 2004; Herman et al., 2013; Sullan et al., 2014). Additionally, some authors reported that the bacteria have also been immobilized by mechanical trapping on porous polycarbonate membrane or aluminum oxide filters with pore sizes lightly smaller than the bacterial cells dimensions for attaching them into the pores when filtering the bacterial suspension (Yao et al., 2002; Vadillo-Rodríguez et al., 2004; Gillis et al., 2012). Within this strategy of mechanical immobilization, Kailas et al. (2009) employed a novel way for anchoring bacterial cells across the entrapment in lithographically patterned substrates (Figure 4.13) that achieve that the AFM cantilever tip will not push the cells off during imaging, while allowing the bacteria to continue with their cellular processes since the cells are not perturbed cause there is not chemical linkage to the surface and they are no trapping in porous membranes.

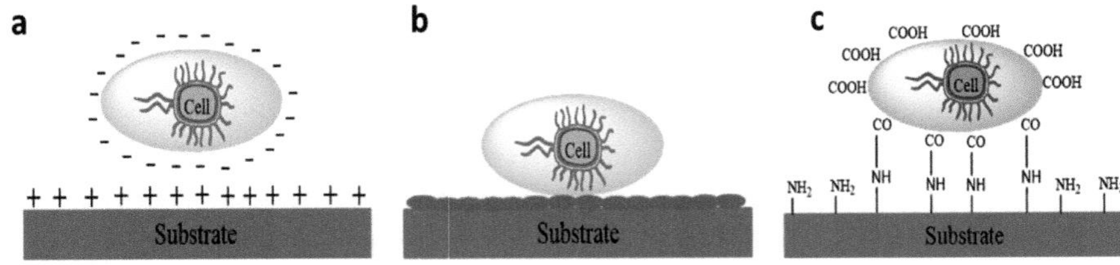

FIGURE 4.12 Schematic representation of the immobilization techniques most commonly used. (a) Attractive electrostatic interactions in positively charged glass surfaces coated by poly-L-lysine or PEI; (b) attachment to surfaces coated by gelatin or glutaraldehyde; and (c) covalent binding between amine groups-functionalized glass surfaces and carboxyl groups present in cell surfaces.

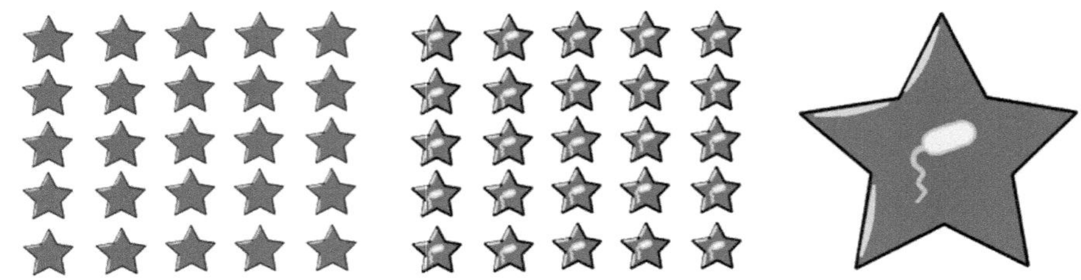

FIGURE 4.13 Diagram of bacterial cells trapped in holes of the lithographically patterned substrate; (a) empty; (b) with bacteria; and (c) zooming.

4.5 TIP SELECTION

The AFM consists of different elements (Figure 4.1(a)) that work together to obtain good resolution images. One of these elements is the cantilever beam in which there is a very sharp point at the free end. This tip interacts with the surface of the samples, which allows it to obtain topographic, phase, deflection, and mechanical images, etc. The cantilevers are disposable since they deteriorate by contact with the surface of the samples, getting dirty or damaged, which reduces the quality of the images obtained, for this reason, should be changed when it is suspected of being damaged. The probes are generally manufactured of silicon nitride or silicon. Nevertheless, they can be made of other materials such as diamond that are used for nanolithography and to indent hard materials.

In the market, there is a wide variety of probes manufactured by different companies that recommend certain cantilever models for each application and AFM operation mode. However, it is necessary to consider some basic issues for a preliminary selection or preparation of the probes. A very important parameter is the cantilever beam stiffness (k), which is related to the range of force that can be applied in a sample. Low values indicate that the range of force that can be applied is low, which is necessary in the case of working with sensitive samples such as bacteria, cells, DNA, and proteins without fixing, etc. On the other hand, high values indicate that it is possible to apply a relatively greater force range, which is useful in the case of rigid samples, such as silicon wafers, polymers, and metals, etc. Another important parameter is the diameter of the tips, which is generally of the order of nanometers. The selection of the diameter of the tip is guided by the size of the details to be observed. As a rule, with a smaller diameter, a better resolution is obtained, however, for soft samples such as bacteria *in vivo*, the small diameter of the tips can tear the surface of the samples reducing the quality of the images obtained. Another important aspect is the AFM is the operation mode because manufacturers have designed specific cantilevers for some of the modes according to their operating principles and the algorithms used in the equipment to ensure good results. Another key issue is the chemical functionalization of the tips to improve or avoid the adhesion of the tip with the sample according to the objective of the investigation.

Thus, to exemplify the types of cantilevers are used in bacteria, some studies are described as follows. Vadillo-Rodríguez et al. (2004) functionalized the cantilever tip with poly-L-lysine (PEI) to create positively charged glass surfaces, improving the adhesion of the bacteria to the tip. The cantilever specifications used in this work were, k 0.06 N/m and tip diameter of 50 nm. Schaer-Zammaretti and Ubbink (2003) studied mechanical characteristics and the adhesion of lactic acid bacteria. The images were obtained in contact mode applied a force below 1 nN, and scan rated between 1 Hz and 2.5 Hz. The adhesion and the elasticity were obtained using force volume mode, matrixes of 32 x 32 force *vs* distance curves were analyzed in each area. A Si_3N_4 cantilever was used with a spring contact of 0.06 N/m. Fantner et al. (2010) obtained images in tapping mode of live bacteria in an aqueous environment using high-speed AFM with a cantilever model NPS, spring constant of 0.06 N/m, 4 V shape cantilevers and tip radius 10 nm.

4.6 RECENT STUDIES IN THE IMAGINING AND NANOMECHANICAL PROPERTIES

There are several studies using AFM imaging and force spectroscopy techniques to study the morphology, structure, and nanomechanical properties of living bacteria and their components such as cell wall and flagella, bacteria interaction with some antibacterial agents, obtaining valuable information at nanometer scales, this information is necessary to comprise the physiological processes and to help elucidate the relationship between cell surface structure and its function, in presence with and without antimicrobials.

4.7 IMAGING

Fernandes et al. (2009) studied the relationship between the MW of chitosan and its antibacterial effect in vegetative cells and spores of *Bacillus cereus* by means of topography images and nanoindentation using AFM. Topography images showed that when chitosan of higher MW at 100 and 628 kDa are applied to vegetative cells and spores of *Bacillus cereus* a polymer layer is formed in the bacteria surfaces that modified drastically the bacterial morphology, which led to the death of the vegetative form, as it did not allow the absorption of nutrients, although this layer has no affected to the spores, because they can survive for long periods without nutrients. While, when chitooligosaccharides (COS) of low MW (< 3 kDa), caused more visible damage to the vegetative form of *B. cereus*, probably due to cell penetration by COS. Using only COS on *B. cereus* spores is not enough to destroy many cells, but it does weaken the structure of the spore and its capacity to contaminate by inducing the loss of exospores. Nanoindentation studies showed that the spores decreased in cell stiffness as COS concentration increased, specifically to a concentration of 0.5% (w/v); however, these are considerably stiffer than vegetative cells under all conditions. This mostly reflects the weakening of the spore wall, due to the loss of the exospores and probably other inner layers of protection. Finally, the average results in vegetative cells of *B. cereus* showed no trend in cell stiffness when COS concentration increase.

In another study, Fantner et al. (2010) analyzed the individual kinetics of bacterial cell death using a high-speed AFM optimized to visualize live cells in real-time. With a higher time resolution (13 s/image), it was possible to characterize the initial stages of the action of an antimicrobial peptide in *E. coli* cells at nanoscale resolution. The results indicated that the death process is a combination of a time variable incubation phase (lasting seconds or minutes to complete) with a faster execution phase. The most logical effect with the addition of the peptide was a change in the surface state of the bacterium, from smooth to rough surfaces, which allowed the inactivation of bacterial cells. The observation of this two-stage process was due to the high-speed AFM capable of recording dynamic changes at the level of a single cell.

While, La Storia et al. (2011) evaluated the antimicrobial activity of carvacrol in some Gram-positive and Gram-negative bacterial strains linked to food (*Brochothrix thermosphacta, Carnobacterium maltaromaticum, Lactobacillus plantarum, Listeria innocua, E. coli, Hafnia alvei, Pseudomonas fragi,* and *Serratia proteamaculans*), and studied its effect on the bacterial cell cove by AFM. The cells were treated with carvacrol, and the images were analyzed to observe the effect of the treatment and to obtain both the roughness values of the cell surface and some morphology parameters of the cell such as length and width, by image analysis. In addition, the results showed that all bacteria used were sensitive to carvacrol in solid and liquid media since cell images of all carvacrol-treated strains had modifications in cell surface structure, therefore, additionality in length and diameter of bacteria decreased upon contact with carvacrol.

Yamashita et al. (2012) also used high-speed AFM in liquid media to visualize the molecular dynamics of isolated proteins at sub-molecular resolution on the surfaces of living *Magnetospirillum magneticum* AMB-1 bacterial cells, which were fixed in mica modified with poly-L-lysine and glutaraldehyde. Surface images of the living cells showed that the bacteria's outer membrane was covered by a net-shape structure that contained holes and the edges of the holes framed them, showing that the holes were slowly diffusing into the cell surface. The nanodissection showed that the porine trimmers form the structure in net-shape. Providing for the first time the direct observation of dynamic molecular architectures on the surface of individual living cells using high-speed AFM.

Gillis et al. (2012) used the AFM to obtain information on the morphology of the surface at nanoscale and on the motility of two wild strains and four mutants of *Bacillus thuringiensis* that showed several levels of flagellation, they used the AFM in air, demonstrating that it is a simple and reliable method to observe the morphological details of the bacteria and to quantify the density and the dimensions of their flagella, allowing to perceive three levels of flagellation: absent, moderate, and abundant. This study provided novel information on flagella in Gram-positive bacteria, demonstrating the scope of AFM in genetic studies for rapid assessment of phenotypic characteristics of altered bacterial strains.

Van Der Hofstadt et al. (2015) focused on nanoscale *in situ* research of the growth and division of individual bacterial cells from two different strains of *E. coli* (MG1655) and enteroaggregative *E. coli* (042) fixed into planar substrates and observed with AFM. For this study, the lateral shear forces responsible for the release of weakly adsorbed bacteria on planar substrates were reduced by using the dynamic jump mode with very soft cantilever probes. These results showed that the dynamic AFM jump mode is a very useful technique for the observation of physiological processes of bacteria that are weakly adhered to coated planar substrates. In addition, soft surface imaging conditions can be maintained for long periods of time, allowing continuous imaging of bacterial cell growth and division, even on planar substrates.

4.8 NANOINDENTATION AND NANOMECHANICAL PROPERTIES

Vadillo-Rodríguez et al. (2004) compared the interaction forces obtained between *Klebsiella terrigena* and silicon nitride, using the methods of mechanical trapping of bacteria in membrane filters, physical adsorption of negatively charged bacteria to a positively charged surface and fixing of glutaraldehyde bacteria to the AFM tip. They demonstrated that the use of different techniques for sample preparation generates different interaction forces. Also, physical adsorption of bacterial cells considering modified substrates promotes rearrangements in their surface structures and treatment with glutaraldehyde showed that it induces physical-chemical and mechanical changes in the surface properties of bacterial cells. Mechanical trapping of each bacterial cell in the filters proved to be the most suitable method for fixation.

In another study Vadillo-Rodriguez et al. (2008) measured the mechanical properties of surface viscoelasticity in individual gram-negative *Pseudomonas aeruginosa* PAO1 cells by AFM, applying to the cell a constant compressive force in aqueous conditions, and measured the time-dependent displacement (creep) of the V-shaped AFM tip. They stated that the cell response is correctly described by a mechanical model that provide an effective cell spring constant, k_1, and an effective time constant, τ, for creep deformation. With the addition of glutaraldehyde, it is being an agent that increases the covalent bond of the cell surface and generates a significant increase of k_1 and a decrease in τ.

Raman et al. (2011) applied a dynamic AFM method to obtain quantitative maps of nanomechanical properties, surface elastic response, and adhesion of living cells, with a performance measured in pixels/minute that is 10–1,000 times greater than that achieved with quasi-static AFM techniques. So, in this study it is presented a new method using multiharmonic channels in the amplitude modulation of the AFM liquid environment to obtain quantitative maps and spatial patterns of the local mechanical properties of living cells at an imaging yield 10–1,000 times better than the standard force-volume method and have demonstrated the method with three completely different cell lines, bacterial cells, red blood cells; and rat fibroblast cells.

Chopinet et al. (2013) compared the quantitative AFM imaging mode, which allows acquisition of high-resolution maps and mechanical properties in soft samples, against the contact imaging mode and the force volume mode, applying six different models of living cell samples: *E. coli*, *Candida albicans*, *Aspergillus fumigatus*, Chinese hamster ovarian cells and their isolated nuclei and human colorectal tumor cells, from which they reported that the quantitative imaging mode allows to visualize and measure simultaneously soft or slightly fixed samples, in a fast way and with high resolution, giving results of their force curves that can be analyzed by any model. The versatility of this method allows it to be the most suitable for kinetic studies in complicated biological samples, for cases where image data, or mechanical/adhesive properties, or both, are required.

Baniasadi et al. (2014) applied nanoindentation with AFM and finite element analysis (FEA) to characterize the mechanical properties of a human-derived bacterial biofilm of *Pseudomonas aeruginosa*. AFM and scanning electron microscopy images showed that the individual bacteria are approximately 1.5–1.6 μm and 800–900 nm, in length and width, respectively. The nanoindentation was performed with a spherical tip, and this showed in its results that the elastic modules of the biofilm vary from 40 to 45 kPa. The heterogeneity presented in the mechanical properties was attributed to the variation of the packing density of the bacteria in different areas of the biofilm and to the existence of water channels in the structure of the biofilm. Using FEA, the effect of the placement of a single bacterium in the mechanical properties under nanoindentation was obtained, showing that the heterogeneity of the mechanical properties is based on the interaction of the probe tip with a few bacteria embedded in the extracellular matrix.

Chang and Liu (2018) studied by AFM the mechanical properties of hyperelastic materials, including some polymers and bacterial cells (*S. aureus* and *P. aeruginosa*), by nanoindentation its elastic modulus (E) was estimated with Hertz's model, and found that the conventional theory of contact mechanism resulted in a high uncertainty of E. They proposed a new equation for the calculation of E by directly adopting the mechanical behavior of the hyperelastic material, PDMS, and in microbial samples (*S. aureus* and *P. aeruginosa*) its equation was applied and showed results consistent with their membrane structures, and the model improved the description of its mechanical behavior and yielded more accurate E values of these materials.

The following tables summarize various studies based on the use of AFM for bacterial cells, which are considered relevant and were previously cited. They also provide relevant information on the types of materials, equipment, type of tips, and the mode of operation used to develop them. Therefore, Table 4.1 highlights the work on topographic images, while Table 4.2 shows the work related to measurements of nanomechanical properties.

4.9 CONCLUSIONS

AFM is a technique widely used in many investigations for its multiple advantages, mainly for its ability to obtain descriptions of the morphological structures of samples at nanometric scales, as well as their characteristics and mechanical properties; in the study of bacterial cells, it has been a very important tool to observe the structural and mechanical properties of their surfaces and follow their behavior under the effect of external stimuli, such as interaction with antimicrobial agents *in vivo* and in real-time. Although AFM techniques are already widely used, their applications and tools continue to grow with the development of new methodologies to acquisition of bacterial imaging at high resolution and force spectroscopy with very fast scanning. Thus, we believe that surely the AFM techniques will continue to be increasingly applied to develop new agents that inhibit or retard the growth of microorganisms by studying their mechanisms of action on cell surfaces. Moreover, nowadays, AFM coupled with Raman spectroscopy and confocal microscopy are interesting tools to study the bacteria at the nanoscale, in the future is foreseen a vertiginous increase of the researches with these powerful analytical devices.

TABLE 4.1 Compilation of Relevant Works about Topographic Imaging Obtained by AFM in Bacterial Cells

Biomaterials	Microscope and Cantilever	Software	Operation Mode	Equation Model	Explanation	References
Bacillus cereus	Multimode IVa (Veeco), scanner j-type Fresh silicon cantilevers (AppNano)	PUNIAS (P. Carl, P. Dalhaimer)	Bacterial morphology and selection of viable bacteria, tapping mode in air. Nanoindentation in the bacteria center, contact mode	Hertz model	They studied the antibacterial effects of chitosans and chitooligosaccharides (COS) on *B. cereus* and its spores by AFM. They observed through the topographic images that chitosans inactivated the cells but did not affect the spores and COS was not enough for the destruction of cells, but it debilitated the spore structure. Additionally, they evaluated the effect of the COS on cell rigidity by nanoindentation.	Fernandes et al. (2009)
Escherichia coli	Multimode with Nanoscope 5 controller (Veeco) Si_3N_4 cantilevers (NPS-D, Veeco)	Image J, image SXM and Gwyddion	Imaging in tapping mode in fluid	n/a	They studied the antimicrobial peptide dynamic activity on single *E. coli* cells using high-speed AFM. They could observe by the imaging that the cells death process is a combination of two-stage process, an incubation phase, and an execution phase.	Fantner et al. (2010)
Brochothrix thermosphacta, *Carnobacterium maltaromaticum*, *Lactobacillus plantarum*, *Listeria innocua*, *Escherichia coli*, *Hafnia alvei*, *Pseudomonas fragi*, and *Serratia proteamaculans*	EasyScan 2 (Nanosurf AG)	Scanning probe image processor (image metrology A/S)	Topographic and phase images in non-contact mode in air	n/a	They studied the effect of carvacrol on the cells of several bacterial strains by means of the AFM, and they reported that the bacteria treated were sensitive to carvacrol since exhibited appreciable changes in their cell structure and dimensions.	La Storia et al. (2011)

TABLE 4.1 *(Continued)*

Biomaterials	Microscope and Cantilever	Software	Operation Mode	Equation Model	Explanation	References
Magnetospirillum magneticum	Microscope ensembled in Laboratory Small cantilevers	SPM simulator (advanced algorithm systems)	Images in tapping mode in liquid medium	Simple hard-sphere model	They employed high-speed AFM to visualize the dynamics of molecular complexes on living *M. magneticum* cell surfaces. The images showed that the bacterial external membrane is covered by a net-like structure and dynamic moves of membrane proteins.	Yamashita et al. (2012)
Bacillus thuringiensis	Nanoscope V multimode and MSCT cantilevers (Veeco)	n/a	Images in contact mode in liquid and air media	n/a	They obtained AFM imaging of six *B. thuringiensis* strain producing different quantity of flagella, and they observed that the flagella quantity expressed by the strains is correlated with their microscopic motility.	Gillis et al. (2012)
Escherichia coli MG1655 and enteroaggregative *Escherichia coli* 042	Cervantes microscope (Nanotec Electronica) Biolevers BL-RC150VB-C1 (Olympus)	WSxM 5.0 Develop 6.5	Bacterial imaging in dynamic jumping mode plus (Nanotec Electronica) in liquid media	n/a	They obtained AFM imaging of live cells of two *E. coli* strains employing the dynamic jumping mode in order to reduce the horizontal shear forces of the tip. They reported that this mode has the advantage to observe living processes of growth and division of bacteria that are weakly attached to planar substrates.	Van Der Hofstadt et al. (2015)

Note: n/a: not applicable.

TABLE 4.2 Compilation of Relevant Works about Nanomechanical Properties Measurements by AFM in Bacterial Cells

Biomaterials	Microscope and Cantilever	Software	Operation Mode	Equation Model	Explanation	References
Klebsiella terrigena	Nanoscope III digital instrument V-shaped Si_3N_4 cantilevers (Park Scientific)	n/a	Individual force curves in the bacteria top center in contact mode	n/a	They obtained the force curves of *K. terrigena* cells employing three commonly immobilization methods, and they demonstrated that different sample preparation techniques result in different interaction forces.	Vadillo-Rodríguez et al. (2004)
Pseudomonas aeruginosa	Asylum MFP-3D (Asylum Research) V-shaped Si_3N_4 cantilevers OTR4 (Veeco)	n/a	Cells imaging in contact mode, force curves in the bacteria top center	Standard solid model	They studied the mechanical properties of individual *P. aeruginosa* cells by applying a constant force to the cell and measured the time-dependent tip displacement to know the cells viscoelastic properties. They reported that the cell response is correctly described by a viscoelastic mechanical model.	Vadillo-Rodriguez et al. (2008)
Escherichia coli, rat fibroblasts and human red blood cells	MFP-3D (Asylum Research) Soft cantilevers TR/RC400 and TR/RC 800 (Olympus)	n/a	Imaging in tapping mode in liquid environment	Hertzian model	They mapped the mechanical properties of living cells employing a new dynamic AFM technique developed by them, and they demonstrated that the method could be used to study the mechanical changes in tumors, cells, and biofilm with a resolution around 10 nm.	Raman et al. (2011)
Escherichia coli, *Candida albicans*, *Aspergillus fumigatus*, Chinese hamster ovary cells and their	NanoWizard 3 (JPK instrument) and catalyst AFM (Bruker) MLCT cantilevers	Data processing (JPK instrument) OpenFovea	Experiments in quantitative imaging, contact, and force volume	Hertz model	They compared the quantitative AFM imaging mode of six different models living cells with two more modes, and they reported that this mode can obtain simultaneously image and measure soft	Chopinet et al. (2013)

TABLE 4.2 *(Continued)*

Biomaterials	Microscope and Cantilever	Software	Operation Mode	Equation Model	Explanation	References
isolated nuclei, and human colorectal tumor cells	(Bruker)		modes in liquid environment		samples or weakly immobilized, by fast and with good resolution obtaining force curve data that can be analyzed by any model.	
Pseudomonas aeruginosa bacterial biofilm	MFP-3D (Asylum Research) Soft triangular SiO_2 cantilever (sQube)	Data analysis by MATLAB finite element analysis by FEA ABAQUS	Nanoindentation in liquid environment	Hertzian contact model	They studied the mechanical properties of *P. aeruginosa* biofilm by AFM nanoindentation. The nanoindentation experiments obtained with a spherical tip exposed that the biofilm elastic module oscillates between 40 kPa and 45 kPa.	Baniasadi et al. (2014)
Staphylococcus aureus and *Pseudomonas aeruginosa*	Dimension icon system and 6 different probe types (Bruker)	n/a	Nanoindentation in contact mode	Hertz and flat-punch model	They compared the Hertz and the flat-punch model proposed by them with the mechanical behavior of PDMS polymer and *S. aureus* and *P. aeruginosa* cells by nanoindentation. Hertz resulted in a high uncertainty of E data, while the second model improved the description of mechanical properties and obtained the correct E for the materials since it shown consistent results with their membrane structures.	Chang and Liu (2018)

Note: n/a: not applicable.

ACKNOWLEDGMENTS

Héctor Hiram Torres-Ventura wishes to thank CONACyT, BEIFI, and Instituto Politécnico Nacional (IPN) in Mexico for the scholarship provided during his PhD studies, and the financial support provided by CONACyT projects (239899, 268660) and Secretaría de Investigación y Posgrado-IPN (20181349, 20180455, 20195428, and 20195198).

KEYWORDS

- **bacterial imaging**
- **cantilever tip**
- **force spectroscopy**
- **high-speed AFM**
- **immobilization**
- **nanoindentation**
- **nanomechanical properties**
- **PeakForce tapping**
- **quantitative imaging**
- **quantitative nanomechanics**

REFERENCES

Aguayo, S., Strange, A., Gadegaard, N., Dalby, M. J., & Bozec, L., (2016). Influence of biomaterial nanotopography on the adhesive and elastic properties of *Staphylococcus aureus* cells. *RSC Advances, 6*(92), 89347–89355. doi: 10.1039/c6ra12504b.

Athanasiadou, D., Jiang, W. G., Goldbaum, D., Saleem, A., Basu, K., Pacella, M. S., & McKee, M. D., (2018). Nanostructure, osteopontin, and mechanical properties of calcitic avian eggshell. *Science Advances, 4*(3). doi: 10.1126/sciadv.aar3219.

Baniasadi, M., Xu, Z., Gandee, L., Du, Y. J., Lu, H. B., Zimmern, P., & Minary-Jolandan, M., (2014). Nanoindentation of *Pseudomonas aeruginosa* bacterial biofilm using atomic force microscopy. *Materials Research Express, 1*(4). doi: 10.1088/2053-1591/1/4/045411.

Boussu, K., Van, D. B. B., Volodin, A., Snauwaert, J., Van, H. C., & Vandecasteele, C., (2005). Roughness and hydrophobicity studies of nanofiltration membranes using different modes of AFM. *Journal of Colloid and Interface Science, 286*(2), 632–638. doi: 10.1016/j.jcis.2005.01.095.

Cardenas-Perez, S., Chanona-Perez, J. J., Mendez-Mendez, J. V., Arzate-Vazquez, I., Hernandez-Varela, J. D., & Vera, N. G., (2019). Recent advances in atomic force microscopy for assessing the nanomechanical properties of food materials. *Trends in Food Science and Technology, 87*, 59–72. doi: 10.1016/j.tifs.2018.04.011.

Cardenas-Perez, S., Chanona-Perez, J. J., Mendez-Mendez, J. V., Calderon-Dominguez, G., Lopez-Santiago, R., & Arzate-Vazquez, I., (2016). Nanoindentation study on apple tissue and isolated cells by atomic force microscopy, image and fractal analysis. *Innovative Food Science and Emerging Technologies, 34*, 234–242. doi: 10.1016/j.ifset.2016.02.004.

Cardenas-Perez, S., Mendez-Mendez, J. V., Chanona-Perez, J. J., Zdunek, A., Guemes-Vera, N., Calderon-Dominguez, G., & Rodriguez-Gonzalez, F., (2017). Prediction of the nanomechanical properties of apple tissue during its ripening process from its firmness, color and microstructural parameters. *Innovative Food Science and Emerging Technologies, 39*, 79–87. doi: 10.1016/j.ifset.2016.11.004.

Chang, A. C., & Liu, B. H., (2018). Modified flat-punch model for hyperelastic polymeric and biological materials in nanoindentation. *Mechanics of Materials, 118*, 17–21. doi: 10.1016/j.mechmat.2017.12.010.

Chopinet, L., Formosa, C., Rols, M. P., Duval, R. E., & Dague, E., (2013). Imaging living cells surface and quantifying its properties at high resolution using AFM in QI (TM) mode. *Micron, 48*, 26–33. doi: 10.1016/j.micron.2013.02.003.

Escamilla-Garcia, M., Alvarado-Gonzalez, J. S., Calderon-Dominguez, G., Chanona-Perez, J. J., Mendez-Mendez, J. V., Perea-Flores, M. D., & Farrera-Rebollo, R. R., (2015). Tools for the study of nanostructures. *Food Nanoscience and Nanotechnology*, 5–38. doi: 10.1007/978-3-319-13596-0_2.

Fantner, G. E., Barbero, R. J., Gray, D. S., & Belcher, A. M., (2010). Kinetics of antimicrobial peptide activity measured on individual bacterial cells using high-speed atomic force microscopy. *Nature Nanotechnology, 5*(4), 280–285. doi: 10.1038/nnano.2010.29.

Fernandes, J. C., Eaton, P., Gomes, A. M., Pintado, M. E., & Malcata, F. X., (2009). Study of the antibacterial effects of chitosans on *Bacillus cereus* (and its spores) by atomic force microscopy imaging and nanoindentation. *Ultramicroscopy, 109*(8), 854–860. doi: 10.1016/j.ultramic.2009.03.015.

Garrett, T. R., Bhakoo, M., & Zhang, Z. B., (2008). Bacterial adhesion and biofilms on surfaces. *Progress in Natural Science-Materials International, 18*(9), 1049–1056. doi: 10.1016/j.pnsc.2008.04.001.

Gillis, A., Dupres, V., Delestrait, G., Mahillon, J., & Dufrene, Y. F., (2012). Nanoscale imaging of *Bacillus thuringiensis* flagella using atomic force microscopy. *Nanoscale, 4*(5), 1585–1591. doi: 10.1039/c1nr11161b.

Herman, P., El-Kirat-Chatel, S., Beaussart, A., Geoghegan, J. A., Vanzieleghem, T., Foster, T. J., Dufrene. Y. F., (2013). Forces driving the attachment of *Staphylococcus epidermidis* to fibrinogen-coated surfaces. *Langmuir, 29*(42), 13018–13022. doi: 10.1021/la4029172.

Imbert, L., Auregan, J. C., Pernelle, K., & Hoc, T., (2014). Mechanical and mineral properties of osteogenesis imperfecta human bones at the tissue level. *Bone, 65*, 18–24. doi: 10.1016/j.bone.2014.04.030.

James, S. A., Hilal, N., & Wright, C. J., (2017). Atomic force microscopy studies of bioprocess engineering surfaces - imaging, interactions and mechanical properties mediating bacterial adhesion. *Biotechnology Journal, 12*(7). doi: 10.1002/biot.201600698.

Jembrek, M. J., Simic, G., Hof, P. R., & Segota, S., (2015). Atomic force microscopy as an advanced tool in neuroscience. *Translational Neuroscience, 6*(1), 117–130. doi: 10.1515/tnsci-2015-0011.

Kailas, L., Ratcliffe, E. C., Hayhurst, E. J., Walker, M. G., Foster, S. J., & Hobbs, J. K., (2009). Immobilizing live bacteria for AFM imaging of cellular processes. *Ultramicroscopy, 109*(7), 775–780. doi: 10.1016/j.ultramic.2009.01.012.

La Storia, A., Ercolini, D., Marinello, F., Di Pasqua, R., Villani, F., & Mauriello, G., (2011). Atomic force microscopy analysis shows surface structure changes in carvacrol-treated bacterial cells. *Research in Microbiology, 162*(2), 164–172. doi: 10.1016/j.resmic.2010.11.006.

Liu, Q., & Yang, H. S., (2019). Application of atomic force microscopy in food microorganisms. *Trends in Food Science and Technology, 87*, 73–83. doi: 10.1016/j.tifs.2018.05.010.

Marin-Bustamante, M. Q., Chanona-Perez, J. J., Guemes-Vera, N., Arzate-Vazquez, I., Perea-Flores, M. J., Mendoza-Perez, J. A., Casarez-Santiago. R. G., (2018). Evaluation of physical, chemical, microstructural and micromechanical properties of nopal spines (*Opuntia ficus-indica*). *Industrial Crops and Products, 123*, 707–718. doi: 10.1016/j.indcrop.2018.07.030.

Martin, Y., Williams, C. C., & Wickramasinghe, H. K., (1987). Atomic force microscope force mapping and profiling on a sub 100-Å scale. *Journal of Applied Physics, 61*(10), 4723–4729. doi: 10.1063/1.338807.

Mendez, M. J. M., (2010). *Mechanical Properties of Single Keloid and Normal Skin Fibroblast Measured Using an Atomic Force Microscope*. Thesis of Doctor of Philosophy. University of Manchester.

Meyer, R. L., Zhou, X. F., Tang, L. N., Arpanaei, A., Kingshott, P., & Besenbacher, F., (2010). Immobilization of living bacteria for AFM imaging under physiological conditions. *Ultramicroscopy, 110*(11), 1349–1357. doi: 10.1016/j.ultramic.2010.06.010.

Nanowerk, (2008). *Atomic Force Microscopy (AFM): A Key Tool for Nanotechnology*. https://www.nanowerk.com/spotlight/spotid=4876.php (accessed on 8 July 2021).

Nicolas-Alvarez, D. E., Andraca-Adame, J. A., Chanona-Perez, J. J., Mendez-Mendez, J. V., Cardenas-Perez, S., & Rodriguez-Pulido, A., (2019). Evaluation of nanomechanical properties of tomato root by atomic force microscopy. *Microscopy and Microanalysis, 25*(4), 989–997. doi: 10.1017/s1431927619014636.

Nicolas-Bermudez, J., Arzate-Vazquez, I., Chanona-Perez, J. J., Mendez-Mendez, J. V., Rodriguez-Castro, G. A., & Martinez-Gutierrez, H., (2018). Morphological and micro-mechanical characterization of calcium oxalate (CaOx) crystals embedded in the pecan nutshell (*Carya illinoinensis*). *Plant Physiology and Biochemistry, 132*, 566–570. doi: 10.1016/j.plaphy.2018.10.008.

Raman, A., Trigueros, S., Cartagena, A., Stevenson, A. P. Z., Susilo, M., Nauman, E., & Contera, S. A., (2011). Mapping nanomechanical properties of live cells using multi-harmonic atomic force microscopy. *Nature Nanotechnology, 6*(12), 809–814. doi: 10.1038/nnano.2011.186.

Rogers, B., Manning, L., Sulchek, T., & Adams, J. D., (2004). Improving tapping mode atomic force microscopy with piezoelectric cantilevers. *Ultramicroscopy, 100*(3, 4), 267–276. doi: 10.1016/j.ultramic.2004.01.016.

Schaer-Zammaretti, P., & Ubbink, J., (2003). Imaging of lactic acid bacteria with AFM-elasticity and adhesion maps and their relationship to biological and structural data. *Ultramicroscopy, 97*(1–4), 199–208. doi: 10.1016/s0304-3991(03)00044-5.

Severa, L., Buchar, J., Nemecek, J., & Nedomova, S., (2010). Nanoindentation as a new tool for evaluation of hen's eggshell local mechanical properties. *Chemicke Listy, 104*, S368–S370.

Sullan, R. M. A., Beaussart, A., Tripathi, P., Derclaye, S., El-Kirat-Chatel, S., Li, J. K., Dufrene. Y. F., (2014). Single-cell force spectroscopy of pili-mediated adhesion. *Nanoscale, 6*(2), 1134–1143. doi: 10.1039/c3nr05462d.

Vadillo-Rodriguez, V., Beveridge, T. J., & Dutcher, J. R., (2008). Surface viscoelasticity of individual gram-negative bacterial cells measured using atomic force microscopy. *Journal of Bacteriology, 190*(12), 4225–4232. doi: 10.1128/jb.00132-08.

Vadillo-Rodríguez, V., Busscher, H. J., Norde, W., De Vries, J., Dijkstra, R. J. B., Stokroos, I., & Van, D. M. H. C., (2004). Comparison of atomic force microscopy interaction forces between bacteria and silicon nitride substrata for three commonly used immobilization methods. *Appl. Environ. Microbiol., 70*(9), 5441–5446.

Van, D. H. M., Huttener, M., Juarez, A., & Gomila, G., (2015). Nanoscale imaging of the growth and division of bacterial cells on planar substrates with the atomic force microscope. *Ultramicroscopy, 154*, 29–36. doi: 10.1016/j.ultramic.2015.02.018.

Yamashita, H., Taoka, A., Uchihashi, T., Asano, T., Ando, T., & Fukumori, Y., (2012). Single-molecule imaging on living bacterial cell surface by high-speed AFM. *Journal of Molecular Biology, 422*(2), 300–309. doi: 10.1016/j.jmb.2012.05.018.

Yao, X., Walter, J., Burke, S., Stewart, S., Jericho, M. H., Pink, D., & Beveridge, T. J., (2002). Atomic force microscopy and theoretical considerations of surface properties and turgor pressures of bacteria. *Colloids and Surfaces B-Biointerfaces, 23*(2, 3), 213–230. doi: 10.1016/s0927-7765(01)00249-1.

CHAPTER 5

NANOSTRUCTURED BIOMATERIALS FOR TARGETED DRUG DELIVERY

SAHER ISLAM,[1] DEVARAJAN THANGADURAI,[2] CHARLES OLUWASEUN ADETUNJI,[3] OLUGBENGA SAMUEL MICHEAL,[4] WILSON NWANKWO,[5] OSENI KADIRI,[6] OSIKEMEKHA ANTHONY ANANI,[7] SAMUEL MAKINDE,[5] and JULIANA BUNMI ADETUNJI[8]

[1]*Institute of Biochemistry and Biotechnology, Faculty of Biosciences, University of Veterinary and Animal Sciences, Lahore–54000, Pakistan*

[2]*Department of Botany, Karnatak University, Dharwad–580003, Karnataka, India*

[3]*Applied Microbiology, Biotechnology, and Nanotechnology Laboratory, Department of Microbiology, Edo University Iyamho, PMB 04, Auchi, Edo State, Nigeria*

[4]*Cardiometabolic Research Unit, Department of Physiology, College of Health Sciences, Bowen University, Iwo, Osun State, Nigeria*

[5]*Informatics and Cyber-Physical Systems Laboratory, Department of Computer Science, Edo University Iyamho, PMB 04, Auchi, Edo State, Nigeria*

[6]*Department of Biochemistry, Faculty of Basic Medical Sciences, Edo University Iyamho, Nigeria*

[7]*Laboratory of Ecotoxicology and Forensic Biology, Department of Biological Science, Faculty of Science, Edo University, Iyamho, Edo State, Nigeria*

[8]*Nutrition and Toxicological Research Laboratory, Department of Biochemistry Sciences, Osun State University, Osogbo, Nigeria*

5.1 INTRODUCTION

The National Nanotechnology Initiative defines nanostructures which range in sizes from 1 nm to 100 nm. The prefix "nano" is however used for particles of several nanometers in size. Typical examples of nanocarriers (NCs) include dendrimers, polymers, magnetic nanoparticles, silicon materials, carbon materials, and liposomes. They all have been examined as targeted drug delivery structures. Researchers have manipulated nanoparticle surface and size properties so as to improve its stability, decrease clearance and optimize its bioavailability. When these characteristics are controlled, drugs can have access to inaccessible tissue. But firstly, the drug needs to be released from its nanoparticle matrix (Mura et al., 2013; Son et al., 2017). Reactivity, surface curvature, appropriate targeting ligands incorporation are some conditions to be considered for creating an optimum nanoparticle which is able to effectively deliver drugs to target tissues. Achieving this optimum system will help to improve system stability, receptor binding, and inhibition of aggregation and which subsequently will improve the drug pharmacological effects (Khanbabaie and Jahanshahi, 2012). Nanotechnology provides a lot of benefits in the cure of many chronic ailments by target-oriented and site-specific delivery of accurate medicine. Such benefits have applied in immunotherapeutic, biological, and chemotherapeutic messengers in the cure of different diseases. Nanomaterials can be utilized to enhance the efficacy of natural (old) and new drugs via selective analysis by malady indicator molecules (Patra et al., 2018). They can be manipulated into smart encasing systems in medicine that can be used as imaging and therapeutic messengers; having stealth healing properties. Advancement of drugs formulations from nano-based materials has yielded several opportunities in addressing and treating issues of diseases. That these systems have the potential to transport drug to exact tissues as well as offer controlled discharged therapy. This pattern of sustained and targeted drug transfer reduces the drug-associated noxiousness and elevates the patient's agreement with less frequency of drug dosages. This recounted that nanotechnology has many applied benefits in medicine. It has been employed in the treatment of some chronic diseases like AIDS, cancer, renal failure and also offers improvement in problem-solving testing (Syed et al., 2018). This chapter is entitled to focus on nanomedicine (NM) and different nanomaterials that could be used in targeted drug delivery systems.

5.2 UNIQUE PROPERTIES OF NANOPARTICLES FOR DRUG DELIVERY SYSTEMS

5.2.1 PARTICLE SIZE

Nanoparticles size and shape affect how body cell relates with it and thereby dictate its targeting strength, distribution, and toxicity. It is pertinent to note that nanoparticles have the ability to cross the blood-brain barrier (BBB) and sustain continual medication for disease with a low response to treatment (McMillan et al., 2011). These techniques have also been manipulated for effective drug control and distribution. Gratton et al. (2008) reported that nanoparticles of 100 nm showed an uptake which was 2.5-fold more than those exhibited by a nanoparticle of 1 μm diameter and uptake which is 6-fold greater than a 10 μm nanoparticle. The closer a drug is to the surface, the more effective is it releases (Bantz et al., 2014). More of a nanoparticle-based drug system is closer to the surface as there is an increase in surface area to volume ratio as particles get smaller. At the size range of nanoparticles which is between 1 nm and 100 nm, particle can easily pass through the BBB, thereby delivering a sufficient amount of blood and bypassing the immediate clearance (Khanbabaie and Jahanshahi, 2012). Clearance issues are addressed through the creation of polymer complexes. However, nanoparticles are easily prone to aggregation due to their small sizes and immensely large surface area. A typical example of nanoparticles prone to aggregations includes micelles, quantum dots, and dendrimers. Li and Kaner (2006) suggested the alteration of the zeta potential of particles. Coating of nanoparticles with capping agents was also suggested as a suitable strategy for prevention the limitation posed by aggregation. In summary, there is a leading idea guiding these theories which are: particle size must high enough so as to prevent its leakage to the blood capillaries but with the appropriate size so as to prevent it susceptible to clearance by the macrophage (Sykes et al., 2016).

5.2.2 THE LOADING AND RELEASE OF PARTICLE

Factors that determine drug release from nanoparticle-based formulation includes temperature, pH, nanoparticle matrix swelling, nanoparticle matrix erosion, desorption of adsorbed drug and of drugs across the nanoparticle matrix (Mura et al., 2013; Son et al., 2017). Drug release depends on nature of nanoparticle been used. Based on composition, polymeric nanoparticles (PNPs) has been sorted as nanospheres or nanocapsules. Drug released in relation to nanospheres is by matrix erosion. In this nanoparticle model, there is a rapid burst followed by a sustained release of drug (Lee and Yeo, 2015). In the case of nanocapsules, drug release is controlled through a diffusion process across the polymeric layer. Deliverability depends on drug diffusibility across the polymer. However, there will be formation of complexes which impedes drug release from the capsule in a situation where ionic interaction exists between the polymer and the drug. The addition of an auxiliary agent like polyethene oxide-propylene oxide can help avoid this limitation. Calvo et al. (1997) noted that this auxiliary agent results in decline interaction between the capsule and the drug matrix thereby improving drug release to the target tissues.

5.2.3 SURFACE PROPERTIES

Surface characteristics manipulation is another procedure for improving the performance of the nanoparticle-based drug, aside from size manipulation (Bantz et al., 2014). An important factor to consider in nanosystem is the clearance by body immune system since they are recognized as foreign bodies. Nanoparticles are more prone to clearance as their hydrophobicity increases (Sushnitha, 2020). Likewise, when nanoparticles are coated with surfactants or polymers, opsonization is lessened. Another factor which diminishes includes poloxamine, polyoxamer, and polyethylene glycol (PEG). PEG is known to inhibit splenic and hepatic localization (Suk et al., 2016). Incorporation of PEG on nanoparticle surface inhibits opsonization, thereby preventing significant loss of the given dose. Creation of polymer complexes resolves the challenges posed by clearance issues though aggregation still posed a problem due to the large surface area of small particles. This problem is however mitigated by coating particles with capping agents. Alteration of zeta potential can also help in preventing aggregation (Li and Kaner, 2006). In summary, the nanoparticle size should be large enough to avoid its entrance into blood capillaries but be non-susceptible to macrophage clearance. Therefore, when the surface of the nanoparticle is manipulated, aggregation or clearance can be monitored and controlled (Sykes et al., 2016).

5.3 NANOPARTICLES AS DRUG CARRIERS

Nanotechnology is an evolving novel science that requires special analytic tools in delivery nanoparticles or nanomaterials for efficient target to specific sites as messengers. This technology has been widely used in the agricultural, environmental, engineering, nutritional, food, and medical sectors. Nanoparticles differ in shapes and sizes from 100–500 nm (Gratton et al., 2008). Wilczewska et al. (2012) in a review looked at the potential of nanoparticles for drug delivery. The authors reported that controlled nanoparticles for drug delivery have been of recent proven efficient and have more advantages and benefits when compared to the traditional types of drugs. Such benefits are that they aid in the transportation of drugs to specific sites of action, thus reducing the impacts on vital tissues as well as side effects faced through the discharge of their physiological functions. More so, the dosages site to therapeutic sites are lower and the compound delivered increased at the site of actions. This novel type of drugs is very vital in terms of incongruity between noxious effects, treatment results, or the level of drug dosages (Bawarski et al., 2008). Different nanostructures such as magnetic nanoparticles, silicon-carbon materials, dendrimers, polymers, and liposomes used in the delivery of specific cell targeting drugs, have been evaluated as major nanoparticles for drug delivery (Figure 5.1). NCs and the related drugs they carry were evaluated and discussed as well as their disadvantages and advantages of the utilization (Hild et al., 2008).

5.3.1 APPLICATION OF SMART NANOPARTICLES FOR DRUG DELIVERY

Lombardo et al. (2019) did a review of the application of smart nanoparticles for drug delivery as a versatile transporter in NM. The authors recounted that in severe microenvironment of unhealthy tissue; nanodrug delivery systems permit the advancement of new platforms for an effective transport as well as regulate the discharge of the treatment molecules in the site-specific regions, thus providing a versatile range of the nano-platforms medicine to act successfully. Examples of such NCs are hydrogels, dendrimers, liposomes, vesicles, and polymeric micelles, which are organic-based and mesoporous silica, gold, and quantum dots nanoparticles which are inorganic based materials (Table 5.1). Notwithstanding the amazing advances of nontechnology, almost all NCs mechanism is linked with unwarranted side effects which moderate their effective utilization in NM and biotechnology. Viewpoints on some important problems in engineering and design of a typical nanocarrier system for effective biotechnology utilization were discussed. Interaction of multiform and environmental complexes of established biological media via utilizing nanomaterials were highlighted and evaluated.

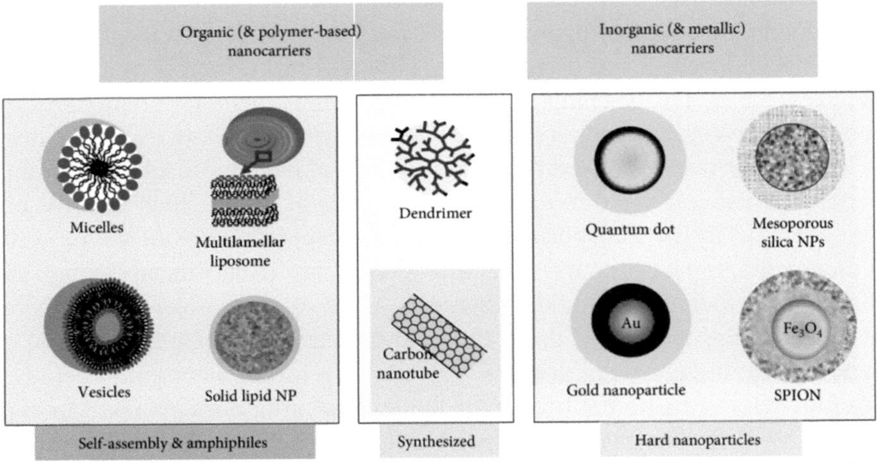

FIGURE 5.1 Promising nanocarriers (organic and inorganic) used in smart drug delivery.

Source: Reprinted from Lombardo et al. (2019); Creative Commons Attribution 4.0 International.

Xiaojiao et al. (2016) in a review looked at the strategy of employing nanomaterial-based transporters for effective delivery of drugs in NM. They stated that NCs serve as a vehicle for the efficient delivery of drugs. That the novel nanoparticles are engineered with computer modeling design for an improved process in the discharge of drugs to the site of tumor or diseased cell regions. Because during the clinical processes (*in vivo* and *in vitro*) testing, several problems might emanate in their delivery process that might result to deleterious influence to animal and living tissues. They further opined those environmental factors can impede the influence of nanoparticles drug delivery efficiency. The authors reported that computational methods are not widely utilized in the pharmaceutical sector because of its lowly probability. However, the models singly lay emphasis on the employment of a specific nanodrug delivery system to a site rather than in the entire targeting phases of cells of the treated patient. Several types of nanoparticles for drug delivery; physical, active, and passive targeting, recent phases of research, limitations, advantages, and methods of targeting diseased cells using biotechnology were discussed. In conclusion, the authors recommended a multi-scale modeling approach at various spatial and temporal scales for an effective drug delivery system in diseased patients.

Li et al. (2017) in a review looked at the potentials of utilizing nanotechnology in the delivery of drugs for the management of cancer. The authors stated that of recent, nanomaterials have been used in the transportation of cancer drugs.

TABLE 5.1 Treatment of Wound and Regeneration of Skin Using Nanodrug Delivery Systems*

Formulation	Drug	Administration	Outcome
Liposomes	bFGF	Smearing wound every 3 days	Accelerated the wound closure of mice with deep second-degree scald, expedited regeneration of vascular vessels
Liposomes	Madecassoside	Smearing wound once a day for 12 days	Enhanced permeation and distribution in skin so as to exhibit superior burn wound healing effect
Liposomal membrane	Usnic acid	Topical treatment applied every 4 days	Enhanced maturation of granulation tissue and better collagen deposition
Deformable liposomes	EGF, PDGF-A, and IGF-1	Topical treatment, once a day for 18 days	Significantly enhanced the healing of chronic wounds due to synergistic effect of complex
Deformable liposomes	Curcumin	Daily topical treatment for 3 days	Shorten inflammatory process, prevent infection, and promoted fibrosis, angiogenesis, re-epithelialization, and wound contraction
Deformable liposomes	Baicalin	Topical treatment	Complete skin restoration and inhibition of inflammatory markers such as edema, TNF-α and IL-1β
Nanoparticles	Thrombin	Topical treatment	Advanced process of healing, improved skin tensile strength, reduced complications in surgery
Nanoparticles	Silver	Topically given with dressing every day	Rapid healing and improved cosmetic appearance were achieved via reduction in wound inflammation, and modulation of fibrogenic cytokines
Nanoparticles	Cerium oxide	Topical treatment once a day for 13 days	Reduced the oxidative stress at wound site and protected regenerative tissue, providing favorable environment for restoration
Nanoparticles	ZnO_2	Topical treatment	Had good antibacterial activity and accelerated wound healing in animal model
Hydrogel loading nanoparticles	Asiatic acid/ZnO/CuO	Topical treatment	Raised DNA, total protein, hexosamine, and hydroxyproline content and rendered superior re-epithelization, collagen fibers arrangement and angiogenesis
Hydrogel loading nanoparticles	Silver oxide	Local injection	Showed excellent antimicrobial activity and burn wound healing in the second burn rat model

TABLE 5.1 *(Continued)*

Formulation	Drug	Administration	Outcome
Nanoparticles	LL37	Intradermally injection	Significantly promoted granulation, collagen deposition, re-epithelialized, and neovascularized composition
Microspheres/scaffold	Mupirocin	Applied topically, covered, and tied	Combated infection and stimulated fibroblast proliferation and dermal collagenization
Nanoparticles	Norfloxacin	Applied topically	Sustained drug release to 24 h, retained antimicrobial efficacy and exhibited good stability
Nanoparticles	Amphotericin B	Applied topically every other day	Equipped with strong anti-fungi capacity and quicker efficiency in fungal clearance
Nanoparticles	hVEGF gene/stem cells	Intramuscularly injection	Facilitated angiogenesis and limb salvage, also reduced muscle degeneration and tissue fibrosis
SLNPs/NLCs	EGF	Administered topically twice a week	Significantly improved wound closure, restoration of the inflammatory phase, and re-epithelialization
SLNPs	LL37 and A1	Applied topically	Promoted wound closure in fibroblast cells and keratinocytes, simultaneously enhanced antibacterial activity
NLCs/scaffold	Andrographolide	Applied every other day 21 days	Enhanced the wound healing with no scar and improved tissue quality
Bioadhesive gel containing SLNPs	Cyclosporine A	Administered topically twice a day	Significantly increased rate of mucosal repair
Nanofibers	Lawsone	Locally delivered every three days over a period of 14 days	Significantly increased in TGF-β1 and Collagen gene expression *in vitro* and promoted re-epithelialization of the wound *in vivo*
Nanofibrous dressing	Astragaloside IV	Locally delivered at the wound site every 2 days	Stimulated wound closure, increased angiogenesis, regulated newly formed types of collagens, and collagen organization
Nanofibrous membranes	Collagen/Zein	Applied topically/single	Exerted antibacterial activity and induced fast tissue regeneration
Nanofibers loaded with nanoparticles	Cefazolin/zinc oxide	Applied topically/single	Showed great anti-bacterial activity, enhanced cell adhesion and epithelial migration, contributed to faster and more efficient collagen synthesis

TABLE 5.1 *(Continued)*

Formulation	Drug	Administration	Outcome
Nanofibrous scaffold	Human bone marrow stem cells	Applied topically	Boosted cell growth rate and accelerated wound recovery
Scaffold loaded with nanoparticles	Silver	Applied topically	Strong anti-microbial capability and excellent biocompatibility with fibroblast cells
Nanofibrous scaffold carrying with nanoparticles	PDGF-BB and VEGF	Topically placed on the wound site	Accelerated tissue regeneration and remodeling, promoted angiogenesis
Hydrosol/scaffold	TiO_2	Applied topically	Strongly inhibited the growth of *Staphylococcus aureus* and induced red blood cells aggregation to stop bleeding
Nanofibrous scaffold carrying with nanoparticles	Fe_3O_4	Applied topically	A suitable scaffold for cell adhesion with favorable magnetic behavior and low cytotoxicity
Hybrid nanostructures/ hydrogel	Ag/Ag-AgCl/ZnO	Applied every 2 days for 14 days	Stimulated the immune function, produced the synergistic antibacterial effects, and accelerated wound healing
Nanohydrogel	Nanosilicates/ VEGF	Applied topically	Enhanced cell adhesion and spreading, reduced blood clotting time, facilitated *in vitro* tissue regeneration, and wound healing
Nanohydrogel	$K_2(SL)6K_2$	Subcutaneous injection	Provoked an inflammatory response, stimulated host cells to secret wide range of cytokines, so as to promote cell recruitment and angiogenesis
Nanohydrogel	Acrylic acid	Applied topically	Maintained the activity and morphology of human dermal fibroblasts, promoted rapid cell proliferation, and affected 9 gene expressions related to wound healing
Nanohydrogel	Baicalin	Topically smeared every day for 4 days	Faster and more complete skin restoration and inhibition of specific inflammatory markers were noticed
Nanohydrogel	Ultrashort aliphatic peptides	Applied topically	Earlier onset and completion of autolytic debridement as well as faster wound closure compared with a commercial product

[*]*Source:* Adapted from: Wang et al. (2019); Creative Commons Attribution 4.0 International.

More so, this technology is also used in the areas of biomarker discovery and cancer diagnosis. They recounted that many nano-antitumor delivery drugs such as nanomicelles, dendritic polymers, polymer nanoparticles and liposomes, are currently developed and tested in both clinical and preclinical trials and have also shown to have promising therapeutics values. Thus, these materials can be optimized and organized for site-specific radiotherapy, photodynamic therapy, and thermotherapy and chemotherapy purposes. However, the issues of toxicology arising from nano-utilization remain life-threatening issues of lately. The issue results from combined therapeutic routines of various cancer forms by the utilization of various mechanisms. In conclusion, the authors recommended that such issues pervading the toxicology and combined therapy of nanomaterials should be addressed. In addition, other challenges as a result of functionalized and modified nanomaterials are corrected with the improvement of the efficacy, biocompatibility, biodistribution, and localization of the nanoformulation via *in vivo* screening in order to meet the basic necessities of cancer therapy and diagnosis precisions.

5.3.2 EFFICIENCY OF UTILIZING MULTIFUNCTIONAL NANOPARTICLES AS DRUG DELIVERY SYSTEMS

Fortuni et al. (2019) tested and evaluated the efficiency of utilizing multifunctional nanoparticles as drug delivery systems. The authors stated that most anticancer drug delivery systems depend on the superficial functionalization of specific NCs ligands that can activate the internalization of in tumor cells through endocytosis mediated receptor. This endocytosis mediated receptor infers that the set-up of the drug delivery systems in lysosomes and endosomes (acidic vesicles) and the final removal by a process called exocytosis. This mechanistic process is the major obstacle of drug delivery systems from the lysosomes and endosomes because it decreases the bioavailability of the drug in the intracellular region of the cell environment. The discharge of drug delivery systems from the lysosomes and endosomes is therefore vital in the improvement as well as the performance of the therapy in a small dosage. The authors were able to establish that multifunctionalized drug delivery systems have high endosomal emission abilities. Those nanoparticles derived from mesoporous silica were able to combine with polyethylenimine to prompt the break of the endosomal and the release of hyaluronic acid, that later combine with CD44 receptors that are widely expressed in cancerous cells. The findings from their study showed that the drug delivery systems were shown to have cytotoxic influence when compared to the traditional drugs showing high specificity to the cancerous cells. More so, hyaluronic acid was able to provide the system with enzymes and target capacity to regulate the release of the drugs. In conclusion, the authors recommended multifunctionality polymeric engineering as a perfect drug delivery systems tool for the improvement and management as well as for the treatment of cancer.

5.3.3 *FUTURE PROSPECTS OF NANOTECHNOLOGY IN DRUG DELIVERY*

In line with this, Banerjee (2018) in a review looked at present as well as future prospects of nanotechnology in drugs delivery. The authors stated that biomedical importance and the biological characteristics of nanosized particles with a broad spectrum of drug delivery characteristics. That this can improve both subcellular and intracellular targeting delivery by accessing sites that are not readily accessible as well as combating physiological and anatomical barriers. However, barriers intended to overcome might be different based on the employed route of administration of the drugs. The authors stated that in as far the delivery system of the nanoparticle is effective, there might be a lacuna in the technology that might impede the translational potentials of the drugs. In conclusion, the authors recommended a novel platform for the delivery of nanodrugs as well as the application of specific issues in the translation of nanodrugs into possible effective nanoproducts.

5.4 NANOMEDICINE (NM) PRODUCTION AND AVAILABILITY

Nanomaterials exhibit potentials and their relatively unlimited sources make them unique substances. Currently, many nanomaterials have been identified, and many are being tested clinically for use in medicine and healthcare. Cardiac disease is one of the primary causes of morbidity and mortality in humans. Every day people are suffering from various heart diseases including arrhythmias, myocardial infarction, ischemic, restenosis, and atherosclerosis (Kang et al., 2010). Systemic and oral drugs administration do not provide the proper level of therapeutic drugs in target arteries for a sufficient period. Though, biomedical engineers are already getting succeeded in advancing micro-scale instruments to treat cardiovascular (CV) diseases by opening the blocked arteries. Nevertheless, these devices are cumbersome, vulnerable to contamination and susceptible to certain disorders. NM currently provides a comprehensive platform in the CV area of science by offering resources for exploring the limits of cardiac research at the cellular level (Weissig et al., 2014).

Even due significant valuable nanotechnology knowledge is available and in use; however, at the moment, it is really incomplete and scattered. Therefore, this implies that many clinicians are unfamiliar with nanotechnology applications during surgical treatment. While, most nanotechnology-using medical devices are in research still, some are at the development stage and few commercially available (Rizzo et al., 2013). Over the previous few years, nanotechnology has evolved rapidly, and its applications have become a significant spin-off in the medicine and surgery field. Given the innumerable advantages of nanostructures in the field of NM, but still due to many drawbacks that these technologies are associated with like toxicity of the materials (Table 5.2), only a handful of products have been able to enter the market. Maintaining a balance between benefits and drawbacks, however, would undoubtedly open up avenues for personalized medicine through counseling, diagnostics, and theranostics (Wang et al., 2013).

5.5 NANOMEDICINE (NM) FORMULATIONS

NM formulations are nanometer-sized carrier materials designed to improve the bio-distribution of chemotherapeutic drugs used systemically. Clinically relevant examples of NM formulations for antivirals are liposomes, polymers, micelles, niosomes, and ethosomes, PNPs, solid-lipid nanoparticles (SLNPs), polymeric micelles, dendrimers, cyclodextrin derivatives, hydrogel-based NCs, micelles, nanoemulsions, nanosuspensions, stimuli-responsive drug delivery systems, and toxicology aspects (Lembo et al., 2018). The goal of NM formulations is to enhance the balance between the effectiveness and toxicity of therapeutic interventions; through more selectively delivering pharmacologically active agents to pathological locations, or by directing them away from potentially threatened healthy tissues.

Remarkably, nanocarrier formulations modify the physicochemical characteristics of incorporated materials, thereby enabling modified pharmacokinetics, controlled release, and targeted action. This can lead to an increase in effectiveness of the drug and a reduction in associated side effects. Intriguingly, NM formulations could represent a new avenue for controlling the amount, dosage frequency, and delivery site of antivirals, as well as for targeting the virus life cycle (Jain et al., 2007). Indeed, the features of nanoparticles, such as sizes, morphology, and surface charge, can be modulated to promote the targeting of the drugs. NCs for oral

TABLE 5.2 Toxicity and Biodistribution of Different Nanocarriers*

SDDS Name	Toxicity		Bio-Distribution of Nanocarrier and Renal Excretion
	Cytotoxicity	Immunogenicity	
Liposome-based SDDS	Cationic liposome affects the *in vitro* growth of different cell lines, such as L 1210, HepG2, A549, etc. *In vivo* study shows DNA damage due to the cationic surface charge.	Positively charged liposome has toxic effect on macrophages and U937 cells.	Majority accumulates in the liver followed by spleen. Rapid clearance with urine.
Micelle-based SDDS	Toxicity of polymeric micelles shows no pathological abnormalities. Many investigations show that polymeric micelles are less toxic.	Polymeric micelle-based drug carriers trigger transient immunogenicity in the MPS system.Polymeric micelles based on poly(ethylene oxide) and α-carbon substituted poly(ε-caprolactone) are found to be non-immunogenic to dendritic cells-the antigen-presenting cell of the mammalian immune system.	The *in vivo* toxicity screening of well-characterized cationic polymeric micelles shows that particles could be found in major organs, such as lung, liver, kidney. Peptide Amphiphile accumulates primarily in the bladder then pass through the urine.
Dendrimer-based SDDS	Dendrimers, such as PPI, PAMAM, and PLL, exert significant *in vitro* cytotoxicity due to their surface catatonic groups, but significantly lowered cytotoxicity is observed with the PEG-modified dendrimer. PAMAM has adverse effects on mammalian cells. Proper surface modification can reduce cytotoxicity.	Dendrimers show no or little immunological response.	They are present in the intracellular compartment of the kidney, liver, and lung.
Meso-porous silica nanoparticle-based SDDS	*In vitro* cytotoxicity is controversial. MCM-41 and two of its functional analogs kill human neuroblastoma (SK-N-SH) cells. Mesoporous silica does not affect cell viability or the plasma membrane.Silica nanoparticle cytotoxicity is size-dependent; smaller particles have higher toxicity.	Functionalized mesoporous silica nanoparticles do not affect the viability of primary immune cells from the spleen in relevant concentrations. Potential adverse effects on the immune system are not clear and need further research.	MSNs mainly distribute in the liver and spleen; minority can be found in the lungs, kidneys, and heart. Silica nanoparticles have a toxic effect on the lliver. PEGylated MSNs with smaller particle sizes possess longer blood circulation and lower gradated products in the urine.

TABLE 5.2 *(Continued)*

SDDS Name	Toxicity		Bio-Distribution of Nanocarrier and Renal Excretion
	Cytotoxicity	Immunogenicity	
Gold nanocarriers-based SDDS	*In vitro* cytotoxicity screening of K562 leukemia cells shows that they do not exhibit an acute toxic effect based on the MTT assay-colorimetric assay for assessing cells' metabolic activity. Experiment on RAW264.7 also shows no considerable cytotoxicity based on the MTT assay. Cationic gold nanocarriers show toxicity.	The immunological study of the RAW264.7 macrophage did not indicate any immunological toxicity. *In vivo* experiment showed size-dependent toxicity; that is, nanoparticles with certain sizes show lethal toxicity while other sizes of nanoparticles show no considerable toxicity.	GSH coated GNP nanocarriers have lower accumulation in the kidneys and liver compared to bare GNPs. Mostly excreted with urine and no systemic toxicity.
SPION-based SDDS	SPIONs are toxic to brain cells with different coatings. Compatible to kidney cells.	The generation of ROS could trigger immunological toxicity.	75% found in spleen. Primarily found in the spleen and liver.
CNT-based SDDS	Interaction of functionalized SWCNTs with CHO and 3T3 cells exhibited no toxicity.	CNTs functionalized with peptides do not trigger anti-peptide antibodies.	Well individualized MWCNTs with shorter lengths and higher degrees of oxidation escape the RES in organs (liver, spleen lungs) and clear through renal excretion.
Quantum dot-based SDDS	QD-induced cytotoxicity is not observed in many *in vivo* and *in vitro* experiments.	Immune response could be suppressed by CdSe/ZnS QDs.	QDs primarily deposit in the lung and atriums of the heart. Not excreted with urine.

**Source:* Reproduced with permission and slight modification from: Hossen et al. (2019); Copyright © Elsevier (2019).

nanoformulations should be resistant to acidic pH of stomach and intestinal enzymes and capable to penetrate the mucus secretion that limits the intestinal presence of these drugs (Lee and Yeo, 2015).

So far, NM formulations have generally struggled to boost the efficacy of chemotherapeutic treatments, despite strong evidence of extended diffusion periods and increased concentrations of tumors. For example, the primary reason for approving the well-known liposomal doxorubicin formulations Doxil and Myocet (PEGylated and non-PEGylated liposomal doxorubicin), was their potential to reduce drug-related toxicity (cardiomyopathy, bone marrow impairment, alopecia, and nausea) rather than improving antitumor effectiveness (Heidari et al., 2013). This can be demonstrated by taking into account the findings of a head-to-head phase III analysis of free doxorubicin versus Myocet in patients that have metastatic breast cancer, where comparable response rates (26%) and progression-free survival periods (4 months) were seen but where the occurrence of cardiac events (29% vs 13%) and congestive heart failure (8% vs 2%) was observed (Milla et al., 2012). A list of FDA-approved nanobiomedicines has been given in Table 5.3.

5.6 NANOPARTICLES USE IN MEDICINE AND DRUG DELIVERY

5.6.1 SOLID-LIPID NANOPARTICLES (SLNPs)

Solid-lipid nanoparticles (SLNPs) are composed mainly of lipid nanoparticles (LNPs) which are either solid or semisolid at ambient temperature. These formulations supply improved physical steadiness, decreased decomposition of drug, diminished toxicity and simplicity and effortless manufacturing process (Kang et al., 2010). Interestingly changing the features of SLNPs through alteration of their lipid constituents, surface charge and size site-specificity, continuous drug delivery can be achieved. Solid-lipid formulations offer better drug durability and improved drug pharmacokinetics management than the liquid lipid preparations. For example, SLNPs have been employed in drug delivery for the management of cancer as noticed in the combination of curcumin an anticancer compound with transferrin-mediated SLNPs for specific delivery to breast cancer cells (Mulik et al., 2010). In addition, another study has reported the SLNPs application for anticancer drugs targeted delivery such as doxorubicin and camptothecin successfully to the brain (Yang et al., 1999; Zara et al., 1999). The encapsulation of drugs into the SLNPs system resulted in increased solubility and permeability several times better than the conventional system.

TABLE 5.3 List of FDA Approved Nanobiomedicines*

Clinical Products	Formulation	Indication	Company	Year
Polymer-based Nanoparticles				
Renagel	Poly(allylamine hydrochloride)	Chronic kidney disease	Sanofi	2000
Eligard	Leuprolide acetate and polymer PLGA (poly(DL-lactide-co-glycolide))	Prostate cancer	Tolmar	2002
Estrasorb	Micellar estradiol	Menopausal therapy	Novavax	2003
Cimzia/ certolizumab pegol	PEGylated antibody fragment (certolizumab)	Crohn's disease, Rheumatoid/psoriatic arthritis, Ankylosing spondylitis	UCB	2008–2013
Genexol-PM	mPEG-PLA micelle loaded with paclitaxel	Metastatic breast cancer	Samyang Corporation, South Korea	2007
Adynovate	Polymer-protein conjugate (PEGylated factor VIII)	Hemophilia	Baxalta	2015
Lipid-based Nanoparticles				
Doxil/Caelyx	Liposomal doxorubicin	Ovarian, breast cancer, Kaposi's sarcoma, and multiple myeloma	Janssen	1995–2008
DaunoXome	Liposomal daunorubicin	AIDS-related Kaposi's sarcoma	Galen	1996
Myocet	Liposomal doxorubicin	Combination therapy with cyclophosphamide in metastatic breast cancer	Elan Pharmaceuticals	2000
Marqibo	Liposomal vincristine	Acute lymphoblastic leukemia	Talon Therapeutics Inc.	2012
AmBisome	Liposomal amphotericin B	Fungal/protozoal infections	Gilead Sciences	–
Visudyne	Liposomal verteporfin	Choroidal neovascularization, macular degeneration, wet age-related, myopia, and ocular histoplasmosis	Bausch and Lomb	2000
Onivyde	Liposomal irinotecan	Pancreatic cancer	Merrimack	2015

TABLE 5.3 *(Continued)*

Clinical Products	Formulation	Indication	Company	Year
		Inorganic and Metallic Nanoparticles		
INFed	Iron dextran (low MW)	Iron deficiency in chronic kidney disease (CKD)	Sanofi Aventis	1957
Feridex/Endorem	SPION coated with dextran	Imaging agent	AMAG Pharmaceuticals	1996–2008
Venofer	Iron sucrose	Iron deficiency in chronic kidney disease (CKD)	Luitpold Pharmaceuticals	2000
GastroMARK, umirem	SPION coated with silicone	Imaging agent	AMAG Pharmaceuticals	2001–2009
NanoTherm	Iron oxide	Glioblastoma	MagForce	2010

Source: Adapted from: Lombardo et al. (2019); Creative Commons Attribution 4.0 International.

5.6.2 LIPOSOME

Liposomes are vesicles of spherical shape and composed of either natural or synthetic phospholipid. Phospholipids are self-assemble and in sizes ranging from micrometers to tens of nanometers (Wang et al., 2012). The aqueous core of the resulting vesicle can accommodate several molecules of hydrophilic or hydrophobic nature for therapeutic purposes. Phospholipids, majorly phosphatidylcholine, are naturally occurring compounds from which liposomes are synthesized. The hydrophilic polymer protects liposomes against macrophage recognition and decreases clearance. Surface charge, bilayer fluidity and size also result in the alteration of liposome delivery kinetics (Lian and Ho, 2001). There is bloodstream diffusion of liposomes into the interstitial space, which is closer to the target site. Liposomes can fuse directly with cell membranes since cell membranes are made of phospholipids and empty their content to the cytosol. Another pathway for this action is through the active transport pathways or phagocytosis.

Liposomes need more study, because of their different forms. Such structures are useful in antibacterial, cancer, immunomodulation, antifungal, ophthalmic, vaccinations, enzymes, genetic, and diagnostics elements for hydrophobic and hydrophilic medicines (Figure 5.2). For these systems, the preparation of liposomes marks in different products. Additionally, the forms of liposomes may be multilamellar and unilamellar based on its preparation methods. However, there are many variables and difficulties influencing the production of liposomes as drug delivery carriers (Alavi et al., 2017). An important problem is the use of clinical trials of various types of liposomal preparations, which is usually more complicated than traditional liposomal forms. In such case, no formulations have a complete framework, and each contains its own limitation, given their performance. Therefore, further studies are required to broaden the positive outlooks of drug supply structures in medical trials (Petrilli et al., 2018).

5.6.3 POLYMERIC NANOPARTICLES (PNPs)

In polymeric nanoparticles (PNPs), drugs molecules can either be infused in the polymerization phase or after polymerization. Some common typical synthetic PNPs include polyacrylate (Turos et al., 2007), polyacrylamide (Bai et al., 2007), and chitosan (Mao et al., 2001). Drugs can bound covalently, conjugated electrostatically, or be encapsulated in a core which is hydrophobic in nature (Figure 5.3) (Reis et al., 2006; Turos et al., 2007).

Electrodropping (Choi et al., 2013), microfluidic approaches (Shim et al., 2013), emulsion-based interfacial polymerization (Song et al., 2013) and high-pressure homogenization are typical procedures for PNPs. Polymers that are biodegradable to be considered when making a choice for the right nanoparticle chemistry. Kumari et al. (2010) reported that NCs are able to hydrolyze to produce smaller molecules which are biocompatible when hydrolyzing. This hydrolysis process results in the production of molecules such as glycolic and lactic acid. Method for creating PNPs include self-assembly or method such as particle replication. The non-wetting template is used later, which allows for the shape, size, and composition to be customized using tiny molds (Wang et al., 2012).

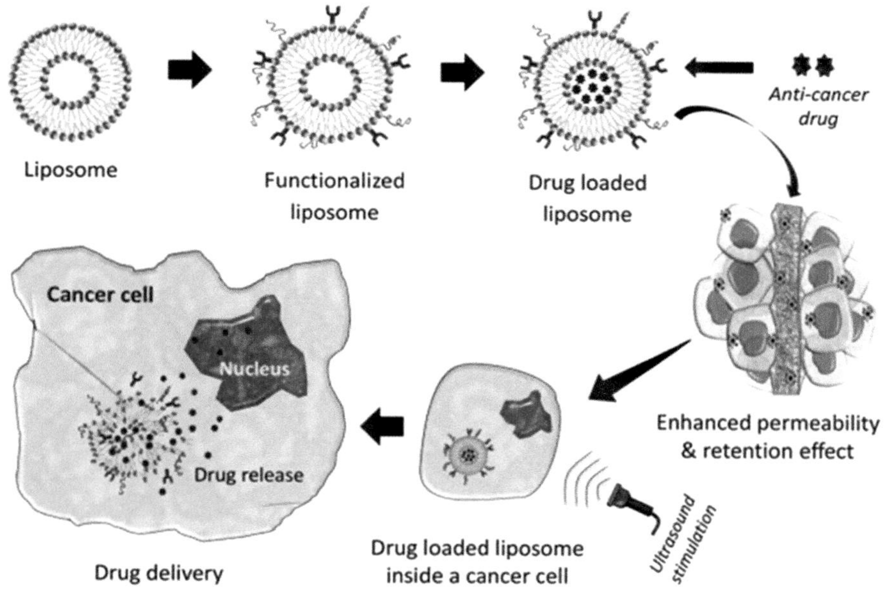

FIGURE 5.2 Liposome-mediated smart drug delivery in cancer therapy.

Source: Reproduced with permission from: Hossen et al. (2019); Copyright © Elsevier Ltd. (2019).

The use of PNPs for the treatment of NM is a program intended to mitigate adverse effects and to maximize therapeutic effects. Nanoformulations such as PNPs have already been used as a drug carrier to transport device with excellent oncological performance. Given the potential for anti-tumor treatment, gene therapy, radiation therapy, and HIV therapy have all been used. It is far more capable of supplying multiple proteins, vaccines, virostatics, and antibiotics. In CNS therapy, it also serves as vesicles for crossing BBB. PNPs thus provide interfaces for a wide variety of potential biological applications (Thakur et al., 2017).

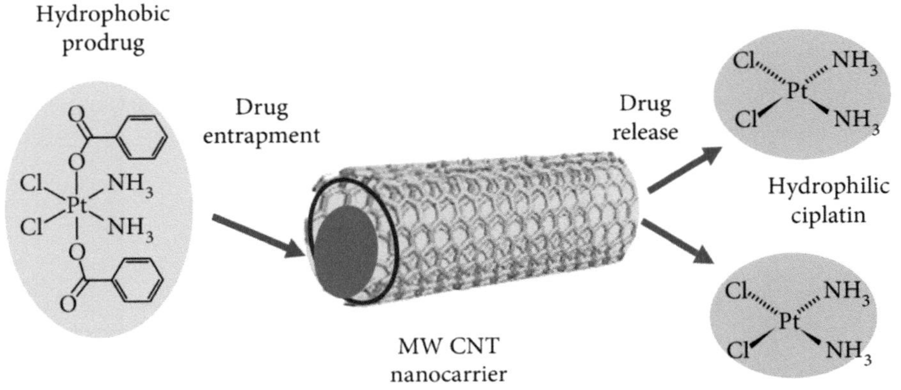

FIGURE 5.3 Entrapment and release of hydrophobic platinum (IV) prodrug in multiwalled carbon nanotubes through chemical reduction and hydrophobic reversal.

Source: Reprinted from Lombardo et al. (2019); Creative Commons Attribution 4.0 International.

5.6.4 METAL NANOPARTICLES AND METAL OXIDE NANOPARTICLES

Due to their unusual optical, magnetic, and electrical properties, metal nanoparticles and metal oxide nanoparticles have aroused some considerable interest. These flexible nanoparticles based on metals have great potential for selective bacterial detection and therapy. Nanoparticles made of metal and metal oxide have a large surface-to-volume ratio and are simple to modulate. Different detection techniques have been designed and tested for use in identifying bacteria, including colorimetric, fluorescence, nonlinear optical, electrochemical, and magnetic methods. Nanoparticles based on metal have often regarded as superior agents for bacterial detection and treatment (Yuan et al., 2018). Nevertheless, several obstacles still remain to be addressed before these nanoparticles could be employed clinics as diagnostic and pharmacotherapeutic agents. Specific surface modification techniques need to boost further the detection selectivity and the stability of the material.

5.6.5 DENDRIMERS

These are unique synthetic polymers which are hyper-branched, well-defined structure, monodispersed size and a terminal surface which is highly functionalized. They are mainly composed of nucleic acid, natural or synthetic amino acids, and carbohydrates molecules. The terminal surface or dendrimers interiors are points by which therapeutics can be loaded via hydrogen bonds, electrostatic interaction, chemical linkages, hydrophobic interactions, or covalent conjugation. Drugs half-life can be elongated through drug-dendrimer conjugation. However, the use of dendrimer in a living host is limited due to toxicity (Jain et al., 2010). There are also reported cases of their method of synthesis (Pooja et al., 2018). A factor which can be said to be responsible for its low yield is its narrow size which is within a range of <15 nm. Dendrimers as sort of artificial polymers contain specific structural characteristics; have been explored extensively for applications in the biomedical field, specifically in a drug delivery. Nonetheless, one major concern regarding the most widely used dendrimers is the non-degradability, which can cause side effects caused by cell or tissue aggregation of synthetic polymers. For that reason, biodegradable dendrimers that combine biodegradability with dendrimers merits, such as abundant internal cavities, distinct architectures as well as surface characteristics are far more favorable for the novel drug carriers' production (Huang and Wu, 2018).

5.6.6 QUANTUM DOTS

Quantum dot is another class which we are hearing increasingly in NM, particularly in multimodal imaging applications. Battle against the cancer requires increasingly sophisticated weapons, including instruments to identify micro-metastasis with sensitive spatiotemporal resolution. For this reason, there can be reliance on this multimodal imaging, as it comes with a combination of several techniques that can go beyond limitations of every single modality, particularly for the deep tissue imaging. Due to their highly small scale, remarkable clarity, photostability, and sufficient light colors emission for an optical detection, quantum dots may be functional components for sophisticated nanodevices (Figure 5.4). Nevertheless, their chemical existence poses significant limitations, since Quantum dots are usually comprised on heavy metals including lead and cadmium, for which durability of quantum dots and the safe elimination from human body is must (Pisanic et al., 2014).

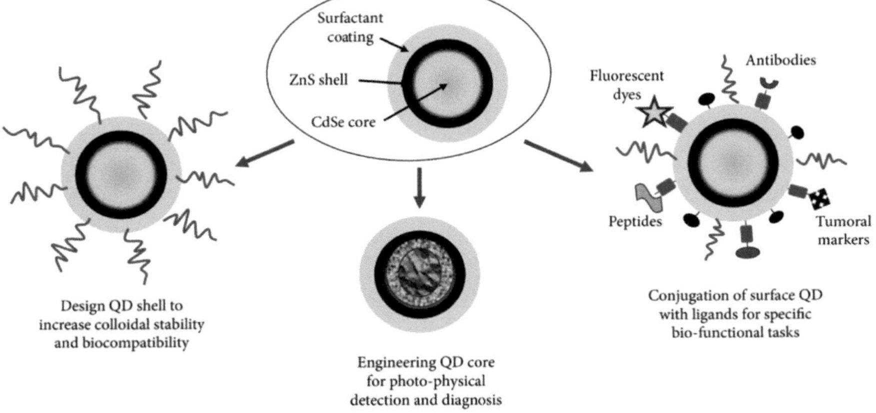

FIGURE 5.4 QD nanocarrier designing for nanobiomedicine applications.

Source: Reprinted from Lombardo et al. (2019); Creative Commons Attribution 4.0 International.

5.6.7 CARBON NANOTUBES (CNTs)

Carbon nanotubes (CNTs) are tremendous candidates because they show many theranostic-related properties. For example, CNTs have relatively resilient near-infrared region absorption that can be employed as photothermal therapy and photoacoustic modality in high-resolution imaging. While biocompatibility and protection of CNTs is still open question. It is essential to remember that CNTs constitute the great heterogeneous materials class for that biocompatibility data cannot be generalized (Bianco et al., 2005).

5.7 CONCLUSION

Of recent, nanotechnology has played immense contribution in the field of medicine and pharmacy by maximizing the effort of natural and novel drugs in the therapeutics of severe diseases like cancer, tumor, and HIV/AIDS. The incorporation of suitable targeting ligands, reactivity, and surface curvature are requirements for the creation of optimized nanoparticle drug delivery system. By manipulation of surface properties and molecular size of nanomaterial, scientists are able to transport drugs for extended time without penetration, and there will be greater accuracy in targeting problematic access to tissues. Deeper understanding and knowledge of the actual connections related to unhealthy tissues is basic for the advancement of a new treatment methods as well as procedures in the engagement of smart nano-based carrier systems.

KEYWORDS

- **carbon nanotubes**
- **dendrimers**
- **liposomes**
- **metal nanoparticles**
- **multifunctional nanomaterials**
- **nanocarriers**
- **nanodrugs**
- **nanomedicine**
- **polymeric nanoparticles**
- **quantum dots**
- **smart drug delivery systems**
- **solid-lipid nanoparticles**

REFERENCES

Alavi, M., Karimi, N., & Safaei, M., (2017). Application of various types of liposomes in drug delivery systems. *Adv. Pharm. Bull., 7*(1), 3–9.

Bai, J., Li, Y., Du, J., Wang, S., Zheng, J., Yang, Q., & Chen, X., (2007). One-pot synthesis of polyacrylamide-gold nanocomposite. *Mater. Chem. Phys., 106*(2, 3), 412–415.

Banerjee, R., (2018). Nanotechnology in drug delivery: Present status and a glimpse into the future. *Ther. Deliv., 9*(4), 231, 232.

Bantz, C., Koshkina, O., Lang, T., Galla, H. J., Kirkpatrick, C. J., Stauber, R. H., & Maskos, M., (2014). The surface properties of nanoparticles determine the agglomeration state and the size of the particles under physiological conditions. *Beilstein. J. Nanotechnol., 5*, 1774–1786.

Bawarski, W. E., Chidlowsky, E., Bharali, D. J., & Mousa, S. A., (2008). Emerging nanopharmaceuticals. *Nanomedicine, 4*(4), 273–282.

Bianco, A., Kostarelos, K., Partidos, C. D., & Prato, M., (2005). Biomedical applications of functionalized carbon nanotubes. *Chem. Commun. (Camb)., 7*(5), 571–577.

Calvo, P., Remuñan-López, C., Vila-Jato, J. L., & Alonso, M. J., (1997). Chitosan and chitosan/ethylene oxide propylene oxide block copolymer nanoparticles as novel carriers for proteins and vaccines. *Pharm. Res., 14*, 1431–1436.

Choi, D. H., Subbiah, R., Kim, I. H., Han, D. K., & Park, K., (2013). Dual growth factor delivery using biocompatible core-shell microcapsules for angiogenesis. *Small, 9*(20), 3468–3476.

Fortuni, B., Tomoko, I., Monica, R., Yasuhiko, F., Indra, V. Z., Akito, M., Eduard, F., et al., (2019). Polymeric engineering of nanoparticles for highly efficient multifunctional drug delivery systems. *Sci. Rep., 9*, 2666. doi: https://doi.org/10.1038/s41598-019-39107-3.
Gratton, S. E., Ropp, P. A., Pohlhaus, P. D., Luft, J. C., Madden, V. J., Napier, M. E., & Desimone, J. M., (2008). The effect of particle design on cellular internalization pathways. *Proc. Natl. Acad. Sci. USA., 105*(33), 11613–11618.
Heidari, M. M., Asgari, D., Barar, J., Valizadeh, H., Kafil, V., Abadpour, A., Moumivand, E., et al., (2013). Tamoxifen loaded folic acid armed PEGylated magnetic nanoparticles for targeted imaging and therapy of cancer. *Colloids Surf. B., 106*, 117–125.
Hild, W., Breunig, M., & Gopferich, A., (2008). Quantum dots-nano-sized probes for the exploration of cellular and intracellular targeting. *Eur. J. Pharm. Biopharm., 68*(2), 153–168.
Hossen, S., Hossain, M. K., Basher, M. K., Mia, M. N. H., Rahman, M. T., & Uddin, M. J., (2019). Smart nano carrier-based drug delivery systems for cancer therapy and toxicity studies: A review. *Journal of Advanced Research, 15*, 1–18. https://doi.org/10.1016/j.jare.2018.06.005.
Huang, D., & Wu, D., (2018). Biodegradable dendrimers for drug delivery. *Mat. Sci. Eng. C-Mater., 90*, 713–727.
Jain, K., Kesharwani, P., Gupta, U., & Jain, N. K., (2010). Dendrimer toxicity: Let's meet the challenge. *Int. J. Pharm., 394*(1, 2), 122–142.
Jain, S., Tiwary, A. K., Sapra, B., & Jain, N., (2007). Formulation and evaluation of ethosomes for transdermal delivery of lamivudine. *AAPS Pharm. Sci. Tech., 8*(4), 249–257.
Kang, K. W., Chun, M. K., & Kim, O., (2010). Doxorubicin-loaded solid lipid nanoparticles to overcome multidrug resistance in cancer therapy. *Nanomedicine, 6*(2), 210–213.
Khanbabaie, R., & Jahanshahi, M., (2012). Revolutionary impact of nanodrug delivery on neuroscience. *Curr. Neuropharmacol., 10*(4), 370–392.
Kumari, A., Yadav, S. K., & Yadav, S. C., (2010). Biodegradable polymeric nanoparticles based drug delivery systems. *Colloids Surf. B., 75*(1), 1–18.
Lee, J. H., & Yeo, Y., (2015). Controlled drug release from pharmaceutical nanocarriers. *Chem. Eng. Sci., 125*, 75–84.
Lembo, D., Donalisio, M., Civra, A., Argenziano, M., & Cavalli, R., (2018). Nanomedicine formulations for the delivery of antiviral drugs: A promising solution for the treatment of viral infections. *Expert Opin. Drug Deliv., 15*(1), 93–114.
Li, D., & Kaner, R. B., (2006). Shape and aggregation control of nanoparticles: Not shaken, not stirred. *J. Am. Chem. Soc., 128*(3), 968–975.
Li, Z., Tan, S., Li, S., Shen, Q., & Wang, K., (2017). Cancer drug delivery in the nano era: An overview and perspectives. *Oncol. Rep., 38*(2), 611–624.
Lian, T., & Ho, R. J., (2001). Trends and developments in liposome drug delivery systems. *J. Pharm. Sci., 90*(6), 667–680.
Lombardo, D., Mikhail, A. K., & Maria, T., (2019). Smart nanoparticles for drug delivery application: Development of versatile nanocarrier platforms in biotechnology and nanomedicine. *J. Nanomater., 3702518*, 1–26. https://doi.org/10.1155/2019/3702518.
Mao, H. Q., Roy, K., Troung-Le, V. L., Janes, K. A., Lin, K. Y., Wang, Y., & Leong, K. W., (2001). Chitosan-DNA nanoparticles as gene carriers: Synthesis, characterization and transfection efficiency. *J. Control Release, 70*(3), 399–421.
McMillan, J., Batrakova, E., & Gendelman, H. E., (2011). Cell delivery of therapeutic nanoparticles. *Prog. Mol. Biol. Transl. Sci., 104*, 563–601.

Milla, P., Dosio, F., & Cattel, L., (2012). PEGylation of proteins and liposomes: A powerful and flexible strategy to improve the drug delivery. *Curr. Drug. Metab., 13*, 105–119.
Mulik, R. S., Mönkkönen, J., Juvonen, R. O., Mahadik, K. R., & Paradkar, A. R., (2010). Transferrin mediated solid lipid nanoparticles containing curcumin: Enhanced *in vitro* anticancer activity by induction of apoptosis. *Int. J. Pharm., 398*(1, 2), 190–203.
Mura, S., Nicolas, J., & Couvreur, P., (2013). Stimuli-responsive nanocarriers for drug delivery. *Nat. Mater., 12*(11), 991–1003.
Patra, J. K., Das, G., Fraceto, L. F., Campos, E., Rodriguez-Torres, M., Acosta-Torres, L. S., Diaz-Torres, L. A., et al., (2018). Nano based drug delivery systems: Recent developments and future prospects. *J. Nanobiotechnol., 16*(1), 71. https://doi.org/10.1186/s12951-018-0392-8.
Petrilli, R., Eloy, J. O., Lee, R. J., & Lopez, R., (2018). Preparation of immunoliposomes by direct coupling of antibodies based on a thioether bond. *Methods Mol. Biol., 1674*, 229–237.
Pisanic, T. R., Zhang, Y., & Wang, T. H., (2014). Quantum dots in diagnostics and detection: Principles and paradigms. *Analyst, 139*(12), 2968–2981.
Pooja, D., Sistla, R., & Kulhari, H., (2018). Dendrimer-drug conjugates: Synthesis strategies, stability and application in anticancer drug delivery. In: Grumezescu, A. M., (ed.), *Design of Nanostructures for Theranostics Applications* (pp. 277–303). William Andrew Publishing.
Reis, C. P., Neufeld, R. J., Ribeiro, A. J., & Veiga, F., (2006). Nanoencapsulation I. Methods for preparation of drug-loaded polymeric nanoparticles. *Nanomedicine, 2*(1), 8–21.
Rizvi, S., & Saleh, A. M., (2018). Applications of nanoparticle systems in drug delivery technology. *Saudi Pharm. J., 26*(1), 64–70.
Rizzo, L. Y., Theek, B., Storm, G., Kiessling, F., & Lammers, T., (2013). Recent progress in nanomedicine: Therapeutic, diagnostic and theranostic applications. *Curr. Opin. Biotechnol., 24*(6), 1159–1166.
Shim, T. S., Kim, S. H., & Yang, S. M., (2013). Elaborate design strategies toward novel microcarriers for controlled encapsulation and release. *Part. Syst. Charact., 30*(1), 9–45.
Son, G. H., Lee, B. J., & Cho, C. W., (2017). Mechanisms of drug release from advanced drug formulations such as polymeric-based drug-delivery systems and lipid nanoparticles. *J. Pharmaceut. Invest., 47*, 287–296.
Song, Y., Fan, J. B., & Wang, S., (2017). Recent progress in interfacial polymerization. *Mater. Chem. Front., 1*(6), 1028–1040.
Suk, J. S., Xu, Q., Kim, N., Hanes, J., & Ensign, L. M., (2016). PEGylation as a strategy for improving nanoparticle-based drug and gene delivery. *Adv. Drug Deliv. Rev., 99*(Pt.: A), 28–51.
Sushnitha, M., Evangelopoulos, M., Tasciotti, E., & Taraballi, F., (2020). Cell membrane-based biomimetic nanoparticles and the immune system: Immunomodulatory interactions to therapeutic applications. *Front. Bioeng. Biotechnol., 8*, 627. https://doi.org/10.3389/fbioe.2020.00627.
Sykes, E. A., Dai, Q., Sarsons, C. D., Chen, J., Rocheleau, J. V., Hwang, D. M., Zheng, G., et al., (2016). Tailoring nanoparticle designs to target cancer based on tumor pathophysiology. *Proc. Natl. Acad. Sci. USA, 113*(9), E1142–E1151. https://doi.org/10.1073/pnas.1521265113.
Thakur, S., Pramod, K., & Malviya, R., (2017). Utilization of polymeric nanoparticle in cancer treatment: A review. *J. Pharm. Care Health Syst., 4*(2), 172. doi: 10.4172/2376-0419.1000172.

Turos, E., Shim, J. Y., Wang, Y., Greenhalgh, K., Reddy, G. S. K., Dickey, S., & Lim, D. V., (2007). Antibiotic-conjugated polyacrylate nanoparticles: New opportunities for development of anti-MRSA agents. *Bioorg. Med. Chem. Lett., 17*(1), 53–56.

Wang, A. Z., Langer, R., & Farokhzad, O. C., (2012). Nanoparticle delivery of cancer drugs. *Annu. Rev. Med., 63*, 185–198.

Wang, R. B., Billone, P. S., & Mullett, W. M., (2013). Nanomedicine in action: An overview of cancer nanomedicine on the market and in clinical trials. *J. Nanomater., 2013*, 12. https://doi.org/10.1155/2013/629681.

Wang, W., Lu, K., Yu, C., Huang, Q. L., & Du, Y. Z., (2019) Nano-drug delivery systems in wound treatment and skin regeneration. *J. Nanobiotechnol., 17*, 82. https://doi.org/10.1186/s12951-019-0514-y.

Weissig, V., Pettinger, T. K., & Murdock, N., (2014). Nanopharmaceuticals (part 1): Products on the market. *Int. J. Nanomedicine, 9*, 4357–4373.

Wilczewska, A. Z., Niemirowicz, K., Markiewicz, K. H., & Car, H., (2012). Nanoparticles as drug delivery systems. *Pharmacol. Rep., 64*(5), 1020–1037.

Yang, S., Zhu, J., Lu, Y., Liang, B., & Yang, C., (1999). Body distribution of camptothecin solid lipid nanoparticles after oral administration. *Pharm. Res., 16*(5), 751–757.

Yu, X., Trase, I., Ren, M., Duval, K., Guo, X., & Chen, Z., (2016). Design of nanoparticle-based carriers for targeted drug delivery. *J. Nanomater., 2016*, 1087250. https://doi.org/10.1155/2016/1087250.

Yuan, P., Ding, X., Yang, Y. Y., & Xu, Q. H., (2018). Metal nanoparticles for diagnosis and therapy of bacterial infection. *Adv. Healthc. Mater., 7*(13), 1701392. https://doi.org/10.1002/adhm.201701392.

Zara, G. P., Cavalli, R., Fundaro, A., Bargoni, A., Caputo, O., & Gasco, M. R., (1999). Pharmacokinetics of doxorubicin incorporated in solid lipid nanospheres (SLN). *Pharmacol. Res., 40*(3), 281–286.

CHAPTER 6

NANOTECHNOLOGY-BASED DELIVERY SYSTEMS FOR TYROSINE KINASES INHIBITORS IN NON-SMALL CELL LUNG CANCER TREATMENT

CATARINA SOUSA,[1] MARIA JACOB,[1] AMANY M. BESHBISHY,[2] GABER E. BATIHA,[3] NATÁLIA CRUZ-MARTINS,[4,5,6] and MARIA GABRIELA O. FERNANDES[1,4]

[1]*Pulmonology Department, Centro Hospitalar e Universitário de São João, Porto, Portugal*

[2]*National Research Center for Protozoan Diseases, Obihiro University of Agriculture and Veterinary Medicine, Nishi 2-13, Inada-Cho, Obihiro–080-8555, Hokkaido, Japan*

[3]*Department of Pharmacology and Therapeutics, Faculty of Veterinary Medicine, Damanhour University, Damanhour–22511, AlBeheira, Egypt*

[4]*Faculty of Medicine, University of Porto, Porto, Portugal*

[5]*Institute for Research and Innovation in Health (i3S), University of Porto, Porto, Portugal*

[6]*Laboratory of Neuropsychophysiology, Faculty of Psychology and Education Sciences, University of Porto, Portugal*

6.1 INTRODUCTION

Lung cancer (LC) is linked to high mortality rates worldwide, with 2.1 and 1.8 million new cases and deaths, respectively, predicted in 2018 (Bray et al., 2018). The impact on patients' lives is far-reaching and depends on a number of factors such as clinical manifestations, comorbidities, staging, treatment side effects, and requirements, as well personal responsibilities.

Based on histological characteristics, LC is broadly divided into small (SCLC) and non-small cell LC (NSCLC), which is further divided into squamous and large cell carcinoma, and adenocarcinoma, with more than 50% of LC cases being adenocarcinomas (Pao and Girard, 2011). In the last decade, a shift towards a further subdivision of NSCLC into molecular-based subtypes has been stated through the identification of oncogenic driver mutations (Travis et al., 2011). Detection of targetable genetic alterations in LC has transformed the therapeutic strategy in recent years, contributing to the improvement of LC outcomes. Driver oncogenic mutations, particularly at *EGFR* or *ALK* gene rearrangements, have led to a paradigm change with the advances in specific molecular treatments, up to the point that LC guidelines are currently incorporating molecular testing and the application of targeted drugs (Reck et al., 2014).

EGFR mutations have been seen in around 15% of pulmonary adenocarcinomas, occurring more often in women, non-smokers, and Asian populations (Dearden et al., 2013; Kawaguchi et al., 2016). Regarding anti-EGFR agents, two classes have shown clinical effects in NSCLC: (i) tyrosine kinase inhibitors (TKIs), triggering the EGFR tyrosine kinase inhibition; and (ii) monoclonal antibodies (mAbs) targeting the EGFR extracellular domain. Currently, EGFR-TKIs, such as Erlotinib, Gefitinib, Afatinib, and Osimertinib, are viewed as the standard of care in EGFR-mutated NSCLC, besides to provide a more favorable prognosis over the standard chemotherapy (Maemondo et al., 2010; Rosell et al., 2012; Mok et al., 2013, 2016; Sequist et al., 2013; Soria et al., 2017). Regarding anti-EGFR mAbs, Cetuximab is an approved drug by FDA for colorectal, and head and neck squamous cell cancers, with some evidence suggesting that it improves the efficacy over standard therapy in locally advanced and metastatic NSCLC (Yang et al., 2014), although its impact on survival is low and, considered non-justifiable so far, given its associated toxicity (Mazzarella et al., 2018).

Similarly, the EML4-ALK (echinoderm microtubule-associated protein-like 4-anaplastic lymphoma kinase) fusion gene has been stated as another independent LC driver. ALK gene rearrangements are present in approximately 4% of NSCLC adenocarcinomas and are more often observed in non-smokers and younger patients (Soda et al., 2007; Shaw et al., 2009). Targeting this abnormality with Crizotinib and other ALK inhibitors has been clinically successful (Shaw et al., 2013; Hida et al., 2017; Soria et al., 2017).

Despite the initial responses stated to drugs targeting EGFR or ALK, tumors end up acquiring drug resistance. In addition, although TKIs are less toxic than traditional systemic chemotherapy, they are commonly associated with specific side effects. Thus, it remains an unmet need to understand the LC' biology and resistance mechanisms and to develop novel therapies with better bioavailability and less toxicity. In this way, nanotechnology appears as a promising strategy to provide unique solutions able to revolutionize the LC-targeted therapy. Thus, this chapter gives an in-depth vision on nanotechnology-based delivery systems for TKI in NSCLC treatment.

6.2 NANOTECHNOLOGY-BASED DELIVERY SYSTEMS CONTAINING TKIs

There have been developed several nanotechnology-based delivery systems containing TKIs for NSCLC therapy (Figure 6.1 and Table 6.1), with Erlotinib, and Erlotinib combined with Doxorubicin; Afatinib; Gefitinib and Gefitinib combined with Doxorubicin, cyclosporin A (CsA) and Simvastatin; Crizotinib and Crizotinib combined with Palbociclib and Sildenafil being those already investigated so far.

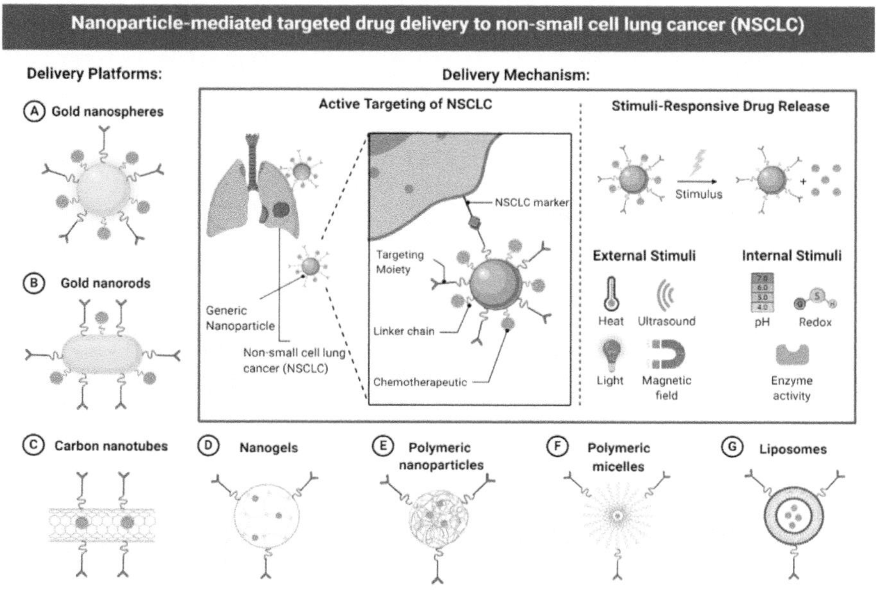

FIGURE 6.1 Nanoparticle-mediated targeted drug delivery to non-small cell lung cancer.

TABLE 6.1 Nano-Based Delivery Systems Containing TKIs for Lung Cancer Therapy

Drug	Delivery System	Outcome	References
Erlotinib	Solid self-emulsifying system	Solubility and bioavailability enhancement	Truong et al. (2016)
	Galactosylated liposomes	Improvement of body distribution and bioavailability enhancement	Xu et al. (2018)
	PEGylation liposomes	Bioavailability enhancement	Zhou et al. (2017)
	SLNP (DPI formulation)	Sustain drug release, with efficacy enhancement	Bakhtiary et al. (2017)
	PLGA nanoparticles	Bioavailability enhancement and less subacute toxicity	Marslin et al. (2009)
	PLGA nanoparticles	Sustain drug release, with efficacy enhancement	Vaidya et al. (2019)
	Gold NP	Bioavailability enhancement	Fathi et al. (2018)
	Aptamer-conjugated chitosan liposomal complex	Bioavailability enhancement	Li et al. (2017)
Erlotinib and doxorubicin	pH-responsive MSN	Efficacy enhancement and less toxicity	He et al. (2016)
Afatinib	Gold NP	Efficacy enhancement	Cryer et al. (2019)
	Colloidal PEGylated gold NP	Efficacy enhancement	Coelho et al. (2016)
Gefitinib	Gold NP	Less toxicity	Lam et al. (2014)
	Tri-block nanoparticle	Overcome EGFR-TKI resistance	Srikar et al. (2016)
Gefitinib and doxorubicin	MSN	Overcome EGFR-TKI resistance	Wang et al. (2016)
Gefitinib and cyclosporin A	PEG-PLA	Reverse multi-drug resistance	Han et al. (2018)
Gefitinib and simvastatin	Liposome	Overcome EGFR-T790M resistance	Yin et al. (2018)
Crizotinib	PLA-TPGS micelles	Sustain drug release, with efficacy enhancement	Jiang et al. (2015)
Crizotinib, Palbociclib, and sildenafil	PLA-TPGS micelles	Efficacy enhancement	Melo-Diogo et al. (2014)

First of all, to raise the oral bioavailability of Erlotinib, Truong et al. (2016) designed a solid self-emulsifying drug delivery system (SEDDS) formulation of Erlotinib, prepared by spray drying method. Initially, Erlotinib was dissolved in mixtures of various oil, surfactant, and co-surfactant at room temperature, and then, spray drying procedure was used to prepare the solid SEDDS formulations (Truong et al., 2016). Pharmacokinetic study in rats indicated that the bioavailability of Erlotinib in these formulations was improved when compared to the pure drug. In addition, several studies testing Erlotinib liposomes have reported prominent changes in the pharmacokinetic behavior of the drug, ultimately improving its targeting and bioavailability (Zhou et al., 2017; Xu et al., 2018). For example, Bakhtiary et al. (2017) prepared microparticles containing solid-lipid nanoparticles (SLNPs) of erlotinib along with mannitol, for pulmonary delivery with a dry powder inhaler, with a sustain drug release profile (Figure 6.1) (Bakhtiary et al., 2017). This method represents a potential system for local drug delivery, effective in the lymphatic system and also avoiding systemic toxicity. The encapsulation of Erlotinib within poly(D,L-lactic-coglycolic acid) (PLGA) have also been investigated, with data obtained revealing an enhanced bioavailability and reduced toxicity (Marslin et al., 2009), besides to a greater efficacy against NSCLC cells with increased apoptosis (Vaidya et al., 2019).

For ALK molecular alterations, a nano-carrier encapsulating Crizotinib was developed, within polylactide tocopheryl polyethylene glycol (PEG) 1000 succinate (PLA-TPGS) (de Melo-Diogo et al., 2014; Jiang et al., 2015), being found a sustained release and an improved cytotoxic effect. On the other side, nano-systems using tumor microenvironment-like pH value to improve the cellular internalization of drugs can be used to treat all tumors kinds. Taking this in view, He et al. (2016) developed a pH-responsive mesoporous silica nanoparticle (MSN) to deliver synergistic Erlotinib-Doxorubicin combination for the treatment of NSCLC (He et al., 2016). As main findings, the authors found *in vivo* marked tumor accumulation and tumor growth suppression, without systemic toxicity, in LC-bearing mice.

Another key challenge is to overcome TKIs resistance, with gold nanoparticles (Au NPs) being viewed as promising TKI-drug delivery candidates, with excellent anticancer therapy response and fewer side effects (Lam et al., 2014; Coelho et al., 2016; Fathi et al., 2018; Li et al., 2018; Cryer et al., 2019) (Table 6.1).

Li et al. (2018) developed liposomes anchored with anti-EGFR aptamer (Apt)-conjugated chitosan, and Erlotinib. *In vitro*, this liposomal complex guided by the anti-EGFR Apt showed specific bind to EGFR TKI-resistant cells with increased cell uptake and better antiproliferative effect. Moreover, Wang et al. (2016) designed a cetuximab-modified MSN as a drug carrier, to simultaneously release Doxorubicin and Gefitinib to target EGFR-mutant

LC cells. The authors stated that co-delivery nanoparticles could overcome EGFR-TKI resistance and held strong potential for an effective EGFR-mutant LC management.

From the other point of view, and given that CsA seems to reverse resistance to multiple drugs, especially EGFR-TKI resistance, Han et al. in a deepen study, showed that the simultaneous encapsulation of CsA with Gefitinib within PEG-PLA, more efficiently suppressed the tumor growth. In addition, *in vivo* data revealed that CsA augments the gefitinib potency in Gef-sensitive and -resistant cell lines (Han et al., 2018). More recently, Yin et al. (2018) generating anti-PD-L1-nanobody-decorated liposomes to deliver Gefitinib-Simvastatin treatment (a combination that can remodel the tumor microenvironment) in PD-L1-expressing cancer cells and macrophages, stated that the combination displayed ability to reverse resistance to gefitinib and to enhance EGFR T790M-mutated NSCLC treatment outcomes.

Lastly, and also worth of note is that, the use of TKIs combined with small interfering RNA (siRNA) has been viewed as an additional strategy for cancer therapy. For example, Srikar et al. (2016) demonstrated that porous gelatin triblock nanoparticles containing Gefitinib and Cetuximab-siRNA conjugate is able to deliver siRNA at cytoplasm level of KRAS mutant NSCLC cells for oncogene knockdown, ultimately sensitizing it to the TKI.

6.3 CONCLUSIONS

Several genetic alterations have been clarified as therapeutic targets, namely those occurring in the EGFR gene and ALK gene rearrangements. Target therapies have significantly improved patients' outcomes with more favorable tolerability compared to systemic chemotherapy. Despite this, they are associated with specific adverse events and, in some cases, tumors acquire resistance to TKIs.

Nanomedicine (NM) is in a phase of growing research and interest, with nanomaterials being viewed as promising tools for the development of efficient delivery systems for TKIs. Data currently available underline the need to develop novel therapies with better bioavailability and less toxicity, where nanotechnology appears as a promissory technique to raise drugs' effectiveness. Nonetheless, although several investigations have been made in this field, in recent years, further, and increasingly deepen *in vivo* studies are needed to confirm their therapeutic benefits and safety profiles.

ACKNOWLEDGMENTS

NM acknowledges the Portuguese Foundation for Science and Technology under the Horizon 2020 Program (PTDC/PSI-GER/28076/2017).

KEYWORDS

- **cancer**
- **mesoporous silica nanoparticle**
- **nanomedicine**
- **solid-lipid nanoparticles**
- **target therapy**
- **tyrosine kinase inhibitors**

REFERENCES

Bakhtiary, Z., Barar, J., Aghanejad, A., Saei, A. A., Nemati, E., Ezzati, N. D. J., & Omidi, Y., (2017). Microparticles containing erlotinib-loaded solid lipid nanoparticles for treatment of non-small cell lung cancer. *Drug Development and Industrial Pharmacy, 43*(8), 1244–1253. https://doi.org/10.1080/03639045.2017.1310223.

Bray, F., Ferlay, J., Soerjomataram, I., Siegel, R. L., Torre, L. A., & Jemal, A., (2018). Global cancer statistics 2018: GLOBOCAN estimates of incidence and mortality worldwide for 36 cancers in 185 countries. *CA: A Cancer Journal for Clinicians, 68*(6), 394–424. https://doi.org/10.3322/caac.21492.

Coelho, S. C., Almeida, G. M., Pereira, M. C., Santos-Silva, F., & Coelho, M. A. N., (2016). Functionalized gold nanoparticles improve afatinib delivery into cancer cells. *Expert Opinion on Drug Delivery, 13*(1), 133–141. https://doi.org/10.1517/17425247.2015.1083973.

Cryer, A. M., Chan, C., Eftychidou, A., Maksoudian, C., Mahesh, M., Tetley, T. D., et al., (2019). Tyrosine kinase inhibitor gold nanoconjugates for the treatment of non-small cell lung cancer. *ACS Applied Materials and Interfaces, 11*(18), 16336–16346. https://doi.org/10.1021/acsami.9b02986.

De Melo-Diogo, D., Gaspar, V. M., Costa, E. C., Moreira, A. F., Oppolzer, D., Gallardo, E., & Correia, I. J., (2014). Combinatorial delivery of crizotinib-palbociclib-sildenafil using TPGS-PLA micelles for improved cancer treatment. *European Journal of Pharmaceutics and Biopharmaceutics, 88*(3), 718–729. https://doi.org/https://doi.org/10.1016/j.ejpb.2014.09.013.

Dearden, S., Stevens, J., Wu, Y. L., & Blowers, D., (2013). Mutation incidence and coincidence in non small-cell lung cancer: Meta-analyses by ethnicity and histology (mutMap). *Annals of Oncology: Official Journal of the European Society for Medical Oncology, 24*(9), 2371–2376. https://doi.org/10.1093/annonc/mdt205.

Fathi, M., Sahandi, Z. P., Barar, J., Aghanejad, A., Erfan-Niya, H., & Omidi, Y., (2018). Thermo-sensitive chitosan copolymer-gold hybrid nanoparticles as a nanocarrier for delivery of erlotinib. *International Journal of Biological Macromolecules, 106*, 266–276. https://doi.org/https://doi.org/10.1016/j.ijbiomac.2017.08.020.

Han, W., Shi, L., Ren, L., Zhou, L., Li, T., Qiao, Y., & Wang, H., (2018). A nanomedicine approach enables co-delivery of cyclosporin A and gefitinib to potentiate the therapeutic efficacy in drug-resistant lung cancer. *Signal Transduction and Targeted Therapy, 3*(1), 16. https://doi.org/10.1038/s41392-018-0019-4.

He, Y., Su, Z., Xue, L., Xu, H., & Zhang, C., (2016). Co-delivery of erlotinib and doxorubicin by pH-sensitive charge conversion nanocarrier for synergistic therapy. *Journal of Controlled Release, 229*, 80–92. https://doi.org/https://doi.org/10.1016/j.jconrel.2016.03.001.

Hida, T., Nokihara, H., Kondo, M., Kim, Y. H., Azuma, K., Seto, T., et al., (2017). Alectinib versus crizotinib in patients with ALK-positive non-small-cell lung cancer (J-ALEX): An open-label, randomized phase 3 trial. *Lancet, 390*(10089), 29–39. https://doi.org/10.1016/s0140-6736(17)30565-2.

Jiang, Z. M., Dai, S. P., Xu, Y. Q., Li, T., Xie, J., Li, C., & Zhang, Z. H., (2015). Crizotinib-loaded polymeric nanoparticles in lung cancer chemotherapy. *Medical Oncology, 32*(7), 193. https://doi.org/10.1007/s12032-015-0636-5.

Kawaguchi, T., Koh, Y., Ando, M., Ito, N., Takeo, S., Adachi, H., et al., (2016). Prospective analysis of oncogenic driver mutations and environmental factors: Japan molecular epidemiology for lung cancer study. *Journal of Clinical Oncology, 34*(19), 2247–2257. https://doi.org/10.1200/JCO.2015.64.2322.

Lam, A. T. N., Yoon, J., Ganbold, E. O., Singh, D. K., Kim, D., Cho, K. H., et al., (2014). Colloidal gold nanoparticle conjugates of gefitinib. *Colloids and Surfaces B: Biointerfaces, 123*, 61–67. https://doi.org/https://doi.org/10.1016/j.colsurfb.2014.08.021.

Li, S., Liu, Y., Rui, Y., Tang, L., Achilefu, S., & Gu, Y., (2018). Dual target gene therapy to EML4-ALK NSCLC by a gold nanoshell-based system. *Theranostics, 8*(10), 2621–2633. https://doi.org/10.7150/thno.24469.

Maemondo, M., Inoue, A., Kobayashi, K., Sugawara, S., Oizumi, S., Isobe, H., et al., (2010). Gefitinib or chemotherapy for non-small-cell lung cancer with mutated EGFR. *New England Journal of Medicine, 362*(25), 2380–2388. https://doi.org/10.1056/NEJMoa0909530.

Marslin, G., Sheeba, C. J., Kalaichelvan, V. K., Manavalan, R., Reddy, P. N., & Franklin, G., (2009). Poly(D,L-lactic-co-glycolic acid) nanoencapsulation reduces Erlotinib-induced subacute toxicity in rat. *Journal of Biomedical Nanotechnology, 5*(5), 464–471. https://doi.org/10.1166/jbn.2009.1075.

Mazzarella, L., Guida, A., & Curigliano, G., (2018). Cetuximab for treating non-small cell lung cancer. *Expert Opinion on Biological Therapy, 18*(4), 483–493. https://doi.org/10.1080/14712598.2018.1452906.

Mok, T. S., Wu, Y. L., Ahn, M. J., Garassino, M. C., Kim, H. R., Ramalingam, S. S., et al., (2016). Osimertinib or platinum-pemetrexed in EGFR T790M–positive lung cancer. *New England Journal of Medicine, 376*(7), 629–640. https://doi.org/10.1056/NEJMoa1612674.

Mok, T., Yang, J. J., & Lam, K. C., (2013). Treating patients with EGFR-sensitizing mutations: First line or second line - is there a difference? *Journal of Clinical Oncology, 31*(8), 1081–1088. https://doi.org/10.1200/JCO.2012.43.0652.

Pao, W., & Girard, N., (2011). New driver mutations in non-small-cell lung cancer. *The Lancet Oncology, 12*(2), 175–180. https://doi.org/https://doi.org/10.1016/S1470-2045(10)70087-5.

Reck, M., Popat, S., Reinmuth, N., De Ruysscher, D., Kerr, K. M., & Peters, S., (2014). Metastatic non-small-cell lung cancer (NSCLC): ESMO clinical practice guidelines for diagnosis, treatment and follow-up. *Annals of Oncology, 25*, iii27–iii39. https://doi.org/https://doi.org/10.1093/annonc/mdu199.

Rosell, R., Carcereny, E., Gervais, R., Vergnenegre, A., Massuti, B., Felip, E., et al., (2012). Erlotinib versus standard chemotherapy as first-line treatment for European patients with advanced EGFR mutation-positive non-small-cell lung cancer (EURTAC): A multicenter, open-label, randomized phase 3 trial. *The Lancet Oncology, 13*(3), 239–246. https://doi.org/10.1016/S1470-2045(11)70393-X.

Sequist, L. V., Yang, J. C. H., Yamamoto, N., O'Byrne, K., Hirsh, V., Mok, T., et al., (2013). Phase III study of afatinib or cisplatin plus pemetrexed in patients with metastatic lung adenocarcinoma with EGFR mutations. *Journal of Clinical Oncology, 31*(27), 3327–3334. https://doi.org/10.1200/JCO.2012.44.2806.

Shaw, A. T., Kim, D. W., Nakagawa, K., Seto, T., Crinó, L., Ahn, M. J., et al., (2013). Crizotinib versus chemotherapy in advanced ALK-positive lung cancer. *New England Journal of Medicine, 368*(25), 2385–2394. https://doi.org/10.1056/NEJMoa1214886.

Shaw, A. T., Yeap, B. Y., Mino-Kenudson, M., Digumarthy, S. R., Costa, D. B., Heist, R. S., et al., (2009). Clinical features and outcome of patients with non-small-cell lung cancer who harbor EML4-ALK. *Journal of Clinical Oncology: Official Journal of the American Society of Clinical Oncology, 27*(26), 4247–4253. https://doi.org/10.1200/JCO.2009.22.6993.

Soda, M., Choi, Y. L., Enomoto, M., Takada, S., Yamashita, Y., Ishikawa, S., et al., (2007). Identification of the transforming EML4-ALK fusion gene in non-small-cell lung cancer. *Nature, 448*(7153), 561–566. https://doi.org/10.1038/nature05945.

Soria, J. C., Ohe, Y., Vansteenkiste, J., Reungwetwattana, T., Chewaskulyong, B., Lee, K. H., et al., (2017). Osimertinib in untreated EGFR-mutated advanced non-small-cell lung cancer. *New England Journal of Medicine, 378*(2), 113–125. https://doi.org/10.1056/NEJMoa1713137.

Soria, J. C., Tan, D. S. W., Chiari, R., Wu, Y. L., Paz-Ares, L., Wolf, J., et al., (2017). First-line ceritinib versus platinum-based chemotherapy in advanced ALK-rearranged non-small-cell lung cancer (ASCEND-4): A randomized, open-label, phase 3 study. *The Lancet, 389*(10072), 917–929. https://doi.org/https://doi.org/10.1016/S0140-6736(17)30123-X.

Srikar, R., Suresh, D., Zambre, A., Taylor, K., Chapman, S., Leevy, M., et al., (2016). Targeted nanoconjugate co-delivering siRNA and tyrosine kinase inhibitor to KRAS mutant NSCLC dissociates GAB1-SHP2 post oncogene knockdown. *Scientific Reports, 6*(1), 30245. https://doi.org/10.1038/srep30245.

Travis, W. D., Brambilla, E., Noguchi, M., Nicholson, A. G., Geisinger, K., Yatabe, Y., et al., (2011). International Association for the study of lung cancer/American thoracic society/European respiratory society: International multidisciplinary classification of lung adenocarcinoma. *Proceedings of the American Thoracic Society, 8*(5), 381–385. https://doi.org/10.1513/pats.201107-042ST.

Truong, D. H., Tran, T. H., Ramasamy, T., Choi, J. Y., Lee, H. H., Moon, C., et al., (2016). Development of solid self-emulsifying formulation for improving the oral bioavailability of erlotinib. *AAPS PharmSciTech, 17*(2), 466–473. https://doi.org/10.1208/s12249-015-0370-5.

Vaidya, B., Parvathaneni, V., Kulkarni, N. S., Shukla, S. K., Damon, J. K., Sarode, A., et al., (2019). Cyclodextrin modified erlotinib loaded PLGA nanoparticles for improved therapeutic efficacy against non-small cell lung cancer. *International Journal of Biological Macromolecules, 122*, 338–347. https://doi.org/https://doi.org/10.1016/j.ijbiomac. 2018.10.181.

Wang, Y., Huang, H. Y., Yang, L., Zhang, Z., & Ji, H., (2016). Cetuximab-modified mesoporous silica nano-medicine specifically targets EGFR-mutant lung cancer and overcomes drug resistance. *Scientific Reports, 6*, 25468. https://doi.org/10.1038/srep25468.

Xu, H., He, C., Liu, Y., Jiang, J., & Ma, T., (2018). Novel therapeutic modalities and drug delivery-erlotinib liposomes modified with galactosylated lipid: *In vitro* and *in vivo* investigations. *Artificial Cells, Nanomedicine, and Biotechnology, 46*(8), 1902–1907. https://doi.org/10.1080/21691401.2017.1396222.

Yin, W., Yu, X., Kang, X., Zhao, Y., Zhao, P., Jin, H., et al., (2018). Remodeling tumor-associated macrophages and neovascularization overcomes EGFRT790M-associated drug resistance by PD-L1 nanobody-mediated codelivery. *Small, 14*(47), e1802372. https://doi.org/10.1002/smll.201802372.

Zhou, X., Tao, H., & Shi, K. H., (2017). Development of a nanoliposomal formulation of erlotinib for lung cancer and *in vitro/in vivo* antitumoral evaluation. *Drug Design, Development and Therapy, 12*, 1–8. https://doi.org/10.2147/DDDT.S146925.

CHAPTER 7

NANOPARTICULATE SYSTEMS FOR LUNG CANCER TARGETED THERAPY

ANA CLÁUDIA PIMENTA,[1] LUÍSA NASCIMENTO,[1] and NATÁLIA CRUZ-MARTINS[2,3]

[1]*Pulmonology Department, Centro Hospitalar de Trás-os-Montes e Alto Douro, Vila Real, Portugal*

[2]*Faculty of Medicine, University of Porto, Porto, Portugal*

[3]*Institute for Research and Innovation in Health (i3S), University of Porto, Porto, Portugal*

7.1 INTRODUCTION

Lung cancer (LC) accounts for most cancer-related deaths in the world, and it is likely to prevail an important disease burden (Ridge et al., 2013; Albaba et al., 2017; Nasim et al., 2019). It represents a public health issue, as most cases have an environmental trigger, mainly from tobacco smoke, but also from radon exposure, asbestos inhalation, environmental pollution, and biomass combustion (Nasim et al., 2019).

The poor overall survival rates are mostly due to late diagnosis. In fact, only 15% of LCs are diagnosed in the early stages; these indicators are expected to improve with the implementation of organized surveillance methods for high-risk populations (Ridge et al., 2013; Oudkerk et al., 2017; Nasim et al., 2019). Smoking cessation is the most powerful tool to decrease LC incidence and LC deaths. Nevertheless, as it represents such a major health problem, efforts are needed to improve the current therapies (Ridge et al., 2013; Mayekar and Bivona, 2017; Oudkerk et al., 2017). The term "lung tumor" comprises a variety of histopathological entities (Figure 7.1) (Manuel et al., 2018; Oberndorfer and Müllauer, 2018). Histological and immunohistochemistry traits allow us to distinguish carcinoid tumor, small cell lung carcinoma (SCLC), and non-small cell lung carcinoma (NSCLC), further divided into adenocarcinoma (invasive or non-invasive), adenosquamous cell carcinoma, squamous-cell carcinoma, and large-cell carcinoma (Oberndorfer and Müllauer, 2018). Importantly, adenocarcinoma represents 40% of all LCs. The anatomical site of the primary lesion also typically varies according to histological subtype (Figure 7.2). Currently, it is vastly recognized that lung tumors present a great heterogeneity at molecular, genetic, and epigenetic levels, and understanding these characteristics is crucial for prognostic and therapeutic guidance (Hirsch et al., 2017; Mayekar and Bivona, 2017; Oberndorfer and Müllauer, 2018). To our days, molecular targeted therapy is only available for adenocarcinoma (Mayekar and Bivona, 2017; Oberndorfer and Müllauer, 2018), but in the last years, a higher interest has been put on the design and use of nanocarriers (NCs) to overcome issues related to delivery, pharmacokinetics, and pharmacodynamics, and safety of certain drugs, namely those targeted to fight cancer. In this sense, the present review aims to provide a brief outline on the advances in nanoparticulate targeted therapies for LC.

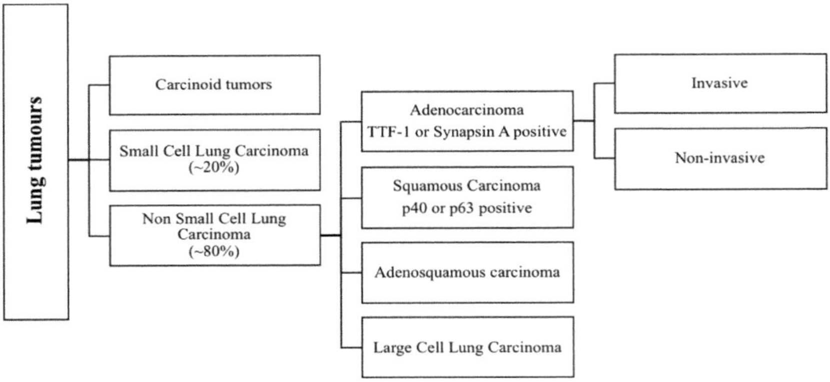

FIGURE 7.1 Histological classification of lung tumors.

Source: Manuel et al. (2018); Oberndorfer and Müllauer (2018).

7.2 LUNG NEOPLASMS AND CURRENT THERAPIES: A BRIEF OVERVIEW

As previously mentioned, the therapeutic approach of lung tumors is guided by specific characteristics that impact prognosis (Goldstraw et al., 2016). Tumor size, invasion degree of adjacent structures, involvement degree of lymph nodes, and presence or absence of distant metastasis are addressed by TNM classification, which is the most widely applied tumor staging method to guide treatment approach (Goldstraw et al., 2016; Nasim et al., 2019). The histologic grade is also important for prognosis prediction, especially for adenocarcinoma (Yasukawa et al., 2018). Even for early stages and small tumors, the low-grade tumors, classified as lepidic, acinar or papillary, are recognized as having a significantly better prognosis than the micropapillary, solid or invasive mucinous adenocarcinomas, considered as high-grade subtypes (Yasukawa et al., 2018). The molecular study is also important, as it can both predict prognosis and determine whether and which immunotherapy targets should be addressed, in advanced stages (Hirsch et al., 2017; Mayekar and Bivona, 2017; Oberndorfer and Müllauer, 2018).

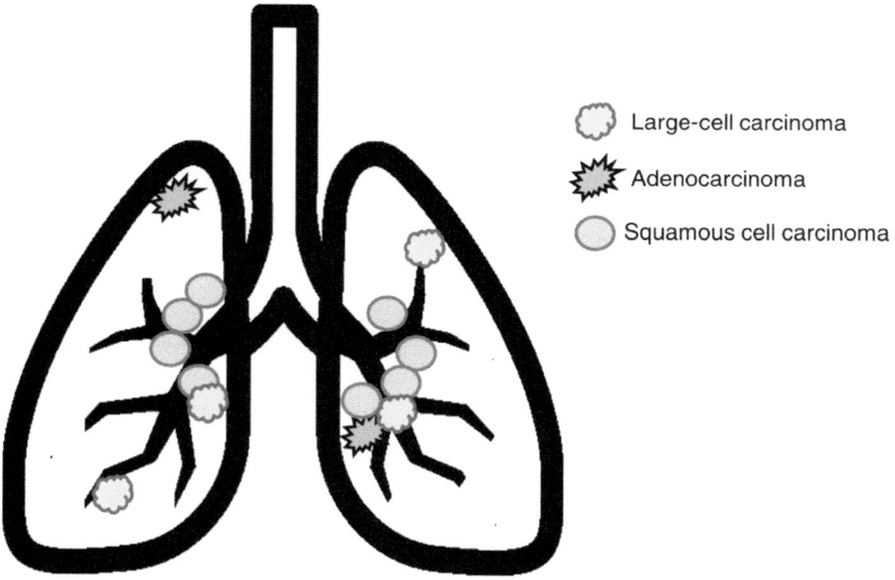

FIGURE 7.2 Pictorial representation of the most common sites of primary lesion, according to histological subtype, for the most common types of lung cancer (NSCLC): squamous cell carcinoma, adenocarcinoma, and large cell carcinoma and most frequent location (Davidson et al., 2013; Lemjabbar-Alaoui et al., 2016). Note that some primary lesions tend to originate at more peripheral sites, arising from alveolar epithelium, and others to develop more centrally, arising from airway epithelium. This observation, along with molecular markers identification, is potentially useful for targeting design.

The TNM classification is the most widely accepted classification system for most types of cancer (Goldstraw et al., 2016). This staging system, revised in 2017, performs well in predicting the prognosis for most histological types in LC. Increases of 1 cm in the tumor size have a significant impact on prognosis. Interestingly, for endobronchial tumors, T2 and T3 share a similar prognosis, even in the presence of total atelectasis and pneumonitis (Goldstraw et al., 2016). However, for SCLC, due to the characteristic aggressiveness of the disease, its applicability is still somewhat limited (Abdel-Rahman, 2018). As a valuable tool for predicting prognosis, along with performance status scales, staging systems also guide therapeutic approaches, illustrated in Table 7.1 (Goldstraw et al., 2016; Hirsch et al., 2017). For earlier stages, therapeutic approaches aim at curative intent; palliative care is reserved for the more advanced disease stages or for patients with low performance, ineligible for chemotherapy, radiotherapy, and/or surgery (Hirsch et al., 2017). Tumor staging methods are also intended to assist in the evaluation of treatment results, in addition to facilitating data shared between centers (Lemjabbar-Alaoui et al., 2016).

7.2.1 SURGICAL TREATMENT

Surgical treatment is used with curative intent, mostly for NSCLC in its early stages, due to its limited role in SCLC (Nasim et al., 2019). It targets the complete resection of the tumor, along with its regional lymph nodes, allowing the correct tumor staging and determining the post-surgical treatment (Nasim et al., 2019). Even for small lesions, lobectomy is usually necessary to obtain R0 resection (disease free margins) owing to the tumor's growth characteristics; wedge resections are associated with higher local recurrence (Lemjabbar-Alaoui et al., 2016; Nasim et al., 2019). Regrettably, lobectomy is only suitable for patients with an estimated postoperative predicted FEV1 >40% or VO_2 max of at least 15 mL/kg/min, criteria that are often not met, due to impaired lung function related to tobacco smoke (Nasim et al., 2019). For stages II and III, even for patients electable for surgery, there is also a need for adjuvant or neoadjuvant chemotherapy (Table 7.1) (Lemjabbar-Alaoui et al., 2016).

7.2.2 PHARMACOLOGICAL TREATMENT

Pharmacological treatment should be offered to all patients with NSCLC stages II, III, or IV as also to stage IB, when the tumor is at least 4 cm, as surgery alone slightly improves survival (Lemjabbar-Alaoui et al., 2016;

TABLE 7.1 Preferred Treatment According to NSCLC Stage*

Stage	Therapeutic Intention	Preferred Treatment	Options and Add-On Treatments to Consider
I	Curative	Surgery	High-dose stereotactic radiation therapy (unfit or refusing surgery) Stage IB-perioperative chemotherapy
II		Surgery	High-dose stereotactic radiation therapy (unfit, unresectable, or refusing surgery) High-dose stereotactic body radiation therapy
III		IIIA resectable-surgery with adjuvant chemotherapy or radiotherapy IIIA unresectable-sequential or concurrent chemotherapy and radiation therapy IIIB-sequential combination of chemotherapy or external radiation therapy IIIC-definitive concurrent chemoradiation or Durvalumab	Heterogeneous group Options include radiotherapy, chemotherapy, and surgical resection in selected cases
IV	Palliative	Targeted therapy, according to driver mutations or PDL-1 expression	Standard combination chemotherapy for patients with performance status 0 to 2 and/or no driver mutations identified. Patients with performance status >2 should receive single-agent chemotherapy and/or palliative care.

**Source:* Lemjabbar-Alaoui et al. (2016); Postmus et al. (2017); Planchard et al. (2018); Ettinger et al. (2019).

Watanabe et al., 2017). Chemotherapy aims to control local disease, as well as microscopic systemic disease (Watanabe et al., 2017). For many years, platinum-based chemotherapy regimens have been considered the standard of care; however, these regimens have demonstrated only 5% to 11% improvement of survival, usually with very significant immediate and long-term toxicity (Michels and Wolf, 2016; Nagasaka and Gadgeel, 2018; Amararathna et al., 2019). As cytotoxic agents, the clinical efficacy of these drugs is conditioned by dosage concerns, which contributes to the development of drug resistance and poor overall prognosis of LC (Pérez-Herrero and Fernández-Medarde, 2015; Mangal et al., 2017; Amararathna et al., 2019). The concentration of the chemotherapeutic agents reached at the solid tumor microenvironment is essential for effectiveness; systemic treatment has failed to achieve the optimal concentrations at these sites (Chatzaki et al., 2012). Strategies are being developed to help improving drug targeting to cancer cells, as to optimize therapeutic effects and limit systemic toxicity (Pérez-Herrero and Fernández-Medarde, 2015; Amararathna et al., 2019). On the other hand, the advances in cell signaling pathways description, responsible for cell proliferation and differentiation has allowed the development of effective drugs targeting these pathways with an improved tolerability profile (Lemjabbar-Alaoui et al., 2016; Mayekar and Bivona, 2017; Amararathna et al., 2019).

7.2.2.1 TARGETED THERAPY

Briefly, cancer cells biology generally involves the overexpression of certain pathways and cell surface receptors, and it is possible to identify such pathways for the various types of LC (Alhajj et al., 2018). Targeted therapy is a designation applied to treatments that aim to block or induce those specific cellular pathways overexpressed or mutated in cancer cells, triggering their apoptosis, stimulating the immune response against them, or aiming at the delivery of cytotoxic agents (Pérez-Herrero and Fernández-Medarde, 2015).

As mentioned, LC presents with a great histological heterogeneity. It is also a disease with pronounced genetic and epigenetic complexity, related, at least in part, to tobacco smoke, which contributes to the discrepant response to certain drugs (Mayekar and Bivona, 2017). Next-generation sequencing (NGS) is a method that has allowed to identify specific genomic regions involved in the disease processes, thus providing a highly sensitive and feasible genetic characterization of LC, ultimately conferring additional inputs and robust data to select the most adequate therapeutic option and improve patients' prognosis (Fernandes et al., 2019). The main drug's specificities, indications, and targets, and routes of administration of the currently used targeted therapy are shown in Table 7.2.

7.2.3 RADIOTHERAPY

Radiotherapy is mostly applied in LC for adjuvant and neoadjuvant purposes as a key driver to improve the patients' outcomes, or even with palliative intent (Fiorino et al., 2020). Initially, radiotherapy was only used in a small number of cases, as surgical resection was largely preferred (Fiorino et al., 2020). However, more recently, radiotherapy has become the local treatment modality with the greatest evidence level and then most usually applied as local treatment (Fiorino et al., 2020). Currently, we have evidences reporting that radiotherapy also presents immunomodulatory potential, as it makes tumors more vulnerable to immune system action (Theelen et al., 2020). Besides the capacity for triggering pro-inflammatory and antitumor effects, it may also act as an immunosuppressive and protumor agent, leading to immunogenic cells death. These scenarios have been closely attributed to the doses applied; thus, low-dose radiation shows an immunomodulatory effect, as it does not damage leucocytes neither local vasculature, while high doses exert the opposite effect (Theelen et al., 2020). In addition, radiotherapy can also incite novel proteins formation or conformational change, thus favoring the neoantigens pool. Also worth of note is that radiotherapy improves the ability of tumor cells recognition through upregulation of major histocompatibility complex (MHC) class I (Theelen et al., 2020). In short, and given these insights, radiotherapy has been increasingly combined with immunotherapy (Theelen et al., 2020), to achieve better patients' outcomes.

7.2.4 OTHER THERAPEUTIC STRATEGIES

Brachytherapy is not yet to be considered a treatment option for LC, but literature data suggest it might be valuable for symptomatic inoperable disease, effectively, and safely relieving symptoms (Skowronek, 2015). In brief, we observe that current pharmacological treatment available is limited to systemic therapy. Also, given the heterogeneity of this group of diseases and the complexity of current practice guidelines, as well as the poor clinical outcomes observed, we may conclude that current therapies still face relevant issues concerning to: (1) Systemic and local side effects, (2) Suboptimal

TABLE 7.2 Drugs Currently Used to Treat Advanced Stage NSCLC Their Specificities*

Drug	Target	Indications	Route
mAbs			
Bevacizumab	VEGF	Non-squamous NSCLC with carboplatin and paclitaxel; unresectable, locally advanced, recurrent, or metastatic disease	Intravenous
Necitumumab	EGFR	Metastatic squamous NSCLC	
Ramucirumab	VEGFR2	Metastatic NSCLC progressing after/under platinum-based chemotherapy with EGFR or ALK genomic tumor aberration	
Atezolizumab	PD-L1	Metastatic NSCLC progressing after/under platinum-based chemotherapy without EGFR or ALK genomic tumor mutations Extensive-stage SCLC	
Durvalumab	PD-L1	Stage IIIC or metastatic NSCLC	
Nivolumab	PD-1	Metastatic NSCLC progressing after/under platinum-based chemotherapy without EGFR or ALK genomic tumor aberration	
Pembrolizumab	PD-1	Used in first-line for metastatic NSCLC with high (>50%) PD-L1 expression without EGFR or ALK tumor mutations, and no prior systemic chemotherapy treatment; Used in second line for Metastatic NSCLC with low PD-L1 expression (≥1%), with disease progression on or after platinum-containing chemotherapy	
TKIs			
Afatinib	EGFR	Advanced or metastatic NSCLC with EGFR exon 19 deletions or exon 21 (L858R) (any histological subtype)	Oral
Erlotinib	EGFR	EGFR+ metastatic NSCLC with exon 19 deletions or exon 21 (L858R)	
Gefitinib	EGFR	Metastatic NSCLC with EGFR exon 19 deletions or exon 21 (L858R)	
Osimertinib	EGFR	Metastatic NSCLC with EGFR T790M mutations that have progressed on or after EGFR TKI therapy	
Crizotinib	ALK	Metastatic NSCLC with ALK+ or ROS1+	

TABLE 7.2 *(Continued)*

Drug	Target	Indications	Route
Brigatinib	ALK	Metastatic NSCLC with ALK+	
Ceritinib	ALK	Metastatic NSCLC with ALK+	
Alectinib	ALK	Metastatic NSCLC with ALK+ that have progressed or are intolerant to crizotinib	
Larotrectinib	TrkA, TrkB, TrkC	NSCLC with NTRK gene fusion+	

Abbreviations: ALK: anaplastic lymphoma kinase; EGFR: epidermal growth factor receptor; mAB: monoclonal antibody; NSCLC: non-small-cell lung cancer; PD-1: programmed cell death protein 1; PDL-1: programmed cell death protein 1-ligand; ROS1: ROS proto-oncogene 1; SCLC: small-cell lung cancer; TKI: tyrosine kinase inhibitor; VEGF: vascular endothelial growth factor; VEGFR2: vascular endothelial growth factor receptor 2.

**Source:* Leung et al. (2018); Ettinger et al. (2019).

tumor tissue drug concentrations raising issues of drug potency and drug resistance, (3) Route and convenience of administration, and (4) Suboptimal cell-specific targeting. In the following sections, strategies currently used in nanotechnology to overcome such problems are briefly discussed.

7.3 NANOPARTICLES AND THE INHALED ROUTE

Aerosol therapy has long been tested and used for many purposes; the respiratory tract offers a large surface area of highly vascular epithelium, which allows rapid absorption and outstrip first-pass hepatic metabolism (Apóstolo et al., 2006). The main transport mechanisms across the lung epithelium are pictured in Figure 7.3. The inhaled route to deliver chemotherapy for LC treatment would hopefully drive drugs' biodistribution and ensure higher concentrations in the target organ (Mangal et al., 2017). However, the normal mechanisms of lung clearance, the presence of disease and the alveolar microenvironment influence drug deposition from inhaled route (Chatzaki et al., 2012). For example, a concentration of CO_2 between 5% and 7% in the inhaled air improves tidal volume and reduces respiratory rate, resulting in greater drug absorption (Chatzaki et al., 2012). In addition, in tumor tissues, the phenomenon of neo-angiogenesis results in dysfunctional vasculature and lymphatics, thus limiting macromolecules distribution (Chatzaki et al., 2012). Furthermore, certain properties of the inhaled suspension may irritate the bronchial tree and cause cough and bronchoconstriction-altered pH and osmolality are likely prone to elicit such responses (Chatzaki et al., 2012). As for particle size, 0.1–3 μm in diameter is the best when seeking alveolar deposition and possibly diffusion and transcytosis (Chatzaki et al., 2012; Amararathna et al., 2019). Nanoparticles (NPs) are colloidal particles with typically <1 μm in diameter, developed from organic and inorganic compounds (Amararathna et al., 2019). In human lung, the pseudostratified architecture of airways epithelium and the thin alveolar epithelium promote both hydrophilic and lipophilic molecules absorption, respectively (Amararathna et al., 2019). Dry-powder inhalation has been described as the optimal route; however, it still raises many technical concerns, regarding particle size (Alhajj et al., 2018).

Nanotechnology allows for optimization of drug carriers, as to enhance drug deposition, targeting, and effectiveness, according to tissue properties and physiologic or pathologic processes and pathways (Hussain, 2016; Amararathna et al., 2019; Cryer and Thorley, 2019). The chemical composition and molecular assembly of a nanoparticle determine the NCs behavior and characteristics, such as shape, size, charge, and deformability can be adjusted, as well as its capacity for sustaining drug delivery, improved dissolution, or delayed degradation (Alhajj et al., 2018; Cryer and Thorley, 2019).

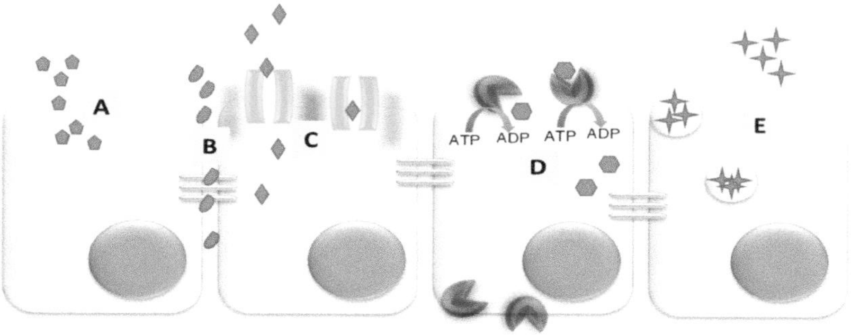

FIGURE 7.3 The most common mechanisms of pulmonary epithelium drug absorption. The different blue figures represent different molecules with different properties. [ATP: adenosine triphosphate; ADP: adenosine-2-phosphate].

Note: A: diffusion-transcellular absorption of lipophilic drug molecules through the plasma membrane; B: paracellular absorption of most hydrophilic molecules through tight intercellular junctions; C: facilitated diffusion; D: active transport, that can be primary or secondary and requires ATP; E: endocytosis of large molecules through membrane-bound receptors.

In particular, solutions involving molecular targeting, with ligand-conjugated NC, and receptor-mediated endocytosis seem an appealing method to refine drug delivery and to further minimize toxicity. Thus, combining molecular targeting with organ-specific delivery of anticancer drugs through nanotechnology is complex and challenging, but it encloses the potential to minimize toxicity and increase therapeutic efficacy, ultimately improving LC outcomes (Alhajj et al., 2018).

7.3.1 NANOCARRIERS (NCS)

Nanoparticle formulations are most often nanocarrier-based, which means applying a system that includes a carrier and a chemotherapeutic agent, and generally also a targeting functional group, bound to the carrier (Hussain, 2016). There is an incredible multiplicity of potential NCs being studied, including lipidic (liposomes, micelles), polymeric (dendrimers, hydrogels), carbon structures, peptide-based, inorganic materials, quantum dots, and viral components (Cryer and Thorley, 2019). The most common types of NC applied so far are liposomes, micelles, dendrimers, polymeric, and peptide-based NCs. Figure 7.4 presents a schematic overview of the different NCs.

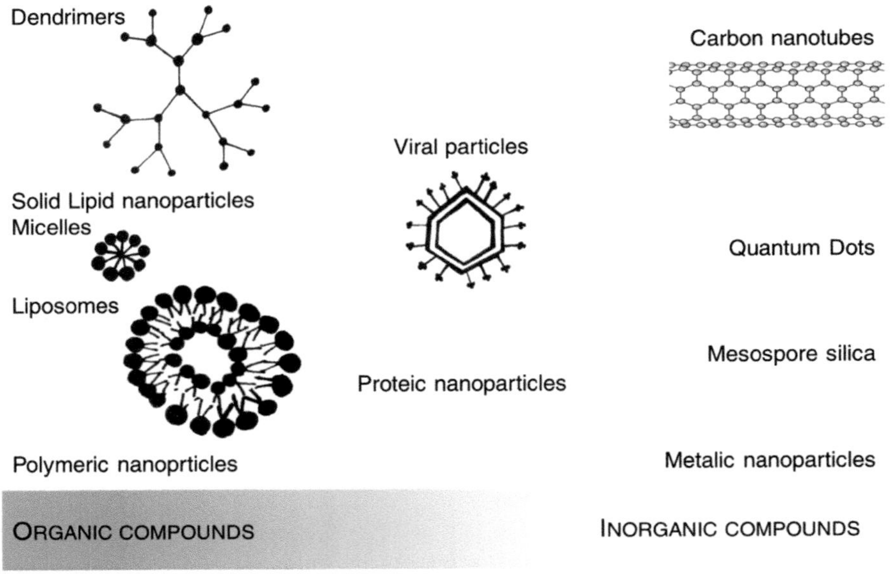

FIGURE 7.4 Spectrum of nanoparticles used in cancer diagnosis and treatment. Nanoparticles composition can generically be defined as organic (polymers, lipids, dendrimers, proteins) or inorganic (metals, rare earth elements, carbonaceous). Although spheres are primarily depicted (and used), nanoparticles can take the form of a variety of shapes, such as rods, wires, and other geometric structures (triangles, diamonds, etc.). The physicochemical properties of nanoparticles determine their in vitro and in vivo behavior, meaning they can be engineered to overcome the body's barriers to effectively target the disease site.

Source: Adapted from Cryer and Thorley (2019).

7.3.1.1 LIPOSOMES

Liposomes are typically sized between 30 nm and 5,000 nm (depending if they are uni- or multi-lamellar in structure), composed of phospholipids, arranged as bilayer vesicles-this structure enables liposomes to carry both hydrophilic and hydrophobic drugs (Amararathna et al., 2019). Hydrophilic molecules remain at the central core, while lipophilic are encapsulated between the amphiphilic layers (Amararathna et al., 2019). Their interest relies on the possibility of co-delivery of drugs to enhance the therapeutic efficacy of chemotherapeutics (Amararathna et al., 2019). According to the liposome's characteristics (size and layers structure), the amount of drug encapsulated and the circulation time may vary (Amararathna et al., 2019). For pulmonary drug delivery, formulations composed of phospholipids and cholesterol are effective and biocompatible, as both are present in plasma membranes and surfactant (Amararathna et al., 2019). Strategies applied to further optimize their properties are PEGylation, which enhances circulation time and the combination with peptides and ligands (Hussain, 2016; Amararathna et al., 2019; Cryer and Thorley, 2019). PEGylation is also considered a passive targeting system, as discussed below. PEG Liposomal encapsulated paclitaxel, for example, passed phase 1 trials; results were promising, with low-plasma-drug concentration and lower systemic toxicity, sustained drug release, and also reduced excretory speed; the drug is delivered intrapleural, a limitation that may conditionate its widespread application (Amararathna et al., 2019).

7.3.1.2 MICELLES

Micelles are simpler structures, composed of amphiphilic polymers and self-assembled in aqueous solution. For micelle assembly, polymers such as polyethylene glycol (PEG) and polyethylene oxide (PEO) are often used and established excipients (Amararathna et al., 2019). The polymers chosen determine the micellar size and shape and, so, their stability and capacity for drug retention (Amararathna et al., 2019). Hydrophobic drugs, such as paclitaxel and tamoxifen, can be incorporated in micelles by acquiring a hydrophilic involucre, which protects these drugs from protein binding and opsonization (Amararathna et al., 2019). Its ability to incorporate both lipophilic and hydrophilic compounds is limited compared to liposomes, but they also allow the co-encapsulation of two different chemotherapeutic agents (Amararathna et al., 2019). *In vitro* studies with models for NSLC showed promising results for micellar co-encapsulation of paclitaxel with β-lapachone (Amararathna et al., 2019). Unimeric micelles, with lower molecular mass, have shown efficient renal clearance (Amararathna et al., 2019). Among other agents successfully incorporated into micelles are cabazitaxel, paclitaxel alone and combined with iRNA, acetylthevetin B and doxorubicin (Hussain, 2016; Amararathna et al., 2019).

7.3.1.3 DENDRIMERS

Dendrimers are multiply branched polymers (i.e., "dendritic"), that organize surrounding a central core (Hussain, 2016; Amararathna et al., 2019).

These structures are advantageous because the tree-like interior facilitates conjugation and/or encapsulation of various drug molecules (Hussain, 2016; Amararathna et al., 2019). The molecular weight (MW) and composition of the terminals influence the properties of the NC (Amararathna et al., 2019). PEGylation or acylation of the active terminal surface, for instance, improve bioavailability (Hussain, 2016; Amararathna et al., 2019). Higher MW results in higher drug retention in lungs, with lower absorption rate (Amararathna et al., 2019). Additionally, we can refine the targeting by attaching peptides to terminal branches; this has been performed with antibodies against EGF, having achieved promising results (Amararathna et al., 2019). Given by inhaled route, small dendrimers seem to be rapidly taken up in the lung and enter the bloodstream, while larger particles are sequestered for longer periods (Hussain, 2016).

7.3.1.4 POLYMER-BASED

Polymeric NCs are those where the therapeutic agent is linked to a soluble backbone, in general by biodegradable or cleavable bridges (Amararathna et al., 2019). There are two types of polymers-based NCs: the so-called nanospheres, in which the drug is uniformly distributed, and the nanocapsules, in which the drug is entrapped into a cavity and surrounded by a polymer layer (Amararathna et al., 2019). A molecule named Abraxano (an albumin-based polymeric NC for paclitaxel) is currently FDA approved for inoperable and metastatic NSCLC in combination with carboplatin (Hussain, 2016).

7.3.1.5 PEPTIDE-BASED

Properties of small peptides (<30 amino acids) have been, to some extent, out of scope for clinical purposes, as they show a lack of stability and short half-lives, being easily eliminated, and degraded (Amararathna et al., 2019). However, these small peptides can be easily internalized in cells and organelles and also show potential for finer interactions with the media, being sensitive to pH, temperature, and other conditions (Amararathna et al., 2019). Thus, they gained interest as targeting elements and as therapeutic agents, and as it is currently possible to modify them to prolong the half-life, with PEGylation, albumin conjugation and other strategies. The inhaled route is auspicious for these agents, since the pulmonary mucosal surface is rich in anti-protease enzymes, which makes it permeable to small peptide molecules (Amararathna et al., 2019). Therefore, pulmonary delivery of peptide-based NPs may result in improved bioavailability in comparison to subcutaneous or intravenous injections routes (Amararathna et al., 2019).

7.3.2 TARGETED THERAPY USING NANOPARTICLES

The main concept of targeted therapy is to further optimize drug delivery, increasing its concentration in the target tumor cells and lowering the uptake and/or effects on healthy cells. The wide variety of compounds available for nanoparticle formulation allows us to think of different strategies to achieve these results.

7.3.2.1 PASSIVE TARGETED THERAPY

It has long been described those larger molecules preferentially accumulate *in vivo* in tumor masses, and this accumulation is a passive targeted therapy (Bertrand et al., 2014). Passive targeted therapy directly reflects the greater permeability and retention effect (EPR). EPR is mainly a consequence of the rapid growth of new and dysfunctional blood vessels and lymphatic drainage system. It has been observed that the disordered endothelium results in fenestrations with an approximate size of 200–2,000 nm and allows the leakage of NPs from the circulation into the tumor. On the other hand, tumor lymphatics are equally inefficient, which allows the particles to interact with the tumor mass for longer periods. In fact, it is a complex phenomenon, encompassing angiogenesis, vascular permeability, hemodynamic regulation, heterogeneities in the tumor's genetic profile, heterogeneities in the tumor microenvironment and lymphangiogenesis (Bertrand et al., 2014). Three distinct, but interrelated phenomena are involved: (i) extravasation of colloids from blood vessels; (ii) their further diffusion through extravascular tissue; and (iii) their interaction with intracellular and/or extracellular targets within the tumor environment (Bertrand et al., 2014; Alhajj et al., 2018).

7.3.2.2 ACTIVE TARGETED THERAPY

Most frequently, LC cells have altered biology, with overexpression and/or activation of certain pathways, typically involving receptor-ligand mechanisms. The overexpressed receptor on cancer cells surface represents an attractive way to achieve cell specificity of drug delivery with nanoparticulate systems, combining them with specific ligands. Mainly, there are two ways of exploiting this possibility: designing NPs to be internalized by endocytosis or designing them to release the drug intracellularly after interaction with the receptor. Table 7.3 shows the described overexpressed surface receptors found to be involved in LC biology, potential targets for nanoparticulate systems.

7.3.2.3 ANTIBODIES

Antibodies and antibody fragments are attractive ligands for active targeted delivery. They exhibit both high binding affinity and high specificity for cell receptors. The use of antibody as a targeting ligand is expensive, its production is time-consuming, storage and stability are of concern, and there are issues in terms of immunogenicity. Cetuximab is an IgG monoclonal antibody that binds to EGFR overexpressed in cancer cells, previously applied to treat advanced NSCLC. About 40% to 80% of NSCLCs show a high EGFR expression, which makes this molecule an interesting target. Conjugates of Cetuximab to paclitaxel, gemcitabine or docetaxel loaded NPs are being tested, showing encouraging *in vivo* results with mice models. Conjugates of cetuximab and liposomes have been tried to deliver siRNA. In general, it has been achieved an improved drug uptake by cancer cells, along with lower toxicity. Matuzumab, another EGFR antibody, is conjugated to magnetic NPs to achieve higher tumor cell capture efficiency. A few studies have shown better outcomes in mice, mainly with smaller tumor mass.

7.3.2.4 APTAMER

Aptamers are small single-stranded DNA or RNA oligonucleotides, lengthening from 30 to 100 base pairs (Zamay et al., 2017). They present great potential to be used in targeted therapy, as they form three-dimensional structures capable of highly selective binding, distinguishing small differences in thousands of proteins (Zamay et al., 2017). *In vitro*, they are capable of targeting ligands, but they also exhibit anticancer activity (Zamay et al., 2017). However, their selectivity depends upon a very sensitive three-dimensional structure, very difficult to maintain under *in vivo* conditions (Zamay et al., 2017).

TABLE 7.3 Molecules with Cell Surface Domains that Commonly Appear with Altered Expression in Lung Cancer Which May be Explored as Targets for Nanoparticulate Systems, According to Their Properties and Lung Cancer Subtype*

Overexpressed Molecule	Functional Specificities	Main Function	Cancer Type/Cell Line
GHR	Extracellular domain with homology to cytokine receptor; activation of JAK2 cascade	Growth and differentiation pathways	A549 NSCLC cell line
EGFR	Tyrosine kinase activity	Facilitates angiogenesis, proliferation, invasion, and metastasis of tumor cells	85% of NSCLC (60% in squamous cell carcinoma, 40% in adenocarcinoma and large cell carcinoma); no EGFR expression was detected for SCLC
VEGF	Tyrosine kinase activity; immunoglob-ulin-like extracellular domains, activated upon ligand-mediated dimerization	Maintenance, differentiation, proliferation, and migration endothelial cells	VEGFR1-squamous cell carcinoma VEGFR2-NSCLC VEGFR3-adenocarcinoma
FGFR	Tyrosine kinase activity; three immunoglobulin-like extracellular domains	Regulation of cell proliferation, differentiation, and angiogenesis	FGFR1-adenocarcinoma (3%) and squamous cell carcinoma (21%)
CD44	Glycoprotein with hyaluronic acid interaction	Adhesion, differentiation, homing, and migration of cancer stem cells	NSCLC CD 44v6-lymph node metastasis in NSCLC
Integrins	Heterodimeric glycoproteins	Cell-cell and cell-substratum adhesion, and help to regulate cellular morphology, differentiation, and proliferation	NSCLC (82%) SCLC (13%); the lack of expression in SCLC has been correlated with higher invasive behavior
CXCR4 receptor	G protein-coupled receptors	Cell migration	80% of NSCLC express CXCL12 (CXCR4 ligand)
Selectin receptor	Calcium-dependent	Control of adhesion to vascular endothelium	Adenocarcinoma (P-selectin receptors)

TABLE 7.3 *(Continued)*

Overexpressed Molecule	Functional Specificities	Main Function	Cancer Type/Cell Line
EpCAM	Transmembrane glycoprotein	Interaction with E-cadherin and cell adhesion promotion	NSCLC (+++ adenocarcinoma)
Transferrin receptor	Dimeric transmembrane glycoprotein with affinity to ferric iron	Promotion of acidic containing endosomes	~88% in NSCLC
TRAIL	Cytokine receptor	Inflammatory pathways. Responsible for cancer-associated	>80% of expression in NSCLC
FRAlpha	Surface glycoprotein with high affinity for folate	Receptor-mediated endocytosis	High expression in NSCLC (+++ adenocarcinoma)
Lectin membrane receptor	Various types (calcium-dependent, thiol dependent and cation dependent); interactions with endogenous carbohydrates	Cellular interactions, cell growth, differentiation, lymphocyte recirculation, and immunomodulation; metastatic spread	Prognostic marker for adenocarcinomas-lectin HPA

Abbreviations: CD44: cluster of differentiation 44; EGFR: epidermal growth factor receptor; EpCAM: epithelial cell adhesion molecule; FGFR: fibroblast growth factor receptor; FRAlpha: folate receptor; GHR: growth hormone receptor; HPA: helix pomatia agglutinin; JAK2: Janus kinase 2; NSCLC: non-small-cell lung cancer; SCLC: small-cell lung cancer; TRAIL: tumor necrosis factor-related apoptosis-inducing ligand; VEGF: vascular endothelial growth factor.

**Source:* Alhajj et al. (2018); Ahmad et al. (2019).

7.3.2.5 CARBOHYDRATES

Carbohydrates also hold a significant potential for targeted chemotherapy, either as attached to NPs or as targets themselves, especially when we intend to target the glycosylated proteins previously mentioned. Complex carbohydrates can be attached to NCs to produce specific interactions with lectins. Another potential application for carbohydrates is glucose coating: glucose-coated magnetic nanoparticles (MNPs) show differential uptake between metabolically active normal cells and cancer cells. Mannan can also be used to target NPs to antigen-presenting cells (APCs) (Zamay et al., 2017).

7.3.2.6 PEPTIDES

Peptides are also able to specifically bind molecules on cell surface and act as targeting elements (Zamay et al., 2017; Alhajj et al., 2018). Peptides have several advantages over antibodies: they are less immunogenic, more chemically stable, easier to synthesize and to conjugate with NCs and also less expensive. All these characteristics make them a highly attractive alternative ligand for active delivery of anticancer therapy targets (Alhajj et al., 2018). For instance, chlorotoxin is a small peptide containing 36 amino acids, and it has been demonstrated to present specific binding affinity to SCLC cells (Zamay et al., 2017). The recombined growth hormone (rGH) has been used as the targeting ligand toward different cancer cells over-expressing growth hormone receptor (GHR), including NSCLC (Alhajj et al., 2018). Arg-Gly-Asp-based peptide (RGD peptide) has been developed as a ligand for targeting integrin receptors (Alhajj et al., 2018). Peptide ligands have been used to actively target drugs to the tumor vasculature in LC; specific targets, such as E-selectin, a transmembrane adhesion protein involved in metastasis, allows drug delivery to the metastatic sites (Alhajj et al., 2018).

7.3.2.7 SMALL MOLECULES

Small molecules, with low MW, are easily modified and combined, in addition to pauci-immunogenic and very stable. The cost of synthesis is also low. Thus, they have attracted significant attention as active targeting ligands. Folic acid-coated particles are the most promising: many combinations have been tested with paclitaxel, docetaxel, and chitosan, and the *in vivo* effective targeting was demonstrated. They successfully target M109 tumors after inhalation, and they are effective in inhibiting tumor growth for A549, HeLa, and M109-HiFR cell lines (Alhajj et al., 2018).

7.3.2.8 OTHER DRUG DELIVERY SYSTEMS

Other notorious and versatile NCs include metal and silica-based NPs. They have been widely and successfully applied to advanced diagnosis techniques, especially gold NPs, which exceed the scope of this review (Amararathna et al., 2019). Gold NPs carrying methotrexate have been shown to improve tumor drug retention and enhance efficacy in a mouse model (Amararathna et al., 2019). Mesoporous silica NPs are primarily taken up by endocytosis after inhalation (Amararathna et al., 2019). They allowed to develop a complex system, that specifically delivers doxorubicin and cisplatin inside cancer cells, using a LHRH peptide for cell targeting, and, combined with two types of siRNA targeting the MRP1 and BCL2 mRNA successfully suppressed cell resistance in NSCLC (Amararathna et al., 2019).

7.3.3 NANOPARTICULATE SYSTEMS IN CURRENT PRACTICE

Currently, there are only two nanoformulations clinically available for treating LC, both with paclitaxel: Abraxane and Genexol-PM (Cynviloq) (Amararathna et al., 2019; Cryer and Thorley, 2019). Abraxane is the only one that has Food and Drug Administration (FDA) approval and is indicated (in combination with platinum therapy) for locally advanced or metastatic NSCLC in non-candidates for surgery or radiotherapy. It is a formulation of about 130 nm in size and, as mentioned, paclitaxel is bound to albumin NPs and shows improved pharmacokinetics and reduced side effects, compared to those of the drug alone (Hussain, 2016; Amararathna et al., 2019; Cryer and Thorley, 2019). Genexol-PM (Cynviloq) is only approved in South Korea for treating NSCLC. It is a polymeric micelle formulation of paclitaxel that seems to show improved tolerability (Cryer and Thorley, 2019).

Being a very hot field, several other nanoparticulate systems are currently undergoing clinical trials. In phase III clinical trials, paclitaxel poliglumex (PPX) showed an improved toxicity profile and similar results compared to gemcitabine, vinorelbine or platinum-based chemotherapy (Cryer and Thorley, 2019). PPX consists of paclitaxel conjugated with biodegradable polymer poly-L-glutamic acid; it is internalized by endocytosis and remains biologically inactive until lysosomal proteases activate the compound (Cryer and Thorley,

2019). Lipoplatin is a PEGylated liposomal formulation of cisplatin that shows lower rates of nephrotoxicity and neutropenia in NSCLC, although data on therapeutic efficacy are still somewhat controversial (Cryer and Thorley, 2019). Doxil (Caelyx) is a PEGylated liposomal formulation of doxorubicin which was the first FDA approved nanoformulation, and it has been investigated both as a monotherapy and as a combination treatment for NSCLC, SCLC, and mesothelioma (Cryer and Thorley, 2019). Prostate-specific membrane antigen is a transmembrane glycoprotein that has been shown to be overexpressed in NSCLC and SCLC tumor vasculature; BIND-014 is a polymeric nanoparticle capsuling docetaxel that targets the mentioned antigen and already reached phase II trials (Cryer and Thorley, 2019). Etirinotecan pegol (NKTR-102) is a polymeric conjugate of irinotecan, and the progressive hydrolysis of the polymer linker results in a slow and sustained release of irinotecan, which is metabolized into the more potent active compound, 7-ethyl-10-hydroxycamptothecin. It is currently undergoing phase II trials for patients with metastatic or recurrent NSCLC who have failed second-line therapy and relapsed SCLC. It demonstrated to be safe along with a more favorable pharmacokinetic profile (Cryer and Thorley, 2019).

7.4 CONCLUSION AND FUTURE PERSPECTIVES

Nanoparticulate systems have the potential to transform the LC treatment. NCs are a valuable resource and seem to aim to overcome the main limitations of current chemotherapy agents, refining their anticancer activity. Different NCs with different properties offer the opportunity to address each specific issue of drugs currently used. So far, paclitaxel and its nanoformulations have shown encouraging clinical results with Abraxane and Genexol-PM. Combining drug and NCs optimization, we may take the tailored therapy much further, hopefully with gains in capacity for disease response, stabilization, and survival. Additionally, NCs have allowed us to explore the advantages of organ specific delivery of anticancer agents, which seems to be a promising approach, even though it is not currently clinically available.

The inhaled route is a highly acceptable, safe, and comfortable (non-invasive) method. Efforts have been made to use it as effective to target LC cells. The current limitations to be surpassed are the lack of stability of liposomal formulations, the low predictability of pulmonary distribution and tissue penetration, as well as the lack of models for structural and functional changes, frequently present in smokers and former smokers. Thus, this route is not yet intended for clinical purposes.

Of course, being a new approach, nanoparticulate systems rise questions of safety, short-, and long-term toxicity, and still largely lack standardized protocols before approval and large-scale applicability. Particularly, the immunogenicity of these structures is of high concern, as well as the ability to cross biological barriers. Probably, the establishment of such safety protocols will be an emerging major challenge for the concerning authorities. So far, *in vivo*, and *in vitro* results, as well as clinical trials are promising. The full potential of nanotechnology in the field of LC therapy is still far underexploited. The field is still maturing and medical community interest is growing, along with the body of evidence. A deeper understanding of the biology of each individual tumor, along with the establishment of crucial dysregulated pathways that drive each cancer, would allow the development of strategies to optimize the delivery systems to overcome several obstacles, including immune reaction, half-lives, drug clearance, tissue penetration and efficiency in targeting and anticancer activity. Take-home messages are as follows:

- Lung cancer is a common condition with a poor overall prognosis and significant burden, which substantiates investment in new therapeutic approaches.
- The term "lung cancer" comprises a wide multiplicity of entities, of which NSCLC is the most common. The prognosis is often compromised by diagnosis at an advanced stage, for which therapy is still very inefficient.
- For each patient, identifying the disease' driver mutation is currently the standard approach.
- Nanotechnology shows great potential to optimize lung cancer therapy, addressing many issues specific to current chemotherapeutic agents, with regard to toxicities and lack of specificity.
- Inhaled chemotherapy is a very promising non-invasive method for drug delivery. NCs allow the optimization of particle size, MW and solubility that determine efficacy and prevent side effects. It is currently not available in clinical practice.
- Active targeting of nanoparticulate systems, especially with monoclonal antibodies (mAbs), shows interesting, although expensive, results.
- Nowadays, for lung cancer treatment, only two paclitaxel nanoformulations are clinically available-Abraxane (FDA approved) and Genexol-PM. Their advantage is, above all, the toxicity profile.
- Many other chemotherapeutic agents' formulations are undergoing clinical trials.

- Although there are many issues to be solved, the authors believe that nanoparticulate drug delivery systems comprise the potential to dramatically improve lung cancer outcomes at advanced stages.

ACKNOWLEDGMENTS

N. Cruz-Martins would like to thank the Portuguese Foundation for Science and Technology (FCT-Portugal) for the Strategic project ref. UID/BIM/04293/2013 and "NORTE2020-Programa Operacional Regional do Norte" (NORTE-01-0145-FEDER-000012).

CONFLICT OF INTERESTS

The authors declare no conflict of interests.

KEYWORDS

- **active targeted therapy**
- **antibodies**
- **clinical outcomes**
- **inhaled route**
- **lung cancer**
- **nanocarriers**
- **nanoparticulate drug delivery systems**
- **passive targeted therapy**
- **pharmacological therapy**
- **radiotherapy**
- **surgical treatment**
- **targeted therapy**
- **toxicity**

REFERENCES

Abdel-Rahman, O., (2018). Validation of the AJCC 8th lung cancer staging system among patients with small cell lung cancer. *Clinical and Translational Oncology, 20*(4), 550–556.

Ahmad, A., Khan, F., Mishra, R. K., & Khan, R., (2019). Precision cancer nanotherapy: Evolving role of multifunctional nanoparticles for cancer active targeting. *J. Med. Chem., 62*(23), 10475–10496.
Albaba, H., Lim, C., & Leighl, N. B., (2017). Economic considerations in the use of novel targeted therapies for lung cancer: Review of current literature. *PharmacoEconomics, 35*(12), 1195–1209.
Alhajj, N., Chee, C. F., Wong, T. W., Rahman, N. A., Kasim, H. A., & Colombo, P., (2018). Lung cancer: Active therapeutic targeting and inhalational nanoproduct design. *Expert Opinion on Drug Delivery, 15*(12), 1223–1247.
Amararathna, M., Goralski, K., & Hoskin, D. W., (2019). Pulmonary nano-drug delivery systems for lung cancer: Current knowledge and prospects. *Journal of Lung Health and Diseases, 3*, 11–28.
Apóstolo, J. L. A., Mendes, A. C., & Azeredo, Z. A., (2006). Adaptation to Portuguese of the depression, anxiety and stress scales (DASS). *Revista Latino-Americana de Enfermagem, 14*(6), 863–871.
Bertrand, N., Wu, J., Xu, X., Kamaly, N., & Farokhzad, O. C., (2014). Cancer nanotechnology: The impact of passive and active targeting in the era of modern cancer biology. *Advanced Drug Delivery Reviews, 66*, 2–25.
Chatzaki, E., Hohenforst, W., Karamanos, N., & Zarogoulidis, K., (2012). Inhaled chemotherapy in lung cancer: Future concept of nanomedicine. *International Journal of Nanomedicine, 7*, 155.
Cryer, A. M., & Thorley, A. J., (2019). Nanotechnology in the diagnosis and treatment of lung cancer. *Pharmacology and Therapeutics, 198*, 189–205.
Davidson, M. R., Gazdar, A. F., & Clarke, B. E., (2013). The pivotal role of pathology in the management of lung cancer. *Journal of Thoracic Disease, 5*(S5), S463–S478.
Ettinger, D. S., Aisner, D. L., Wood, D. E., Akerley, W., Bauman, J., Chang, J. Y., Chirieac, L. R., et al., (2019). NCCN guidelines insights: Non-small cell lung cancer, version 5.2018. *J. Natl. Compr. Canc. Netw., 16*(7), 807–821. doi: 10.6004/jnccn.2018.0062.
Fernandes, M. G. O., Jacob, M., Martins, N., Moura, C. S., Guimarães, S., Reis, J. P., Justino, A., et al., (2019). Targeted gene next-generation sequencing panel in patients with advanced lung adenocarcinoma: Paving the way for clinical implementation. *Cancers, 11*(9), 1229.
Fiorino, C., Guckemberger, M., Schwarz, M., Van, D. H. U., & Heijmen, B., (2020). Technology-driven research for radiotherapy innovation. *Mol. Oncol., 14*(7), 1500–1513.
Goldstraw, P., Chansky, K., Crowley, J., Rami-Porta, R., Asamura, H., Eberhardt, W. E. E., Nicholson, A. G., et al., (2016). The IASLC lung cancer staging project: Proposals for revision of the TNM stage groupings in the forthcoming (eighth) edition of the TNM Classification for lung cancer. *Journal of Thoracic Oncology, 11*(1), 39–51.
Hirsch, F. R., Scagliotti, G. V., Mulshine, J. L., Kwon, R., Curran, W. J., Wu, Y. L., & Paz-Ares, L., (2017). Lung cancer: Current therapies and new targeted treatments. *The Lancet, 389*(10066), 299–311.
Hussain, S., (2016). Nanomedicine for treatment of lung cancer. In: Ahmad, A., & Gadgeel, S. M., (eds.), *Lung Cancer and Personalized Medicine: Novel Therapies and Clinical Management* (pp. 137–147.). Springer International Publishing, Switzerland.
Lemjabbar-Alaoui, H., Hassan, O., Yang, Y. W., & Buchanan, P., (2016). Lung cancer: Biology and treatment options. *Biochim Biophys Acta, 118*(24), 6072–6078.
Leung, V., Hughes, E., Cambareri, G., Rubin, J., & Eaby-Sandy, M. S. N., (2018). Systemic treatments for lung cancer patients receiving hemodialysis. *Journal of the Advanced Practitioner in Oncology, 9*(6), 614–629.

Mangal, S., Gao, W., Li, T., & Zhou, Q. T., (2017). Pulmonary delivery of nanoparticle chemotherapy for the treatment of lung cancers: Challenges and opportunities. *Acta Pharmacol. Sin., 38*(6), 782–797.

Manuel, V., Sousa, L. D., & Carvalho, L., (2018). Heterogeneity in lung cancer. *Pathobiology, 85*(1, 2), 96–107. doi: 10.1159/000487440.

Mayekar, M. K., & Bivona, T. G., (2017). Current landscape of targeted therapy in lung cancer. *Clinical Pharmacology and Therapeutics, 102*(5), 757–764.

Michels, S., & Wolf, J., (2016). Stratified treatment in lung cancer. *Oncology Research and Treatment, 39*(12), 760–766.

Nagasaka, M., & Gadgeel, S. M., (2018). Role of chemotherapy and targeted therapy in early-stage non-small-cell lung cancer. *Expert Review of Anticancer Therapy, 18*(1), 63–70.

Nasim, F., Sabath, B. F., & Eapen, G. A., (2019). Lung cancer. *Medical Clinics of North America, 103*(3), 463–473.

Oberndorfer, F., & Müllauer, L., (2018). Molecular pathology of lung cancer: Current status and perspectives. *Current Opinion in Oncology, 30*(2), 69–76.

Oudkerk, M., Devaraj, A., Vliegenthart, R., Henzler, T., Prosch, H., Heussel, C. P., Bastarrika, G., et al., (2017). European position statement on lung cancer screening. *The Lancet Oncology, 18*(12), e754–e766.

Pérez-Herrero, E., & Fernández-Medarde, A., (2015). Advanced targeted therapies in cancer: Drug nanocarriers, the future of chemotherapy. *Eur. J. Pharm. Biopharm., 93*, 52–79. doi: 10.1016/j.ejpb.2015.03.018.

Planchard, D., Popat, S., Kerr, K., Novello, S., Smit, E. F., Faivre-Finn, C., Mok, T. S., et al., (2018). Metastatic non-small-cell lung cancer: ESMO clinical practice guidelines for diagnosis, treatment and follow-up. *Annals of Oncology, 29*(4), 192–237.

Postmus, P. E., Kerr, K. M., Oudkerk, M., Senan, S., Waller, D. A., Vansteenkiste, J., Escriu, C., & Peters, S., (2017). Early and locally advanced non-small-cell lung cancer (NSCLC): ESMO clinical practice guidelines for diagnosis, treatment and follow-up. *Annals of Oncology, 28*(4), 1–21.

Ridge, C. A., Mcerlean, A. M., & Ginsberg, M. S., (2013). Epidemiology of lung cancer. *Seminars in Interventional Radiology, 30*, 93–98.

Skowronek, J., (2015). Brachytherapy in the treatment of lung cancer: A valuable solution. *Journal of Contemporary Brachytherapy, 7*(4), 297–311.

Theelen, W., De Jong, M., & Baas, P., (2020). Synergizing systemic responses by combining immunotherapy with radiotherapy in metastatic non-small cell lung cancer: The potential of the abscopal effect. *Lung Cancer, 142*, 106–113.

Watanabe, S. I., Nakagawa, K., Suzuki, K., Takamochi, K., Ito, H., Okami, J., Aokage, K., et al., (2017). Neoadjuvant and adjuvant therapy for stage III non-small cell lung cancer. *Japanese Journal of Clinical Oncology, 47*(12), 1112–1118.

Yasukawa, M., Sawabata, N., Kawaguchi, T., Kawai, N., Nakai, T., Ohbayashi, C., & Taniguchi, S., (2018). Histological grade: Analysis of prognosis of non-small. *In Vivo, 1512*, 1505–1512.

Zamay, T. N., Zamay, G. S., Kolovskaya, O. S., Zukov, R. A., Petrova, M. M., Gargaun, A., Berezovski, M. V., & Kichkailo, A. S., (2017). Current and prospective protein biomarkers of lung cancer. *Cancers, 9*(11), 1–22.

CHAPTER 8

NANOPARTICLE-BASED THERAPY IN CHRONIC OBSTRUCTIVE AND INFECTIOUS LUNG DISEASES: PAST, PRESENT, AND FUTURE PERSPECTIVES

ANA CATARINA MOREIRA,[1] GABER E. BATIHA,[2] NOURA H. ABDELLAH,[3] and NATÁLIA CRUZ-MARTINS[4,5]

[1]*Hospital Garcia de Orta, E.P.E Almada, Lisboa, Portugal*

[2]*Department of Pharmacology and Therapeutics, Faculty of Veterinary Medicine, Damanhour University, Damanhour–22511, AlBeheira, Egypt*

[3]*Department of Pharmaceutics, Faculty of Pharmacy, Assiut University, Assiut–71526, Egypt*

[4]*Faculty of Medicine, University of Porto, Portugal*

[5]*Institute for Research and Innovation on Health (i3S), University of Porto, Portugal*

8.1 INTRODUCTION

Nanomedicine (NM) has opened new pathways for lung diseases diagnosis and treatment over the last 20 years. In 1995, the first nanotherapeutics' approval pioneered the NM use on a human set and had been shown potential to have a major impact on human health (Fymat, 2016). Nanotechnology applied to respiratory medicine refers to highly specific molecularly-mediated medical interventions for prevention, diagnosis, and treatment purposes (Hua et al., 2018). It relies on the use of molecular tools and molecular knowledge of the human body to ameliorate human biological systems, using engineered devices and nanostructures for medical applications (Duncan, 2005). In this line, a new paradigm in pharmacotherapy has been introduced. One particular aspect is the design of nanoparticulate nanomedicines (NNMs) for drug delivery, intended to a sustained release of the target drug within the lung tissue, thus allowing a reduction in dose frequency and improving patient adherence through various mechanisms. Inhalable NNMs has appeared as a promisor strategy for oral and/or intravenous administration, once it improves the therapeutic efficacy by decreasing the effective therapeutic dose needed and consequently reducing the risk of systemic side effects associated with high serum doses (Upadhyay and Ganguly, 2015).

In this way, NM has appeared as an emerging science based on matter manipulation at nanoscale (<100 nm) (Silva et al., 2013), that can be extended to multiple medical areas. Specifically addressing pulmonary disorders, lung cancer (LC), chronic obstructive lung diseases, and pulmonary infections appear as excellent examples. Recent studies have revealed that nanoparticles (NPs) can also affect immunological mechanisms, trigger oxidative stress and even genotoxicity (Omlor et al., 2015).

More than 25,000 publications have been released in the past years, with NNMs being one of the most studied drug delivery systems' classes (Anselmo and Mitragotri, 2014), with unique properties that can be used to improve traditional drug delivery. In fact, over the past few years, there has been a great deal of efforts concentrating on the development of advanced nanotherapeutic delivery systems to improve efficacy, increase patient compliance, and ensuring optimal treatment safety (Peer et al., 2007).

Considering lung physiology and lung parenchyma characteristics, NNMs should be designed to ensure that they are resistant to a variety of changes, such as temperature, barrier to mucus hyperproduction and its clearance, subepithelial fibrosis, local cell activation and changes in local pH, to achieve an efficient and safe way for lung delivery (Omlor et al., 2015). Thus, given the above referred aspects, the present chapter aims to provide a brief overview on NM use for chronic obstructive and infectious lung diseases, making an evolutionary analysis over time and pointing out future trajectories. A special emphasis is particularly given to asthma, chronic obstructive pulmonary disease (COPD), and tuberculosis (TB).

8.2 NANOTECHNOLOGY-BASED APPROACHES IN RESPIRATORY MEDICINE

Chronic lung diseases, like asthma and COPD are leading causes of death worldwide, but the lack of cell specific targeting has limited their therapeutic options (Upadhyay and Ganguly, 2015). Thus, alternative approaches are needed to overcome the current gaps, and nanotechnology may be conceived as a key therapeutic strategy. Indeed, the use of NNMs for drug delivery directly to the lung may improve lung deposition with sustained therapeutic effects and less toxicity, while also improving clinical efficacy. Several studies and experiments have progressively assessed a variety of NNMs in the lung (van Rijt et al., 2014; Silva et al., 2017), although several physiological barriers have made their application extremely challenging (Schneider et al., 2017). In the next sections, a brief description of the disease pathophysiology and current clinical data is provided, also highlighting the current insights and upcoming perspectives on NNMs use for therapeutic intervention.

8.2.1 NANOCARRIERS (NCS) FOR RESPIRATORY SYSTEM: AN OVERVIEW

Lung has a wide epithelial surface area (over 100 m^2) for air conduction and gas-exchange, added to an ample vascularization, making them one of the most efficient delivery ways for inhaled xenobiotics (Figure 8.1), such as NNMs (Gehr et al., 1978; Gehr et al., 2010). The vast internal surface area of the adult human lung allows an extremely effective local action of the drug, favored by a faster systemic absorption, although several pharmacokinetic issues need to be overcome to raise the optimum serum therapeutic doses until an ideal time window, long clearance time and low compounds solubility, resulting in undesirable side effects (Torchilin, 2006; Azarmi et al., 2008). In this way, NM, and its recent developments have been crucial to overcome these gaps and constrains, considering the NNMs' biological properties (e.g., liposomes, micelles, and polymeric NPs), highly efficient for pharmacological and therapeutic purposes (Bailey and Berkland, 2009; Mansour et al., 2009).

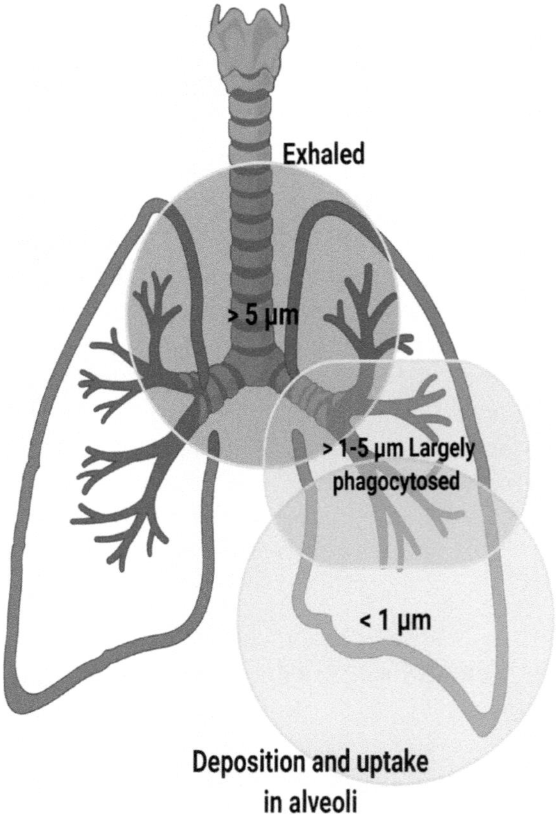

FIGURE 8.1 Schematic diagram showing the size-dependent particles deposition in the lungs.

Over the past years, various studies have revealed that several factors are associated with NPs characteristics, such as the particle size, surface charge, and its modification that determine the NPs pathway through the respiratory tract. Primarily, NPs could target distal lung compartments for prolonged persistence, allowing an interaction between NPs and target cells for longer times, improving local or systemic efficacy, thus staying in the pulmonary compartment, or crossing the air-blood barrier, respectively. Pulmonary administration has advantages over the oral or intravenous route (Figure 8.2), as it allows directly local delivery, avoids the first-pass metabolism, and decreases systemic side effects while increasing the onsite drug levels (Blank et al., 2017).

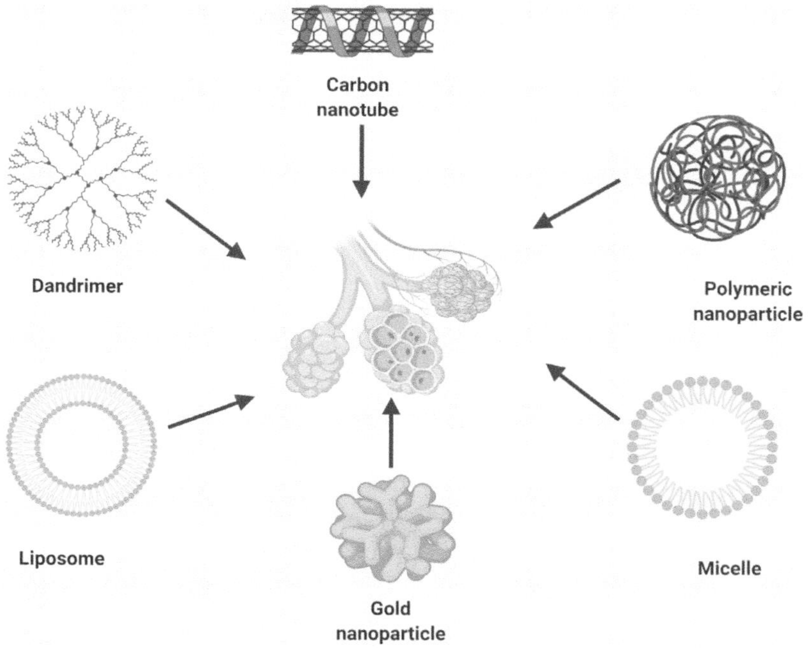

FIGURE 8.2 Nanoparticles drug delivery.

Among NNMs, liposomes are sphere-shaped vesicles with one or more phospholipid bilayers, capable of trapping hydrophobic and lipophilic drugs (Moreno-Sastre et al., 2015). Polymeric NPs have a higher stability when compared to liposomes, explained by the structural integrity in relation to the polymeric matrix, but due to the difficulties on water-soluble drugs encapsulation, they generally require organic solvents use for polymers dissolution (Moreno-Sastre et al., 2015).

Poly(lactic-co-glycolic acid) (PLGA), an FDA-approved polymer, is an excellent option for NPs formulation, considering its minimal toxicity and good biodegradability and biocompatibility. Chitosan, dextran, alginates, polyvinyl alcohol (PVA), polyethylene glycol (PEG) are other biodegradable polymers currently being widely explored (Moreno-Sastre et al., 2015).

8.2.2 NANOCARRIERS (NCS) FOR ASTHMA INTERVENTION

According to Global Initiative for Asthma (2019), asthma affects around 300 million people worldwide (http://ginasthma.org), with a burden of negative consequences on quality of life, work productivity and daily living (Peters et al., 2006; Nelsen et al., 2017). Depending on several aspects, such as clinical features (e.g., onset' age, disease severity), specific triggers and inflammation type (e.g., eosinophilic, neutrophilic, paucigranulocytic, and mixed granulocytic), there are different asthma phenotypes to be considered (Tarlo and Malo, 2006; Geiser et al., 2013). Regarding its complex pathogenesis, no specific treatment is yet available, although long-term standardized treatment seems to be effective in controlling symptoms, preventing exacerbations, stabilizing lung function, and thus improving the overall prognosis. However, it is worth of note that asthma control is difficult in many asthmatic patients (Wenzel, 2012).

The NNMs currently available are briefly splitted into two groups, the first one being an improvement over traditional molecular drugs, and the other one includes brand-new nano-drugs developed themselves as drugs. Asthma treatment usually requires long-term drugs, mostly administered intravenously, *per os*, or by inhalation. Inhalable targeted drug delivery is the mainstream treatment, with improved bioavailability and efficacy in comparison to systemic administration (Wang et al., 2019). Inhaled glucocorticoids, due to their potent anti-inflammatory effects, are the leading drug, being highly effective in controlling asthma-induced airway inflammation, although their pharmacological activity tends to be short (Wang et al., 2019). Direct inhaled drugs delivery in the lungs allows a direct-action respiratory tract, thus requiring lower doses and reducing the likelihood of systemic adverse effects occurrence. However, the long-term use of high-dose corticosteroids will inevitably be associated with adverse effects, such as adrenal axis inhibition, fungal infections in the mouth cavity, and osteoporosis. Thus, in order to reduce the corticosteroid inhalation-derived side effects occurrence and to further improve their bioavailability, PEGylated poly(amidoamine) (PAMAM) dendrimer, a typical dendrimer, has been investigated. Used as a carrier for beclomethasone dipropionate (BDP) and other insoluble drugs, PAMAM improves the solubility of the drug and increases the capacity of pulmonary accumulation, leading to improved bioavailability, dose, and frequency reduction, and avoiding toxic and side effects occurrence (Agarwal et al., 2015). According to Nasr et al. (2014) PAMAM can also act as a key nanocarrier, helping to obtain higher loading capacity and better stability than micelles, reported in more than 6 months (Nasr et al., 2014). Other studies have reported that this nanocarrier is also capable of delivering hydrophobic drugs (e.g., dexamethasone) directly into the lungs, acting on allergic inflammation mechanisms, thus reducing airway hyperresponsiveness by acting on eosinophils count and inflammatory cytokines (Jackson et al., 2004).

Compared to free steroids, nanocarrier-encapsulated steroids achieve better and more lasting therapeutic effects at airway inflammation site, as reported by Matsuo et al. (2009). Furthermore, NPs compacted with salbutamol yield a stronger interaction with pleura in relation to greater peripheral deposition and mucociliar movement back to the tracheobronchial region, thus maintaining a higher and effective dose for a long period, leading to a higher and more sustained bronchospasm relief and better symptoms control (Bhavna et al., 2009). It was also found that liposomes are able to increase the salbutamol sulfate retention in the lungs, thus achieving a significantly higher efficacy than the non-encapsulated drug, maintaining an effective drug concentration for >10 h (Bhavna et al., 2009). Drug-loaded NPs maintained greater stability through the human oropharynx passage and higher peripheral deposition when compared to micronized salbutamol sulfate, related to the smaller size and higher topical NPs bioavailability (Chen et al., 2012).

On the other side, gene therapy and molecularly targeted therapy have also been targeted of extreme interest in NM (Singh et al., 2019). For instance, chitosan through the form of NPs has been used with increasing interest in gene transfer, considering its ability to deliver plasmids (Kumar et al., 2003). Kumar et al. (2003) showed that intranasal chitosan interferon (IFN)-gamma-pDNA NPs (CIN) can be extremely useful in asthma prophylaxis and treatment. CIN was able to markedly reduce both functional and immunological abnormalities associated with sensitization and challenge to allergens, significantly reducing airway hyperresponsiveness and pulmonary histopathology in BALB/c mice with ovalbumin (OVA)-induced allergic asthma (Kumar et al., 2003). In another study, Kong et al. (2008) also stated that CIN treatment effectively inhibited cytokine production in the lung OVA-specific CD8+ T lymphocyte population, thus decreasing dendritic cells activation (Kong et al., 2008). This effect was STAT4 signaling pathway-mediated and the effective regulation of T helper cells was anticipated, which might explain the ability to reverse allergic asthma in humans (Kong et al., 2008). Increased levels of IFN-gamma would also promote Th1 cytokines expression up-regulation by blocking the Th2 cytokines expression (Kong et al., 2008).

The promotion of local drug retention and anti-inflammatory effect in the respiratory tract can also be achieved by thiolated chitosan NPs and NPs containing chitosan and cyclodextrin (Oyarzun-Ampuero et al., 2009). In addition, according to Patil-Gadhe et al. (2014), montelukast incorporated into a nanostructured lipid carrier presented a greater systemic bioavailability and ability to bypass hepatic metabolism, with a consequent reduction in hepatocellular toxicity (Patil-Gadhe et al., 2014). Other studies have tested highly packaged DNA NPs with block copolymers of poly-L-lysine and PEG linked by a cysteine residue (CK30PEG), without evidence of lung toxicity and immunogenicity in mice and humans (Boylan et al., 2012). DNA NPs instillation carrying thymulin, an anti-inflammatory and anti-fibrotic nonapeptide, has been shown to have several effects on OVA-challenged allergic asthma in BALB/c mice, particularly preventing inflammation, collagen deposition, and smooth muscle hypertrophy in murine lungs up to 27 days post-administration, thus improving pulmonary mechanics (Silva et al., 2014). More recently, Mastorakos et al. (2015) using a highly compacted and biodegradable DNA NP proved its efficacy in overcoming the mucus barrier for inhaled lung gene therapy, without toxicity signs after intratracheal administration (Mastorakos et al., 2015). Salem et al. (2014) presented a cytosine-phosphate-guanine (CpG) adjuvant-loaded, biodegradable NP-based vaccine for dust-mite allergies treatment, used as an immunomodulatory agent on T cells immune responses level (shifting Th1 type and suppressing Th2-triggered asthma responses) (Salem, 2014).

Other innovative immunomodulatory therapies for NPs' pulmonary administration have been studied, and promising achievements have been stated in asthma treatment. Considering their specific physicochemical properties that permit deposition and cell uptake, inhalable NPs have been shown to modulate immune responses, by allowing resident antigen-presenting cells (APCs) interactions (Blank et al., 2017). These APCs may represent the ideal target for specific inhaled immunomodulators; however, a comprehensive analysis of their immunomodulatory potential is necessary to develop new therapeutic strategies, in particular, for allergic asthma (Blank et al., 2017).

8.2.3 NANOCARRIERS (NCS) FOR COPD INTERVENTION

COPD is an often, avoidable and treatable disease characterized by persistent respiratory symptoms and airflow limitation related to airway and/or alveolar abnormalities. COPD represents a major cause of death worldwide, and is expected to be ranked third by 2030 (Vij, 2011). Its pathophysiology is marked by chronic inflammation of the airways and destruction of the lung parenchyma, where alveolar macrophages exert a crucial role, releasing inflammatory moderators, such as interleukin (IL)-6, IL-8, leukotriene LTB4, tumor necrosis factor-α (TNF-α), monocyte chemotactic peptide (MCP)-1, and reactive oxygen species (ROS) (Silva et al., 2017).

The currently available therapeutic options for COPD symptoms control include anticholinergics, β2-agonists, and inhaled corticosteroids. However, the airways immune response represents one of the main challenges faced by NNMs (Maurya and Singh, 2019). Thus, considering the COPD physio-pathology, airways defense, severe inflammation, and mucus hypersecretion are the main barriers for NNMs delivery and efficacy. Although, in the last 20 years, there has been an increase in nanotechnology research, also with a growing interest in new therapeutic approaches development for COPD, very few have been tested so far. Vij et al. (2011) described the potential of a multifunctional polymeric vesicle formed by a mixture of poly(ethylene glycol) and PLGA (PLGAPEG) to obtain a combined drug delivery that is prednisolone and/or theophylline. More recently, Muralidharan et al. (2016) reported the anti-inflammatory efficacy of dimethyl fumarate-containing advanced inhalable dry powders, using an *in vitro* predictive lung deposition model, and found that the aerosol deposition patterns of these particles were able to reach the lower airways.

Some authors have also described metallic NPs, previously used as contrast elements, as efficient in the delivery of drugs and genes, given their effective penetration into cells and tissues, which are also promising in COPD treatment. However, its wide application has been limited by toxic effects that are still not completely understood (Sadeghi et al., 2015). A study by Geiser et al. (2013) proved to be a successful pathway based on gold NP delivery to target alveolar epithelial cells and macrophages. Roulet et al. contradicted the previous experiments proving that titanium dioxide (TiO_2) and carbon black (CB) NPs did not impair elastase-induced emphysema besides to be a promisor metallic NCs for COPD treatment (Roulet et al., 2012).

In fact, research that aims to revolutionize lung diseases treatment, namely COPD, through NM application is an ongoing process. More recently, Malaysian researchers from the Malaysia Institute for Innovative Technology are collaborating with experts from the Harvard University on a 5-year project to develop "smart" NPs capable of delivering appropriate drugs to the deepest level of the lung, in a potentially assisted process through using magnetic fields (Semedo, 2016). On the other side, stem cell-based therapies have also been an emerging area in COPD treatment, with clinical infusions of mesenchymal stem cells (MSCs) showing to be safe and leading to a decrease in serum C-reactive protein, although without functional improvement (Weiss et al., 2013). Studies with type II alveolar epithelial cells (AEC2s) have also revealed their ability to produce surfactant proteins (Sugihara et al., 1993; Kotton and Morrisey, 2014). Another promising investigation includes basal, luminal, and Clara cells, which support the lung repair and growth (Kotton, 2012).

8.2.4 NANOCARRIERS (NCS) FOR TUBERCULOSIS (TB) INTERVENTION

Tuberculosis (TB) is an infectious disease triggered by *Mycobacterium tuberculosis* (MTB), with major public health impact (Cruz-Knight and Blake-Gumbs, 2013). MTB mostly affects the lung (75%), although it can also be found in lymph nodes, central nervous system (CNS), liver, bones, genitourinary, and gastrointestinal (GI) system. As it presents high transmission rates through respiratory droplets, the identified cases should be reported as soon as possible to reduce the public health impact (Cruz-Knight and Blake-Gumbs, 2013). In addition, MTB easily acquires resistance to one or more drugs used in the treatment, being thus necessary to increase the doses used. It is noteworthy that the treatment is itself prolonged, which associated with an increase of the doses due to the development of drug resistance, potentiates the occurrence of adverse effects and toxicity. Thus, and considering the high incidence of TB, there is a great need to develop new drug delivery systems that can improve TB treatment intervention (Ahmad et al., 2005; Cruz-Knight and Blake-Gumbs, 2013).

Over the last 20 years, studies have shown that the development of NP-based formulations is a great strategy to achieve higher pulmonary drug levels, directly delivered via aerosol (Rajaonarivony et al., 1993). The encapsulation of ATDs into NCs appears to be a powerful tool to enhance therapeutic effectiveness, minimizing side effects, while also allowing a tight control over the NPs' morphologic characteristics, resulting in aerosols capable of reaching the deep-lung (Melo, 2014). This has been an area of high interest and involves several studies focused on the development of new formulations and highly potent antitubercular drugs (ATDs) to overcome the low drug efficacy and resistance to multiple drugs. Guo proposed a targeting mechanism for multiple subunit complexes crucial for viruses and bacteria function, thus reducing or possibly eliminating their resistance to targeted drugs (Guo, 2010). Another experiment, conducted by Pandey et al. showed that ATDs-loaded aerosolized poly(DL-lactideco-glycolide) NPs by single inhalation kept plasma drug levels at therapeutic levels for >6 days in guinea pigs (Pandey et al., 2003). Repeated ATDs inhalations proved to be as effective as oral administration in the treatment of experimental TB. Another study by Pandey and Khuller showed that a single subcutaneous injection of ATD-containing NPs in mice maintained the drug plasma levels for >32 days, and was also more effective than the oral administration of free drug in reducing bacterial counts in lung and spleen (Pandey and Khuller, 2004). These authors also assessed the solid lipid NPs potential, incorporating rifampicin (RIF), isoniazid (INH) and pyrazinamide (PZA), in TB treatment and stated both *in vitro* and *in vivo* a tardily and held drugs release from solid lipid NPs (Pandey and Khuller, 2004). The pharmacokinetic and chemotherapeutic potential of aerosolized alginate NPs encapsulating ATDs (INH, PZA, and RIF) was also documented, and it was found that inhalable alginate NPs showed to be excellent carriers for ATDs release, resulting in significantly higher bioavailability compared to free oral drugs, with the efficacy of 3 doses of drug-loaded alginate NPs being nebulized 15 days apart, equivalent to 45 daily doses of free oral ones (Ahmad et al., 2005). The aerosolized poly(DL-lactide-co-glycolide) NPs with the three ATDs also showed superior chemotherapeutic effects when compared to oral free drugs (Pandey et al., 2003). The bio-adhesive drug delivery systems were also proposed by Sharma et al. in a study with lectin-functionalized poly(lactide-co-glycolide) NPs (PLG-NPs) as carriers of ATDs drugs orally or in aerosol, as a way to optimize local absorption, thus enhancing drug bioavailability, reducing the drug doses frequency, and improving patient compliance (Sharma et al., 2004). The authors observed that PLG-NPs administration led to a significant raise in ATDs relative bioavailability (Sharma et al., 2004).

8.3 NANOPARTICLES BIODISTRIBUTION IN RESPIRATORY SYSTEM

8.3.1 RATIONALES FOR USE OF NANOTECHNOLOGY IN RESPIRATORY MEDICINE

The design and development of NCs for drug delivery at respiratory system has been at the center of research in recent decades and is one of the most debated topics by the scientific community. NPs as nanocarrier systems are an attractive concept for drug delivery, given all of their properties (Azarmi et al., 2008). However, FDA suggest not to adopt a strict definition of nanomaterials, with pharmacological field' definition being still unclear. Scientific evidence has shown that particles <1 μm in the three dimensions may be viewed as NPs, thus membrane permeability and absorption are enhanced at this level (Kreuter, 2004).

NCs are involved in several phases of pulmonary drug delivery, with the ultimate goal to what concerns the NPs pulmonary delivery being the improvement of powder dispersion and aerosol performance towards to drug delivery to the deep lung (Sahu and Casciano, 2014). In the same line, and as previously referred, it has been shown that drug delivery at lung is more effective than oral or intravenously (Patton and Byron, 2007; Seydoux et al., 2016), and that of the distinct characteristics that affect particles fate, its size, surface charge and modification have a crucial role (Weber et al., 2014). Thus, considering the size issue, NPs can be designed to reach the more distal pulmonary compartments, so guaranteeing prolonged drug latency, given that in these areas the performance of alveolar macrophages is much lower when compared to proximal compartments. In this way, a longer period of activity and permanence between NPs and target cells will be achieved, thus increasing local and/or systemic efficiency, thus managing to cross the air-blood barrier. In this regard, the interaction between NPs and pulmonary immune cells is of extreme interest, and it has even been demonstrated by several studies that lung immune cells provide valuable information for the development of new biomedical tools for pulmonary delivery (Galli et al., 2011; Desai, 2012).

Another important advantage of NCs over conventional inhaled drug delivery includes slower drug degradation and controlled drug release, besides having protective functions against rapid degradation or immune activation and contribute to slower endolysosomal degradation (Kumari et al., 2010; Beck-Broichsitter et al., 2011). Moreover, NPs drug release can be adjusted to a target therapeutic window in order to meet the intended objectives, thereby decreasing the total drug dose required to achieve the desired therapeutic effects (Iyer et al., 2015).

8.3.2 PRESENT CHALLENGES: NANOFORMULATIONS FOR PULMONARY DRUG DELIVERY AND TOXIC HAZARDS

Respiratory system, although readily available, imposes some barriers to an optimal delivery and therapeutic action of inhaled NPs through the distal lung. Literature data emphasize, as most important, an optimal definition of the NPs' spatial placement and the understanding of toxic risks and their mechanisms in order to achieve an adequate control of the nanomaterials structure (Iyer et al., 2015).

NPs-mediated drug delivery relies on customized pharmacokinetic and pharmacodynamic properties, with important therapeutic applications on respiratory disorders being achieved by controlled and slow drug absorption, minimal side/toxic effects, longer retention time and more specific biodistribution patterns (Chung et al., 2007; Sung et al., 2007). Although knowledge on NPs-mediated drug delivery has increased in recent years, the ideal dose to achieve higher delivery or uptake in the target organ/cells with less toxicity is still uncertain (Beck-Broichsitter et al., 2011; Loira-Pastoriza et al., 2014).

From the beginning, nanotechnology applications have raised concerns over risks to human health and guidelines have not been established to determine the NNMs toxic potential (Vlachogianni et al., 2013). Individual studies that rely on this theme, lack specificity, being thus extremely challenging to interpret data and virtually impossible to do a comparative data analysis, even if the same NP has been investigated. Therefore, it is challenging to identify parameters that can influence toxicity (Card et al., 2008).

The NCs deposition or entrapment in the mononuclear phagocytic system can increase local and systemic toxicity due to extra-pulmonary organs involvement. After NPs delivery in the lung by inhalation or intra-tracheal installation, a portion of NPs is translocated to extra-pulmonary tissues, which can lead to local and systemic toxicity (Patton and Byron, 2007; Paranjpe and Müller-Goymann, 2014). Data from *in vitro* and *in vivo* studies suggest possible NPs toxicities through different mechanisms, but specifically *in vitro* toxicity cannot predict *in vivo* effects (Card et al., 2008).

8.4 CONCLUSIONS

Among the multiple pulmonary disorders, chronic respiratory diseases, like asthma and COPD have been the most leading causes of death worldwide, where the lack of cell specific targeting has limited the therapeutic options available. Although close monitoring has proved useful in controlling symptoms, preventing exacerbations, and stabilizing lung function, no specific treatment is already available. NM has been the subject of major advances in recent years, where the unique physicochemical properties of NPs allow for site-specific delivery and appears as a way to overcome the current limitations found, as well as a new paradigm for both treating the disease and preventing side effects occurrence, ultimately improving clinical efficacy. Anyway, although some evidence points to a great potential for the use of NPs in pulmonary drug delivery, more in-depth studies are needed in order to define testing strategies that provide a better understanding of the mechanisms underlying adverse effects and potential hazards related to NPs exposure. So, much still need to be done before nano-therapy can be applied in clinical practice.

ACKNOWLEDGMENTS

N. Cruz-Martins would like to thank the Portuguese Foundation for Science and Technology (FCT-Portugal) for the Strategic project ref. UID/BIM/04293/2013 and "NORTE2020-Programa Operacional Regional do Norte" (NORTE-01-0145-FEDER-000012).

CONFLICT OF INTERESTS

The authors declare no conflict of interests.

KEYWORDS

- **chronic lung diseases**
- **chronic obstructive pulmonary disease**
- **clinical efficacy**
- **drug-delivery systems**
- **infectious lung diseases**
- **local and systemic effects**
- **lung tissue**
- **nanocarriers**
- **nanoparticles biodistribution**
- **nanoparticulate medicines**
- **patients' adherence**
- **toxic hazards**
- **tuberculosis**

REFERENCES

Agarwal, R., Dhooria, S., Aggarwal, A. N., Maturu, V. N., Sehgal, I. S., Muthu, V., Prasad, K. T., et al., (2015). Guidelines for diagnosis and management of bronchial asthma: Joint ICS/NCCP (I) recommendations. *Lung India, 32*(Suppl 1), S3–S42.

Ahmad, Z., Sharma, S., & Khuller, G., (2005). Inhalable alginate nanoparticles as antitubercular drug carriers against experimental tuberculosis. *International Journal of Antimicrobial Agents, 26*(4), 298–303.

Anselmo, A., & Mitragotri, S., (2014). An overview of clinical and commercial impact of drug delivery systems. *Journal of Controlled Release, 190*, 15–28.

Azarmi, S., Roa, W., & Löbenberg, R., (2008). Targeted delivery of nanoparticles for the treatment of lung diseases. *Advanced Drug Delivery Reviews, 60*(8), 863–875.

Bailey, M., & Berkland, C., (2009). Nanoparticle formulations in pulmonary drug delivery. *Medicinal Research Reviews, 29*(1), 196–212.

Beck-Broichsitter, M., Schmehl, T., Seeger, W., & Gessler, T., (2011). Evaluating the controlled release properties of inhaled nanoparticles using isolated, perfused, and ventilated lung models. *Journal of Nanomaterials, 2011*, 163791.

Bhavna, Ahmad, F., Mittal, G., Jain, G., Malhotra, G., Khar, R., & Bhatnagar, A., (2009). Nano-salbutamol dry powder inhalation: A new approach for treating broncho-constrictive conditions. *European Journal of Pharmaceutics and Biopharmaceutics, 71*(2), 282–291.

Blank, F., Fytianos, K., Seydoux, E., Rodriguez-Lorenzo, L., Petri-Fink, A., Von, G. C., & Rothen-Rutishauser, B., (2017). Interaction of biomedical nanoparticles with the pulmonary immune system. *Journal of Nanobiotechnology, 15*(6).

Boylan, N. J., Suk, J. S., Lai, S. K., Jelinek, R., Boyle, M. P., Cooper, M. J., & Hanes, J., (2012). Highly compacted DNA nanoparticles with low MW PEG coatings: *In vitro, ex vivo* and *in vivo* evaluation. *Journal of Controlled Release, 157*(1), 72–79.

Card, J., Zeldin, D., Bonner, J., & Nestmann, E., (2008). Pulmonary applications and toxicity of engineered nanoparticles. *American Journal of Physiology-Lung Cellular and Molecular Physiology, 295*(3), L400–L411.

Chen, X., Huang, W., Wong, B., Yin, L., Wong, Y., Xu, M., & Yang, Z., (2012). Liposomes prolong the therapeutic effect of anti-asthmatic medication via pulmonary delivery. *International Journal of Nanomedicine, 7*, 1139–1148.

Chung, Y., Ahn, K., Jeon, S., Lee, S., Lee, J., & Tae, G., (2007). Enhanced bone regeneration with BMP-2 loaded functional nanoparticle-hydrogel complex. *Journal of Controlled Release, 121*(1, 2), 91–99.

Cruz-Knight, W., & Blake-Gumbs, L., (2013). Tuberculosis: An overview. *Primary Care: Clinics in Office Practice, 40*(3), 743–756.

Da Silva, A., Cruz, F., Rocco, P., & Morales, M., (2017). New perspectives in nanotherapeutics for chronic respiratory diseases. *Biophysical Reviews, 9*(5), 793–803.

Da Silva, A., Martini, S., Abreu, S., Samary, C. S., Diaz, B., Fernezlian, S., De Sá, V. K., et al., (2014). DNA nanoparticle-mediated thymulin gene therapy prevents airway remodeling in experimental allergic asthma. *Journal of Controlled Release, 180*, 125–133.

Desai, N., (2012). Challenges in the development of nanoparticle-based therapeutics. *AAPS Journal, 14*(2), 282–295.

Duncan, R., (2005). ESF Scientific forward look on nanomedicine. *European Science Foundation Policy Briefing, 23*, 1–6.

Fymat, A., (2016). Recent developments in nanomedicine research. *Journal of Nanomedicine Research, 4*(4), 1–12.

Galli, S., Borregaard, N., & Wynn, T., (2011). Phenotypic and functional plasticity of cells of innate immunity: Macrophages, mast cells and neutrophils. *Nature Immunology, 12*(11), 1035–1044.

Gehr, P., Bachofen, M., & Weibel, E. R., (1978). The normal human lung: Ultrastructure and morphometric estimation of diffusion capacity. *Respiration Physiology, 32*(2), 121–140.

Gehr, P., Mühlfeld, C., Rothen-Rutishauser, B., & Blank, F., (2010). *Particle-Lung Interactions*. Informa Healthcare, New York.

Geiser, M., Quaile, O., Wenk, A., Wigge, C., Eigeldinger-Berthou, S., Hirn, S., Schäffler, M., et al., (2013). Cellular uptake and localization of inhaled gold nanoparticles in lungs of mice with chronic obstructive pulmonary disease. *Particle and Fibre Toxicity, 10*(19).

Guo, P., (2010). The emerging field of RNA nanotechnology. *Nature Nanotechnology, 5*(12), 833–842.

Hua, S., De Matos, M., Metselaar, J., & Storm, G., (2018). Current trends and challenges in the clinical translation of nanoparticulate nanomedicines: Pathways for translational development and commercialization. *Frontiers in Pharmacology, 9*(790).

Iyer, R., Hsia, C., & Nguyen, K., (2015). Nano-therapeutics for the lung: State-of-the-art and future perspectives. *Current Pharmaceutical Design, 21*(36), 5233–5244.

Jackson, J., Zhang, X., Llewellen, S., Hunter, W., & Burt, H., (2004). The characterization of novel polymeric paste formulations for intratumoral delivery. *International Journal of Pharmaceutics, 270*(1, 2), 185–198.

Kong, X., Hellermann, G., Zhang, W., Jena, P., Kumar, M., Behera, A., Behera, S., et al., (2008). Chitosan interferon-gamma nanogene therapy for lung disease: Modulation of t-cell and dendritic cell immune responses. *Allergy, Asthma and Clinical Immunology, 4*(3), 95–105.

Kotton, D., & Morrisey, E., (2014). Lung regeneration: Mechanisms, applications and emerging stem cell populations. *Nature Medicine, 20*(8), 822–832.

Kotton, D., (2012). Next-generation regeneration: The hope and hype of lung stem cell research. *American Journal of Respiratory and Critical Care Medicine, 185*(12), 1255–1260.

Kreuter, J., (2004). Influence of the surface properties on nanoparticle-mediated transport of drugs to the brain. *Journal of Nanoscience and Nanotechnology, 4*(5), 484–488.

Kumar, M., Kong, X., Behera, A., Hellermann, G., Lockey, R., & Mohapatra, S., (2003). Chitosan IFN-gamma-pDNA nanoparticle (CIN) therapy for allergic asthma. *Genetic Vaccines and Therapy, 1*(1), 3.

Kumari, A., Yadav, S., & Yadav, S., (2010). Biodegradable polymeric nanoparticles-based drug delivery systems. *Colloids and Surfaces B: Biointerfaces, 75*(1), 1–18.

Kuzmov, A., & Minko, T., (2015). Nanotechnology approaches for inhalation treatment of lung diseases. *Journal of Controlled Release, 219*, 500–518.

Loira-Pastoriza, C., Todoroff, J., & Vanbever, R., (2014). Delivery strategies for sustained drug release in the lungs. *Advanced Drug Delivery Reviews, 75*, 81–91.

Mansour, H., Rhee, Y., & Wu, X., (2009). Nanomedicine in pulmonary delivery. *International Journal of Nanomedicine, 4*, 299–319.

Mastorakos, P., Da Silva, A., Chisholm, J., Song, E., Choi, W., Boyle, M. P., Morales, M. M., et al., (2015). Highly compacted biodegradable DNA nanoparticles capable of overcoming the mucus barrier for inhaled lung gene therapy. *Proc. Natl. Acad. Sci. USA., 112*(28), 8720–8725.

Matsuo, Y., Ishihara, T., Ishizaki, J., Miyamoto, K., Higaki, M., & Yamashita, N., (2009). Effect of betamethasone phosphate loaded polymeric nanoparticles on a murine asthma model. *Cellular Immunology, 260*(1), 33–38.

Maurya, P. K., & Singh, S., (2019). *Nanotechnology in Modern Animal Biotechnology: Concepts and Applications*. Elsevier, St. Louis.

Melo, G. F., (2014). Nanomedicine and therapy of lung diseases. *Einstein (São Paulo), 12*(4), 531–533.

Moreno-Sastre, M., Pastor, M., Salomon, C., Esquisabel, A., & Pedraz, J., (2015). Pulmonary drug delivery: A review on nanocarriers for antibacterial chemotherapy. *Journal of Antimicrobial Chemotherapy, 70*(11), 2945–2955.

Muralidharan, P., Hayes, D., Black, S. M., & Mansour, H. M., (2016). Microparticulate/nanoparticulate powders of a novel Nrf2 activator and an aerosol performance enhancer for pulmonary delivery targeting the lung Nrf2/Keep-1 pathway. *Molecular Systems Design and Engineering, 1*(1), 48–65.

Nasr, M., Najlah, M., D'Emanuele, A., & Elhissi, A., (2014). PAMAM dendrimers as aerosol drug nanocarriers for pulmonary delivery via nebulization. *International Journal of Pharmaceutics, 461*(1, 2), 242–250.

Nelsen, L. M., Vernon, M., Ortega, H., Cockle, S. M., Yancey, S. W., Brusselle, G., Albers, F. C., & Jones, P. W., (2017). Evaluation of the psychometric properties of the St George's respiratory questionnaire in patients with severe asthma. *Respiratory Medicine, 128*, 42–49.

Omlor, A., Nguyen, J., Bals, R., & Dinh, Q., (2015). Nanotechnology in respiratory medicine. *Respiratory Research, 29*(16), 24.

Oyarzun-Ampuero, F., Brea, J., Loza, M., Torres, D., & Alonso, M., (2009). Chitosan-hyaluronic acid nanoparticles loaded with heparin for the treatment of asthma. *International Journal of Pharmaceutics, 381*(2), 122–129.

Pandey, R., & Khuller, G., (2004). Subcutaneous nanoparticle-based antitubercular chemotherapy in an experimental model. *Journal of Antimicrobial Chemotherapy, 54*(1), 266–268.

Pandey, R., Sharma, A., Zahoor, A., Sharma, S., Khuller, G., & Prasad, B., (2003). Poly (D,L-lactide-co-glycolide) nanoparticle-based inhalable sustained drug delivery system for experimental tuberculosis. *Journal of Antimicrobial Chemotherapy, 52*(6), 981–986.

Paranjpe, M., & Müller-Goymann, C., (2014). Nanoparticle-mediated pulmonary drug delivery: A review. *International Journal of Molecular Sciences, 15*(4), 5852–5873.

Patil-Gadhe, A., & Pokharkar, V., (2014). Montelukast-loaded nanostructured lipid carriers: Part I oral bioavailability improvement. *European Journal of Pharmaceutics and Biopharmaceutics, 88*(1), 160–168.

Patton, J., & Byron, P., (2007). Inhaling medicines: Delivering drugs to the body through the lungs. *Nature Reviews Drug Discovery, 6*(1), 67–74.

Peer, D., Karp, J., Hong, S., Farokhzad, O., Margalit, R., & Langer, R., (2007). Nanocarriers as an emerging platform for cancer therapy. *Nature Nanotechnology, 2*(12), 751–760.

Peters, S., Ferguson, G., Deniz, Y., & Reisner, C., (2006). Uncontrolled asthma: A review of the prevalence, disease burden and options for treatment. *Respiratory Medicine, 100*(7), 1139–1151.

Rajaonarivony, M., Vauthier, C., Couarraze, G., Puisieux, F., & Couvreur, P., (1993). Development of a new drug carrier made from alginate. *Journal of Pharmaceutical Sciences, 82*(9), 912–917.

Roulet, A., Armand, L., Dagouassat, M., Rogerieux, F., Simon-Deckers, A., Belade, E., Van, N. J. T., et al., (2012). Intratracheally administered titanium dioxide or carbon black nanoparticles do not aggravate elastase-induced pulmonary emphysema in rats. *BMC Pulmonary Medicine, 12*(38).

Sadeghi, L., Yousefi, B. V., & Espanani, H., (2015). Toxic effects of the Fe_2O_3 nanoparticles on the liver and lung tissue. *Bratisl Lek Listy, 116*(6), 373–378.

Sahu, S. C., & Casciano, D. A., (2014). *Handbook of Nanotoxicology, Nanomedicine and Stem Cell Use in Toxicology*. Wiley, USA.

Salem, A., (2014). A promising CpG adjuvant-loaded nanoparticle-based vaccine for treatment of dust mite allergies. *Immunotherapy, 6*(11), 1161–1163.

Schneider, C. S., Xu, Q., Boylan, N. J., Chisholm, J., Tang, B. C., Schuster, B. S., Henning, A., et al., (2017). Nanoparticles that do not adhere to mucus provide uniform and long-lasting drug delivery to airways following inhalation. *Science Advances, 3*(4), e1601556.

Semedo. D., (2016). *Nanomedicine for Lung Cancer, COPD to be Developed in Joint Malaysian-Harvard Venture* (p. 2). Lung Disease News.

Seydoux, E., Rodriguez-Lorenzo, L., Blom, R. A., Stumbles, P. A., Petri-Fink, A., Rothen-Rutishauser, B. M., Blank, F., & Von, G. C., (2016). Pulmonary delivery of cationic gold nanoparticles boost antigen-specific CD_4^+ T cell proliferation. *Nanomedicine, 12*(7), 1815–1826.

Sharma, A., Sharma, S., & Khuller, G., (2004). Lectin-functionalized poly (lactide-co-glycolide) nanoparticles as oral/aerosolized antitubercular drug carriers for treatment of tuberculosis. *Journal of Antimicrobial Chemotherapy, 54*(4), 761–766.

Silva, A. L., Santos, R. S., Xisto, D. G., Alonso, S., Del, V., Morales, M. M., & Rocco, P. R. M., (2013). Nanoparticle-based therapy for respiratory diseases. *Annals of the Brazilian Academy of Sciences, 85*(1), 137–146.

Singh, A. P., Biswas, A., Shukla, A., & Maiti, P., (2019). Targeted therapy in chronic diseases using nanomaterial-based drug delivery vehicles. *Signal Transduction and Targeted Therapy, 4*(1), 33.

Sugihara, H., Toda, S., Miyabara, S., Fujiyama, C., & Yonemitsu, N., (1993). Reconstruction of alveolus-like structure from alveolar type II epithelial cells in three-dimensional collagen gel matrix culture. *American Journal of Pathology, 142*(3), 783–792.

Sung, J. C., Pulliam, B. L., & Edwards, D. A., (2007). Nanoparticles for drug delivery to the lungs. *Trends in Biotechnology, 25*(12), 563–570.

Tarlo, S., & Malo, J., (2006). An ATS/ERS report: 100 key questions and needs in occupational asthma. *European Respiratory Journal, 27*(3), 607–614.

Torchilin, V., (2006). Multifunctional nanocarriers. *Advanced Drug Delivery Reviews, 58*(14), 1532–1555.

Upadhyay, S., & Ganguly, K., (2015). Wonders of nanotechnology in the treatment for chronic lung diseases. *Journal of Nanomedicine and Nanotechnology, 6*.

Van, R. S. H., Bein, T., & Meiners, S., (2014). Medical nanoparticles for next generation drug delivery to the lungs. *European Respiratory Journal, 44*(3), 765–774.

Vij, N., (2011). Nano-based theranostics for chronic obstructive lung diseases: Challenges and therapeutic potential. *Expert Opinion on Drug Delivery, 8*(9), 1105–1109.

Vlachogianni, T., Fiotakis, K., Loridas, S., Perdicaris, S., & Valavanidis, A., (2013). Potential toxicity and safety evaluation of nanomaterials for the respiratory system and lung cancer. *Lung Cancer, 4*, 71–82.

Wang, L., Feng, M., Li, Q., Qiu, C., & Chen, R., (2019). Advances in nanotechnology and asthma. *Annals of Translational Medicine, 7*(8), 180.

Weber, S., Zimmer, A., & Pardeike, J., (2014). Solid lipid nanoparticles (SLN) and nanostructured lipid carriers (NLC) for pulmonary application: A review of the state of the art. *European Journal of Pharmaceutics and Biopharmaceutics, 86*(1), 7–22.

Weiss, D., Casaburi, R., Flannery, R., LeRoux-Williams, M., & Tashkin, D., (2013). A placebo-controlled, randomized trial of mesenchymal stem cells in COPD. *Chest, 143*(6), 1590–1598.

Wenzel, S., (2012). Asthma phenotypes: The evolution from clinical to molecular approaches. *Nature Medicine, 18*(5), 716–725.

CHAPTER 9

NANOTECHNOLOGY: AN APPROACH FOR ENHANCEMENT OF PLANT SYSTEM IN TERMS OF TISSUE CULTURE

SNEHA BHANDARI,[1] SWATI SINHA,[1] TAPAN K. NAILWAL,[1] and DEVARAJAN THANGADURAI[2]

[1]*Department of Biotechnology, Kumaun University, Nainital, Bhimtal Campus, Bhimta–263136, Uttarakhand, India*

[2]*Department of Botany, Karnatak University, Dharwad–580003, Karnataka, India*

9.1 INTRODUCTION

Plant tissue culture is a technology which involves the proliferation of plant cells or tissues on a culture medium provided with growth supplements and nourishing constituents in terms of minerals, vitamins, and sometimes microbial antibiotics. *In vitro* regeneration is widely used for the conservation of endangered or near to extinct floral species, for large scale multiplication. This system is also used for crop improvement by implementing molecular approaches for transmission of trans-genes of desired characteristics, for enhancement of primary and secondary metabolites in plant system in order to produce pharmaceutically active products (Jeong and Sivanesan, 2015).

Regardless of extensive and crucial developments, plant tissue culture was solely an investigating and analytical tool for fundamental studies related to botany earlier, and only a few pedagogical laboratories had proficiency of it. Its massive application in horticulture and other interconnected fields was first acknowledged in 1964 when Morel first reported the usage of shoot apex culture for fast clonal propagation of orchids (Ioannou, 1989). Plant tissue culture has accredited mass multiplication of even those floral species for which conventional methods have failed.

Bio-products which are in high demand produced via plant tissue culture for commercial purposes. With the development and emergence of novel approaches and technologies, tissue culture can also be modernized and remodeled. The inventions of emerging and compelling fields like nanotechnology can amalgamate with plant tissue culture so as to intensify its potential (Siddiqui et al., 2015; Kim et al., 2017).

Nanotechnology in a narrower sense, may be seen as an archetype of physical sciences. However, the potential of nanotechnology cannot be trivialized for future reference. Modern and progressive nanotechnological science has turned out to be pertinent to the field of life science and thus, plays a pivotal role in ameliorating the status of human life in upcoming years. Nanoparticles have acquired notable recognition due to their distinctive physical and chemical attributes. Special physicochemical properties of these amazing particles can be a breakthrough in delivery systems in the life science field, especially in the case of plants (Galbraith, 2007; Wang et al., 2016).

Plant tissue culture is an effective means for the proliferation and multiplication of important plant species. Both agricultural crops and endangered medicinal plants can be mass propagated via *in vitro* regeneration procedure. Role of nanotechnology is not limited to the field of medicine but has flair to revolutionize agriculture and food industry as well. Certain benefits include; generation of novel tools for molecular management of diseases, rapid detection and diagnosis of disease, and improving plant's ability to absorb nutrients. On the other hand, the biology of numerous cereal crops can be better understood by utilizing nanobiotechnology principles (Giraldo et al., 2014). It can thus, possibly elevate nourishing values, as well as help in developing upgraded systems for commercial use. The capability of plants to absorb nutrients and pesticides can also be intensified (Yunlong and Smit, 1994; DeRosa et al., 2010; Tarafdar et al., 2013). In the present scenario, biotechnology has also been very much influenced by nanotechnology. Some instances of it are:

- Gene or DNA transfers utilizing nanoparticles in case of plants for production of insect pest-resistant varieties;
- Development of various types of biosensors utilizing nanomaterials which would be beneficial in remote sensing devices prerequisite for precision farming (Rai and Ingle, 2012);

- Development in microfabrication using nanotechnology for enhancing detection limit and cost-effectiveness in viral diagnostics (Cheng et al., 2009).

The concept of nanotechnology is not only used to enhance desirable characteristics in plants but it also widely employed in other different sectors, which are described in Figure 9.1.

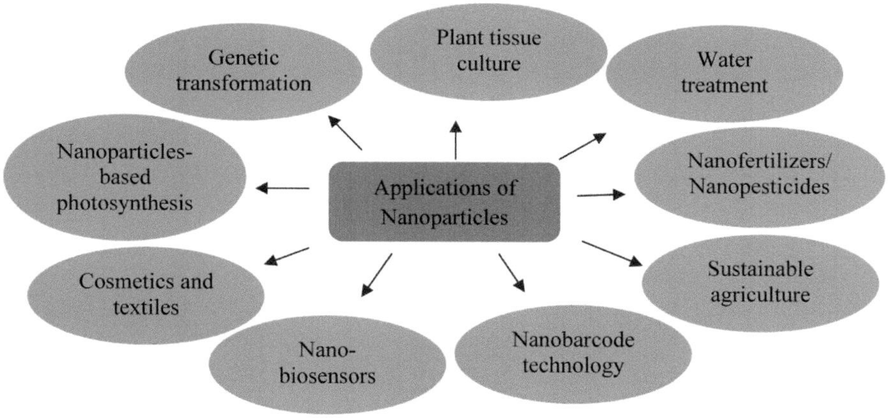

FIGURE 9.1 Applications of nanoparticles in various sectors.

9.2 NANOTECHNOLOGY: AN IDEA AND PRACTICE

In a very fundamental sense "Nanotechnology refers to engineering of functional system which deals particles at atomic and molecular scale." In a more elaborated way, "Nanotechnology deals with miscellaneous tools and techniques which are projected in a way to construct things *from bottom up* to make them into complete and highly valued performance product." Technology when deals with particles ranging in diameter of nanoscale and implementing those particles in other field with determined objectives forms nanosciences.

It all started when a Nobel Prize winner physicist Dr. Richard Phillips Feynman, presented a talk entitled "There's Plenty of Room at the Bottom" at an American Physical Society meeting at California Institute of Technology (CalTech) on December 29, 1959, long before the term nanotechnology came into existence. In the presentation, he talked about how future and upcoming scientists would be able to modify and manipulate individual atoms and molecules for their own need (Feynman, 1960). At that point of time and phase, hardly anybody would have thought that this is possible. But, soon over a decade Professor Norio Taniguchi of Tokyo University of Science coined the term 'nanotechnology' and made this thought of nanoscale molecule a reality.

Substantial development in nanotechnology started in the 1980s when K. Eric Drexler gave a thought of designing and building machines, robots, and even computers at molecular scale much simpler and smaller than a cell. He spent 10 years for growth and development of this idea to come into existence. He is often described as "the founding father of nanotechnology." He introduced his concepts and theories in *Proceedings of the National Academy of Sciences* in the year 1981, where he tried to establish and relate molecular engineering fundamentals to advanced field of nanotechnology in a paper entitled 'Molecular engineering: An approach for the development of general capabilities of molecular manipulation' (Drexler, 1981). In various publications, he reported the implementation of nanotechnology and explained how these nanomaterials can solve trivial issues to large scale problems such as global warming.

Nanosciences has applications in various fields like food packaging, clothing, disinfectant, cosmetics, water purification, environment clean up, machinery, biomedical, and diagnostics. It is important to discuss types of nanoparticles, their characteristics, so that it will be easier to utilize them in various biological systems.

9.3 NANOPARTICLES TYPES

Nanoparticles (NPs) can be engineered to unique parameters such as size, shapes, compositions, and surface chemistries which enable their use in a broad spectrum of biological applications. Unique attributes of nanomaterials and their behavior in biological environment open up wide ranges of compelling and integrative function and advancement in investigating fundamentals of biological systems. These molecules are composed of three layers which are (a) surface layer, formed by various small functionalized molecules like metal ions, polymers, or surfactants. (b) shell layer, a different chemical material layer which is entirely different from core in every aspect, and (c) core, an essential layer which is the central part of nanoparticles and in general referred as NPs itself (Khan et al., 2019). Some important NPs used in plant system are as given in subsections.

9.3.1 LIPOSOMES

In cell membrane model, liposomes were first described in 1965 (Bangham, 1993). Since then, it has become a noteworthy candidate in biophysical investigation and was the first-ever NPs to be utilized for targeted drug and gene delivery. These spherical vesicles are composed of single or multi-bilayered lipid structure which assembles themselves in aqueous environment (Torchilin, 2005). Diameter of these molecules ranges from 10–1,000 nm according to need. Some important characteristic features of liposomes are biocompatibility; versatility and efficiency of entrapment. Liposomes have been widely used to deliver DNA to the cell via lipofection process. Another prime application is targeted drug delivery by using liposome as therapeutic carrier, since it can entrap hydrophilic component centrally and, hydrophobic component on its surface (Felgner, 1987).

9.3.2 CARBON-BASED NANOMATERIALS (CBNs)

Fullerenes and carbon nanotubes (CNTs) are two prime groups of carbon-based NPs. Fullerenes are composed of globular hollow cage such as allotropic forms of carbon (Khan et al., 2019). CNTs is other important carbon-based NPs. These have a diameter ranging from 0.5–3 nm and length varying from 20–1,000 nm, they have high solubility and ability to penetrate inside cell and thus, serve as a good carrier for gene and peptide delivery. They are either present as single layer (single-walled nanotubes: SWNT) or multiple layer (multi-walled nanotubes: MWNT) (Nahar et al., 2006).

9.3.3 METAL NPs

Metal precursors form metal-based NPs. Because of localized surface plasmon resonance (LSPR) feature, these particles have distinctive opto-electrical properties (Khan et al., 2019). In the visible region of the electromagnetic spectrum, nanomaterials are made up of noble elements like Cu, Ag, Au, etc., and have a wide range of absorption bands. In the present-day, cutting-edge materials, size, shape, and facet governed metal NPs synthesis is very important (Dreaden et al., 2012). Since these particles possess optical properties so have a wide range of research applications. Different types of elements which can be used to construct NPs are:

1. **Gold (Au):** Gold NPs provide several shape and size-dependent properties, optical activity, superficial modification of surface and biological compatibility (Daniel and Astruc, 2004). Gold nanomaterials are of less than 100 nm in diameter and are capable of magnifying optical processes like fluorescence, absorption of light, scattering, and surface-enhanced Raman scattering (SERS). This ability is due to distinct reaction between free electrons of NPs with light (Huang et al., 2007). Due to these properties, gold NPs find various applications like biological imaging, gene transporting, diagnostics, and therapeutics.
2. **Silver (Ag):** Due to unique physicochemical properties such as electrical, thermal, optical, and biological properties, silver NPs (Ag NPs) have been used chiefly in pharmaceutics, food industry, textiles, biomedical diagnostics, drug delivery, and gene transfection. In biological system, these particles are used as antiviral, antifungal, antibacterial, anticancer, anti-angiogenic, and anti-inflammatory agent. Size and shape decide their toxicity against microorganisms. Small-sized NPs confer more toxicity than larger ones since the former have more surface area than the latter ones. These come in a size ranging from 5–100 nm according to use (Agnihotri et al., 2014).
3. **Iron (Fe) and Oxides:** Some of the distinctive magnetic properties of iron oxide NPs are low Curie temperature, high coercivity, and high magnetic susceptibility superparamagnetic. Ferromagnetic materials form the basis of these NPs with diameter ranging from less than 10 nm to 20 nm. These magnetic NPs are of substantial importance and have broad ranges of applications in biology. Some common bioapplications are biological entities detection (protein, enzymes, virus, nucleic acids, etc.), clinic diagnosis, bioseparation, clinical therapy (like MRI: magnetic resonance imaging), targeted drug delivery (Wu et al., 2008).
4. **Aluminum (Al) and Oxides:** Colloidal alumina nanoscale particles/fibers usually have a size range of 10–30 nm but sometimes may also occur in diameter of 70–100 nm, with high surface area. They are considered to be highly effective catalysts in many industrial applications. These are extensively applied in drug delivery, heat transfer fluids, water-resistant additives, nano-composites, and material surface coatings. In several plants' species, supplementation of Al-based NPs showed phytotoxicity. The general effect was stunted plant growth with reduced root length.

5. **Titanium (Ti) and Oxides:** These NPs are sometimes referred as ultrafine titanium dioxide, which normally have diameter of less than 100 nm, while pigment-grade titanium oxide particle has a mean diameter of 200–300 nm. Ultrafine TiO_2 is the third most-produced nanoparticles after silicon dioxide and zinc dioxide NPs (Zhang et al., 2015). These have high stability, anticorrosive, and photo-catalytic properties (Riu et al., 2006). These particles are used extensively in paint, plastics, cosmetics, pharmaceuticals, and food products. It has been reported to improve plant growth, promote antioxidant stress, and increase light-harvesting complex II content (Yang et al., 2006; Lei et al., 2007, 2008).

9.3.4 QUANTUM DOTS

These are small nanocrystals of semiconducting material having a size diameter range of 2–10 nm. They show unique electronic properties, with high surface-to-volume ratios. Three main types of quantum dots are (a) Core-type: It consists of single-component material like chalcogenides of various elements such as zinc, cadmium, or lead. (b) Core-shell type: These are made up of two materials in which a tiny region of one material is embedded in another with a broad bandgap, for instance, ZnS in shell covers while CdSe in the core. (c) Alloyed type: These are formed via alloying two semiconductors which have different bandgap energies so that it exhibits exciting features of main parent semiconductors and bulk counterparts.

Quantum dots can produce bright fluorescence, broad UV excitation, and high photostability. These characteristics find applications in various biological aspects like DNA hybridization, receptor-mediated endocytosis, surface labeling of cancerous cells, and immunoassay. These can also be utilized extensively in plant-based research due to their small size. Cell wall porosity of plant is usually around 13 nm which limits other NPs to penetrate through it (Albersheim et al., 2010). Real time monitoring can be done with quantum dots systems for drug delivery in order to find particular disease site, improvement in drug circulation rate and enhanced drug stability, *in vivo* (Zhao and Zhu, 2016).

9.3.5 POLYMERIC NPs

These particles have a size of 10–1,000 nm. Its chief features are biocompatibility, biodegradability, providing overall protection of drugs. These organic-based NPs are basically nanosphere or nanocapsular in shape (Mansha et al., 2017). In nanospheric type of polymeric NPs, overall matrix particle is solid mass with other molecules adsorbed on its spherical surface. In the case of nanocapsular NPs, overall solid mass is entrapped inside the particle completely (Rao and Geckeler, 2011). They can be utilized as drug delivery carriers in a controlled and sustained way. Since these NPs have modified surface, so they can be advantageous for delivery of biologically active molecules actively and passively.

To perform and develop working mechanisms under favorable or adverse conditions is one of the noteworthy abilities of higher plants. In order to boost the native functions of plants, scientists, and researchers are constantly focusing on developing novel techniques. It is important to discuss the issues and challenges while working with plants and how nanosciences will solve these hurdles so that we get a clear vision about nanoscale research for application in plant tissue culture.

9.4 NANOTECHNOLOGY: A NEW PLATFORM FOR DEVELOPMENT OF PLANT TISSUE CULTURE PROTOCOLS OF MEDICINAL FLORA

When a matter is manipulated and altered at atomic, subatomic, molecular, and supramolecular levels, then the technology is termed as nanotechnology (Eric, 1986; Drexler, 1992). This technology deals with nanostructured materials with the dimensions of 1 to 100 nm. NPs in general have various distinctive structural, electrical, chemical, and magnetic attributes. These properties often enable them to be utilized in diverse novel applications such as biosensors, information storage, and biomedical engineering (Leslie-Pelecky et al., 1996; Williard et al., 2004). The magnetic properties of these NPs are substantial hence give an opportunity to be utilized in various procedures. Due to the magnetic properties of NPs, they can selectively attach to a functional molecule; provide magnetic attributes to a target, by controlling magnetic field produced by an electromagnet or permanent magnet, NPs provide modification and transportation to particular molecules at the required location (Nalwa, 2004).

Nanoscale dimensions, structure, and other properties of NPs make them a worthy candidate for diverse biotechnological applications. In a hybrid system, NPs make great companion with many common biomolecules as size ranges are quite similar to both of them. Combining functions of NPs with biologically derived molecules is a boon to research and development as combined functionalities is becoming popular in applications like biomarkers, biosensors, bioimaging, gene therapy, and masking of immunogenic moieties to targeted drug delivery (Bruchez et al., 1998).

Apart from biomedical and pharmaceutical applications, NPs are playing a crucial role in plant tissue culture. Tissue culture provides mass propagation of plant in a short span of time by culturing plant parts on nutrient medium in a controlled sterilized environment. In the present scenario, tissue culture is widely used for several other basic and applied plant biology such as nutrition, cytology, morphology, genetic manipulation, morphogenesis, pathology, large scale propagation of disease-free plants, germplasm conservation, and most importantly production of pharmaceutically active metabolites (Thorpe, 2007). In this era, where nanotechnology has influenced every field then how can the field of plant tissue culture remain untouched with its charm. There are several reports that indicated positive outcomes by combining these two fields. Nanomaterials have been often used to augment plant growth and yield, allow genetic manipulation in plants, enhance seed germination, intensify production of biologically active metabolites, and finally stimulating defense mechanism of plants (Ruttkay-Nedecky et al., 2017). NPs improved seed germination percentage and seedling growth when silicon dioxides (SiO_2) treatment was given to tomato seeds (Siddiqui and Al-Whaibi, 2014). The number of seeds per pod and content of seed protein in black-eyed peas was significantly increased on application of iron and magnesium-based nano-fertilizers (Delfani et al., 2014). DNA was delivered into protoplasts, cells, and leaves of tobacco with the application of gold (Au) capped mesoporous silica NPs (Torney et al., 2007). Content of various cellular constituents such as flavonoids, tannins, anthocyanins, phenolic compounds, and glycyrrhizin increased in licorice seedlings upon copper oxide (CuO) and zinc oxide (ZnO) treatment (Oloumi et al., 2015). Antimicrobial activity against various plant pathogens were reported by utilizing silica-Ag NPs. It was found out that silica-Ag NPs are advantageous against powdery mildew control in infected green squash plants (Sarmast and Salehi, 2016). This chapter deals with an aim to unify all ongoing and contemporary accomplishments made via implementing nanotechnology into plant tissue culture. It will bring positive attributes of NPs in the spotlight for its amalgamation into plant tissue culture. Although the field of tissue culture is itself quite prominent but it still has some drawbacks.

Integration of nanotechnology will help us to find novel prospects and help to minimize problems associated with tissue culture. Some detailed applications of NPs as an aid to plant tissue culture are described below and pictorially represented in Figure 9.2.

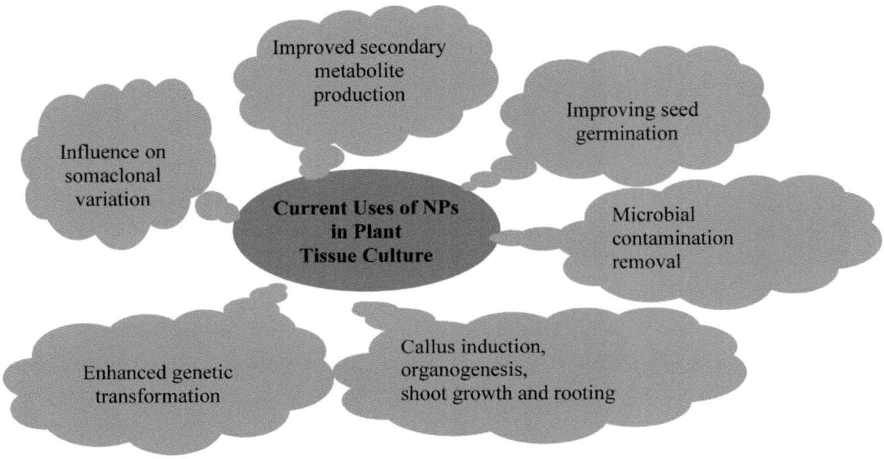

FIGURE 9.2 Various areas of application of nanoparticles in plant tissue culture.

9.4.1 *MICROBIAL CONTAMINATION REMOVAL*

One of the major problems in tissue culture procedures is microbial contamination which intensively affects the plant growth and overall efficiency of propagation system. Plant tissue culture is the best medium for mass propagation of plant variety free from insect-pest-pathogen. Laboratory environment and explants themselves can be a source of contamination (Leifert et al., 1994). 3 to 15% subculture is lost due to microbial contamination in major commercial and scientific laboratories of tissue culture. Major microbial contaminants are bacteria and fungi (Boxus and Terzi, 1987; Leifert et al., 1990; Danby et al., 1994). There are various sterilizing agents present in order to kill microorganisms like ethanol, sodium hypochlorite, calcium hypochlorite, silver nitrate, fungicides, and antibiotics (Leifert et al., 1990). However, supplementation of antibiotics to nutrient media causes phytotoxicity in some plant species (Qin et al., 2011; Tambarussi et al., 2015). For these reasons, metal and oxides of metals as NPs have been used for eliminating microorganisms (Wang et al., 2017). A wide variety of NPs like aluminum oxide, gold, silicon, silver, titanium, ZnO, magnesium oxide (MgO), copper oxide, and iron oxide have been utilized against numerous microorganisms as they possess antimicrobial activities (Beyth et al., 2015). Some of the NPs used for various species for their activity against microorganisms is listed in Table 9.1.

TABLE 9.1 Application of NPs to Remove Microbial Contamination in Plant Tissue Culture

Plant Species	NPs	Size (nm)	Concentration	Phytotoxicity	References
Olive	Ag	–	100–400 mg/L	+	Rostami and Shahsavar (2009)
Banana	Zn, ZnO	35	50–200 mg/L	–	Helaly et al. (2014)
Bacopa monnieri	Ag	–	16×10^{-3} to 16×10^{-10}	+	Kalsaitkar et al. (2014)
Cynodon dactylon	Ag + thymol	–	100 and 200 mg/L	+	Taghizadeh and Solgi (2014)
Gerbera jamesonii	Ag	–	25–200 mg/L	–	Fakhrfeshani et al. (2012)
Nicotinia tabacum cv. Xanthi	TiO_2	10	–	–	Safavi et al. (2011)
Rosa hybrida	Ag	–	100–400 mg/L	–	Shokri et al. (2013)
Valeriana officinalis L.	Ag	30	25–100 mg/L	–	Abdi et al. (2008)
Vitis vinifera L.	Ag	–	100–1,000 mg/L	+	Gouran et al. (2014)

9.4.2 IMPROVING SEED GERMINATION

Several reports have suggested the application of NPs for enhancing seed germination rate in various plant species in plant tissue culture. It has been reported that nanomaterials have the potential to penetrate plant cell wall. According to one research, Au-capped mesoporous silica NPs penetrated the cell wall, and thus, by utilizing particle bombardment method, DNA was delivered into plant cells (Lin et al., 2009). As seed coat covers the seed, it is somewhat difficult to penetrate plant cells when compared to mammalian cell membranes (Serrato-Valenti et al., 2000). However, heavy metals like Lead (Pb_2) and Barium (Ba_2) may selectively penetrate seed coats of various plant species (Wierzbicka and Obidzińska, 1998). List of plant species in which NPs were used to improve seed germination is listed in Table 9.2.

TABLE 9.2 Application of NPs to Improve Seed Germination in Plant Tissue Culture

Plant Species	NPs	Size (nm)	Concentration	Phyto-toxicity	References
Boswellia ovalifoliolata	Ag	30–40	10 to 30 μg/ml	–	Savithramma et al. (2012)
Brassica juncea	Ag	29	25–400 mg/L	+	Sharma et al. (2012)
Gloriosa superba	Au	25	1000 μM	–	Gopinath et al. (2014)
Larix olgensis	SiO_2	–	500 μL/L	–	Bao-Shan et al. (2004)
Lolium perenne	AlO_2	100	50 mg/L	–	Doshi et al. (2008)
Nicotiana tabacum	Activated carbon	–	10–40 μg/mL	–	Khodakovskaya et al. (2012)

9.4.3 CALLUS INDUCTION, ORGANOGENESIS, SHOOT GROWTH, AND ROOTING

It is suggested by various reports that NPs have positive effects on shoot regeneration, callus induction, rooting, and growth in several plant species. For any external agent which also includes NPs, plant cell wall acts as a barrier. Pore diameter of cell wall ranges from 5 nm to 20 nm, which determines the sieving properties (Choat et al., 2004). Therefore, NPs which have a diameter less than that of the pore diameter of the cell wall can pass through and ultimately reach the plasma membrane (Moore, 2006; Navarro et al., 2008). Application of engineered NPs may enlarge the pore or induce new cell wall pore, which could improve nanomaterial uptake. Plasma membrane form a cavity-like structure around the NPs and that is how, further internalization takes place during endocytosis. With the support of ion channels or embedded transport proteins, these particles may cross the membrane. NPs can bind to various organelles in the cytoplasm and thus interfere with metabolic processes either in favor or against the plant cell (Jia et al., 2005). The effects of NPs on various plant species for callus induction, organogenesis, shoot growth, and rooting is explained in Table 9.3.

TABLE 9.3 Application of NPs to Improve Callus Induction, Organogenesis, Shoot Growth and Rooting in Plant Tissue Culture

Plant Species	NPs	Size (nm)	Concentration	Phytotoxicity	References
Arabidopsis thaliana	Ag	8–47	1–100 mM	–	Syu et al. (2014)
Lilium ledebourii	ZnO	–	10–100 mg/L	+	Chamani et al. (2015a)
Mentha longifolia	Cu Co	– –	0.5 mg/L 0.8 mg/L	– –	Talankova-Sereda et al. (2016)
Prunella vulgaris	Au	25–35	30 μg/L	–	Fazal et al. (2016)
Satureja khuzestanica	Carbon nanotubes	–	25–500 μg/mL	–	Ghorbanpour and Hadian (2015)
Stevia rebaudiana L.	ZnO	34	0.1–1000 mg/L	+	Javed et al. (2017)
Tecomella undulata	Ag	18.2	5–80 mg/L	–	Aghdaei et al. (2012)
Verbena bipinnatifida Nutt.	$CuSO_4$	–	5–15 μg/L	+	Genady et al. (2016)

9.4.4 IMPROVED SECONDARY METABOLITES PRODUCTION

Plants are rich source of numerous biologically active secondary metabolites. These metabolites play crucial roles in plants defense mechanism and are important for their survival during stress conditions. Supplementation of precursors, elicitors, optimization of nutrient medium composition, and other culture conditions may enhance secondary metabolite production in plant cell and organ culture (Sivanesan et al., 2016). An efficient means for enhancing the low yields of secondary metabolite is by use of elicitors. Some biological or chemical agents through which physiological and morphological changes takes place, may improve the bioactive components of plant cell (Hussain et al., 2012). Hairy root culture, a type of plant tissue culture, serves as one of the best mediums for the production of secondary metabolites. Biotic or abiotic elicitors may enhance productivity of metabolites (Moharrami et al., 2017). In *in vitro* cultures, NPs may act as an elicitor or nutrient source thus enhance the production of bioactive phytoconstituents.

It is supposed that NPs supplementation during plant tissue culture procedures may induce reactive oxygen species (ROS), which causes oxidative stress. ROS also leads to disruption of the structure of biological molecules like proteins, enzymes, carbohydrates, and lipids. These highly active oxygen entities may cause DNA mutation which is likely to enhance secondary metabolite production rate (Marslin et al., 2017).

It was reported that incorporation of aluminum oxide (Al_2O_3) at a concentration of 10–100 μg/ml in suspension cultures of *Nicotiana tabacum* significantly improved phenolic content in the medium (Poborilova et al., 2013). Some of the plant species in which production of secondary metabolites was enhanced is mentioned in Table 9.4. Possible mechanism action of NPs for enhancing the production of secondary metabolites in plants is explained in Figure 9.3.

9.4.5 ENHANCED GENETIC TRANSFORMATION

Genetic transformation is a robust and crucial procedure that is popularly used in functional genomics of plants, such as novel cognizance into gene function, gene determination, discovery, and exploration of genetically governed characteristics and traits. Additionally, by utilizing functional complementation via genetic transformation gene functions has also been confirmed. Furthermore, with the support of genetic transformation, desirable genes have been introduced in important species which include agricultural crops, and medicinal flora. There are many methods for performing genetic transformation such as electroporation, particle bombardment, liposome-mediated, and *Agrobacterium*-mediated gene transfer. But every method has benefits and limitations of their own.

With the introduction of nanobiotechnology, novel tools such as nanofibers, NPs, and nanocapsules can be incorporated in order to modify genes (McKnight et al., 2003; Roy et al., 2005; Lin et al., 2009). Nanotechnology serves as a platform through which functionalized NPs linked with desirable genes or chemicals may trigger gene expression in plants. It has been reported that carbon nanofibers integrated with plasmid DNA at its surface has been applied for controlled biochemical changes in cells (McKnight et al., 2003, 2004). Incorporation of fluorescent-labeled starch-NPs acting as plant transgenic vehicle was created in such a way that it attaches and transfers genes into plant cells. This transfer takes place by prompt induction of pore channels in cell wall, cell membrane, and nuclear membrane by ultrasound (Liu et al., 2008). In the present scenario, several genes are integrated on fluorescent nanomaterials so that it is possible to

TABLE 9.4 Application of NPs for Improved Secondary Metabolites Production in Plant Tissue Culture

Plant Species	NPs	Size (nm)	Concentration	Type of Secondary Metabolite Produced	References
Aloe vera L.	Ag TiO_2	–	0.63-120 mg/L	Aloin	Raei et al. (2014)
Artemisia annua	Ag-SiO_2	101.8 ± 8.9	900 mg/L	Artemisinin	Zhang et al. (2013)
Artemisia annua	Co	10	0.25–5 mg/L	Artemisinin	Ghasemi et al. (2015)
Calendula officinalis L.	$AgNO_3$	–	0.3–1.2 mg/L	Essential oils	Al-Oubaidi et al. (2014)
Capsicum frutescens	Ag	–	50 mg/L	Capsaicin	Bhat and Bhat (2016)
Corylus avellana	Ag	–	2.5–10 mg/L	Taxol	Jamshidi et al. (2016)
Datura metel	Ag	50–60	20 mg/L	Atropine	Shakeran et al. (2015)
Hyoscyamus reticulatus	Fe_3O_4	–	450–3600 mg/L	Hyoscyamine and scopolamine	Moharrami et al. (2017)
Lilium ledebourii	ZnO	–	10–100 mg/L	Chlorophyll, anthocyanin, flavonoids, phenolics	Chamani et al. (2015b)
Prunella vulgaris	Au, Ag, Au-Ag	25–35	30 μg/L	Flavonoids and Phenolics	Fazal et al. (2016)
Rosa hybrida	Ag	–	100–400 mg/L	Phenolics	Shokri et al. (2013)
Satureja khuzestanica	Carbon nanotube	–	25–500 μg/ml	Secondary metabolite	Ghorbanpour and Hadian (2015)
Stevia rebaudiana	ZnO	34	0.1–1000 mg/L	Steviol glycosides, flavonoid and phenolics	Javed et al. (2017)
Verbena bipinnatifida	$CuSO_4$	–	5–15 μg/L	Phenolics	Genady et al. (2016)
Arabidopsis thaliana	CeO_2 In_2O_3	10–30 to 20–70	0–3000 mg/L	Anthocyanin	Ma et al. (2013)

take image of gene transfer with the help of a fluorescent microscope. In regenerative Calli and tissues, NPs-mediated gene transfer through pores on cell wall and membrane are quite successful with appropriate agent (Nair et al., 2010).

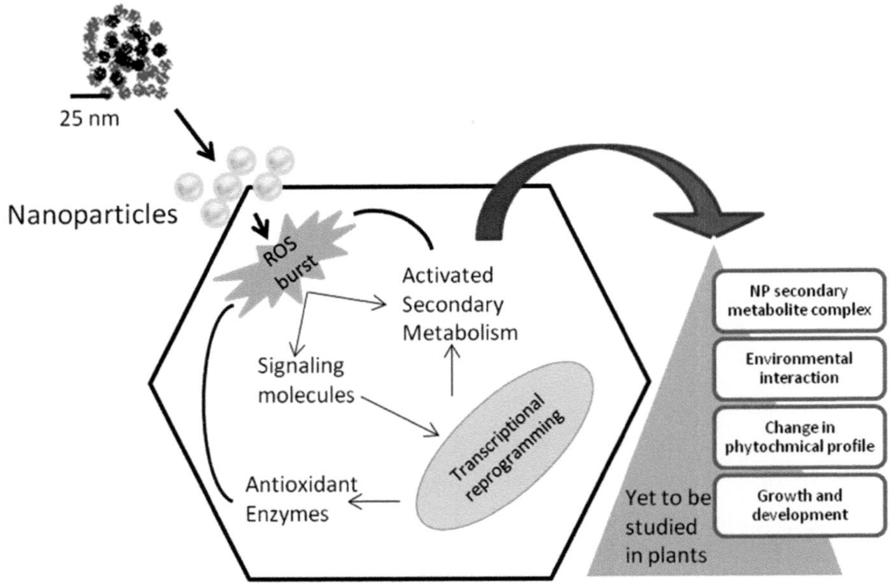

FIGURE 9.3 Possible activation of secondary metabolism pathway mediated by NPs.

It is practically impossible to insert desirable genes in each and every cell. As seeds and plant parts carry billions of cells so, genetic engineers use plant tissue culture in order to produce desired inserted gene in *in vitro* regenerated plantlets. Utilizing tissue culture, transformed cells then regenerate into whole new plant with every cell containing desired transgene. Gene will become a permanent part of plant's genome and thus, will be transferred to next generation of progeny. Some of the important gene transfers via NPs using tissue culture are listed in Table 9.5.

9.4.6 INFLUENCE ON SOMACLONAL VARIATION

Variations that develop in *in-vitro* generated plants in tissue culture are popularly known as somaclonal variations. Changes in chromosome structure,

TABLE 9.5 Application of NPs for Enhanced Genetic Transformation in Plant Tissue Culture

Plant Species	NPs	Size (nm)	Concentration	Gene Transfer Type	References
Agrostis stolonifera L.	Poly(amidoamine) dendrimer	4.5	–	GFP gene	Pasupathy et al. (2008)
Arabidopsis thaliana, *Nicotiana tabacum*	Mesoporous silica	40–50	–	Red fluorescent protein gene	Chang et al. (2013)
Brassica juncea	CaP	20–50	0.5 mg/ml	GUS gene	Naqvi et al. (20120
Nicotiana tabacum, *Oryza sativa*, *Leucaena leucocephala*	Sharp carbon-supported Au	33 ± 4	200– 2000 ng	GUS gene	Vijaykumar et al. (2010)

number, DNA methylation, mitotic crossing over, and activation of transposable elements are some of the common changes associated with somaclonal variations (Bairu et al., 2011; Sivanesan and Jeong, 2012). Advantages and disadvantages both occur due to somaclonal variation. Somaclonal mutants may offer various features and traits such as leaf variegation, flower color, secondary metabolite production, resistance to biotic and abiotic stress (Sivanesan et al., 2016). Treatment via NPs may affect the integrity of DNA, ultimately leading to alteration of DNA and protein expression in plants (Atha et al., 2012; Landa et al., 2015; Tripathi et al., 2017). Treatment of NPs causing somaclonal variation in some flora is listed in Table 9.6).

9.5 LIMITATIONS OF NANOTECHNOLOGY

Nanotechnology and related science have extended their range to various sectors like medical, food, industry, and agriculture. Although exact figures are not available but it is estimated that a massive quantity of synthetic nanotechnology products are being manufactured every year. According to data, the production of various NPs such as TiO_2, CeO_2, Fe_2O_3, Al_2O_3, and ZnO have been extended up to 1,000 of tons (Piccinno et al., 2012) per year. This production level is expected to rise and thus, result in greater discharge of nanomaterials to the environment (Gottschalk and Nowack, 2011). This discharge may be from point sources or from diffuse sources that contaminate water and soil (Colman et al., 2013).

Although NPs have proven exceptional potential and hold diverse promises but, handling, and working with such nanoscale components can be challenging at the same time. Quantum size effects and large surface area to volume are some unique attributes of NPs but, in parallel, they also possess some toxicological features (Zhao and Castranova, 2011). If we talk about inert elements like Gold, Silver, etc., at nano-dimensions, they become highly active and may show negative aspects in plant system.

Nanotechnology has been widely used in the development of plant tissue culture procedures; however, some adverse effects have also been resulted by its incorporation. Some applications in tissue culture, for which NPs have been used, are enhancement in seed germination, removal of microbial growth, improved production of secondary metabolites, and induction of somaclonal variation for better output of traits. Though these positive aspects have been suggested by several reports but simultaneously negative effects have also been reported. Following are some limitations and adverse effects caused by nanoparticles integration in plant system:

TABLE 9.6 Application of NPs to Influence Somaclonal Variation in Plant Tissue Culture

Plant Species	NPs	Size (nm)	Concentration	Somaclonal Variation	References
Linum usitatissimum	Carbon	5–20	1–3 g/L	Decreased percentage of callus induction and somatic embryogenesis, increased number of tetraploid cells and methylation level	Kokina et al. (2012)
Linum usitatissimum L.	Au Ag	– –	1 mg/L 1 mg/L	Au particles increased somatic embryogenesis compared to Ag	Kokina et al. (2013)
Calendula officinalis L.	$AgNO_3$	–	0.3–1.2 mg/L	Increased fresh and dry weights of Calli and secondary metabolite production	Al-Oubaidi et al. (2014)
Prosopis julifora-velutina	ZnO	10	500– 4000 mg/L	Increased level of catalase and ascorbate peroxidase enzyme due to increase in oxidative stress	Hernandez-Viezcas et al. (2011)
Papaver somniferum, *Chelidonium majus*	Petroleum magnetic liquid (iron oxides system, Fe^{2+} and Fe^{3+})	< 100 A^0	5–20 µl/L	Influenced chromosome number thus increases quality and quantity of callus	Pavel et al. (1999)

1. **Decreased Cell Viability:** Supplementation of nanomaterials to plant growth medium has reported cell mortality in some plant species. When CNTs (9.5 nm) were used at a concentration of 10–600 mg/L in *Arabidopsis thaliana* cultures, then, they showed decreased cell viability and biomass. Superoxide dismutase (SOD) activity was also decreased along with chlorophyll content on CNTs application (Lin et al., 2009). In another report, it was found that exposure of *Arabidopsis thaliana* to Ag NPs concentration more than 300 mg L^{-1}, inhibited root elongation and leaf expansion (Sosan et al., 2016).
2. **Inhibition of Seed Germination:** In several plant species, NPs have been used as agents inducing seed germination like *Larix olgensis*, *Boswellia ovalifoliolata*, *Gloriosa superba*, etc., though the fact cannot be neglected that in some other species these particles have inhibited seed germination process. When Al_2O_3 (<50 nm) was used in suspension cultures of *Nicotiana tobacum*, at a concentration of 10–100 μg/mL, phenolic content of the cell increased but simultaneously cell viability decreased (Poborilova et al., 2013). In *Brassica nigra*, when ZnO (100 nm) was used at a concentration of 500–1500 mg/L it inhibited seedling growth and germination process (Zafar et al., 2016). Nanoscale zinc at a concentration of 2,000 mg/L drastically affected seed germination and root elongation in *Lolium perenne* (ryegrass) (Lin et al., 2007).
3. **Genotoxicity:** Application of various NPs at some defined concentration may be genotoxic to plant life cycle. Genotoxicity refers to the property of some chemical agents which damages genetic balance and causes mutation within a cell which may lead to tumor formation, chromosomal aberration, variation in gene, and protein function, etc. β-cyclodextrin ($C_{42}H_{70}O_{35}$) (12.3 nm) coated with magnetic NPs (made up from iron salts) when used at a various concentration (10 and 250 μl/L) in *Zea mays* cultures, induced genotoxicity. Chromosomal aberrations such as retarded chromosomes, chromosome fragments, inter-chromatin bridges, and ring chromosomes were observed (Răcuciu et al., 2009).
4. **Formation of ROS:** ROS accumulation in large scale causes oxidative stress within plant cell. These reactive particles may be formed as by-product during the metabolism of oxygen and play a significant role in cell signaling and homeostasis (Kroneck and Torres, 2015). But, higher amounts of these reactive particles may be lethal

to cells, as these may damage genetic material, oxidize amino acids in proteins, damage polyunsaturated fatty acids in lipids and oxidize co-factors of vital enzymes, making them non-functional (Brooker, 2011). Phytotoxicity due to NPs leading to induction of oxidative stress has been reported by several studies. Treatment of cerium oxide (CeO_2, 10–30 nm) and indium oxide (In_2O_3, 20–70 nm) at a concentration of 250 and 1,000 mg/L causes high accumulation of ROS in roots of *Arabidopsis thaliana* (Ma et al., 2016).

5. **Reduced Secondary Metabolite Production:** Secondary metabolites are bioactive components which are not directly involved in growth, development, and reproduction but used as defensive agents in plant. These biologically active molecules are of therapeutic value for humans, but natural production of these metabolites is quite low in plants, so their level is increased by hairy root culture, suspension culture by providing various elicitors. NPs have been widely used as elicitors for improved production of secondary metabolites in plants. In many plant species like *Calendula officinalis* L., *Artemisia annua*, *Aloe vera* L., *Lilium ledebourii*, production of secondary metabolite was enhanced using several types of NPs as elicitor. But the case is not same in every plant type. In some plant species on application of NPs, there has been a reduced production of secondary metabolite (Poborilova et al., 2013). For instance, when zinc (<100 nm) was used as an elicitor in culture of *Stevia rebaudiana* at a concentration of 50–1000 mg/L, it was shown that it acted as suppressing agent and caused lower production of stevioside, a glycoside of alkaloid origin (Desai et al., 2015).

9.6 FUTURE SCENARIO OF PLANT TISSUE CULTURE AMALGAMATED WITH NPs

From various reports quoting the role and efficiency of NPs in plant biotechnology, we can say that nanotechnology has great potential in improving various aspects of plant research. Right from the first stage of explants disinfection to better response of plants *in vitro*, better secondary metabolite synthesis, NPs have proved their stance as an evolutionary science with many novel applications. Recent progress in understanding, synthesis, and varied handling of NPs undeniably has led to its phenomenal growth, with successive research on nanomaterials encompassed applications and products.

Owing to their distinctive properties, a lot of research has been performed on the toxic effect of NPs in plants; however, research nucleating on understanding the valuable effects of NPs in plant remains deficient. It is apparent from assembled information that effects of NPs varies from plant to plant and depends on type, size, concentration, and their mode of action (MOA). Research on NPs is essential for plants, and it is in a premature phase; in view of this, more rigorous work needs to be done to understand the physiological, biochemical, and molecular mechanisms of plants under the effect of NPs. More studies are needed to understand the effect of NPs on callus induction, organogenesis, shoot, and root proliferation, regulation of gene expressions in plants and their interaction with biomolecules. Though, it is supposed that oxidative stress caused by ROS induced by NPs in the plant system, is the reason behind various effects in plants but more research needs to be done to justify the supposition. Manipulation induced by NPs on secondary metabolite production in several important plant species, and the effect on chemical, physical, and biological activities of compounds obtained from NPs treated plant cultures should be scrutinized.

Nanotoxicity is the bottleneck in freely using nanomaterials in a living system. Hence, it should be thoroughly scrutinized and contemplated for their consequences, without which no genuine progress can be made.

9.7 CONCLUSION

Nanotechnology has promising prospects in assorted fields of science, but one must be very cautious about any new technology for its unlikely associated risks that may be associated along with its positive usage. On the other hand, it is important for the future of a nation to train workforce in nanotechnology and make the public aware of its unparalleled usage and associated risks. Nevertheless, with the anticipated utilities of NPs based products, it is required that methodical study should be done so as to better understand the consequences of NPs in crop plants and their environmental outcomes. Even with all of these unrevealed intense intimidating backgrounds, plant nanobiotechnology unquestionably has shown great potential with flamboyant applications for plant improvement. The need of the hour is to shed light and simplify the process to only utilize the favorable side without the exposure of negative effects.

KEYWORDS

- **callus induction**
- **carbon nanotubes**
- **genetic transformations**
- **genotoxicity**
- **molecular engineering**
- **nanobarcode technology**
- **nanobiosensors**
- **nanofertilizers**
- **nanoparticles**
- **nanotechnology**
- **plant tissue culture**
- **quantum dots**
- **reactive oxygen species**
- **secondary metabolites**
- **somaclonal variations**

REFERENCES

Abdi, G., Salehi, H., & Khosh-Khui, M., (2008). Nanosilver: A novel nanomaterial for removal of bacterial contaminants in valerian (*Valeriana officinalis* L.) tissue culture. *Acta Physiol. Plant., 30*(5), 709–714.

Aghdaei, M., Salehi, H., & Sarmast, M. K., (2012). Effects of silver nanoparticles on *Tecomella undulata* (Roxb.) Seem. micropropagation. *Adv. Hortic. Sci., 26*(1), 21–24.

Agnihotri, S., Mukherji, S., & Mukherji, S., (2014). Size-controlled silver nanoparticles synthesized over the range 5–100 nm using the same protocol and their antibacterial efficacy. *RSC Adv., 4*(8), 3974–3983.

Albersheim, P., Darvill, A., Roberts, K., Sederoff, R., & Staehelin, A., (2010). Cell walls and plant anatomy. In: *Plant Cell Walls* (pp. 19–60). Garland Science, New York.

Al-Oubaidi, H. K. M., & Mohammed-Ameen, A. S., (2014). The effect of ($AgNO_3$) NPs on increasing of secondary metabolites of *Calendula officinalis* L. *in vitro*. *Int. J. Pharm. Pract., 5*(4), 267–272.

Atha, D. H., Wang, H., Petersen, E. J., Cleveland, D., Holbrook, R. D., Jaruga, P., et al., (2014). Correction to copper oxide nanoparticle mediated DNA damage in terrestrial plant models. *Environ. Sci. Technol., 48*(20), 12473.

Bairu, M. W., Aremu, A. O., & Van, S. J., (2011). Somaclonal variation in plants: Causes and detection methods. *Plant Growth Regul., 63*(2), 147–173.
Bangham, A. D., (1993). Liposomes: The Babraham connection. *Chem. Phys. Lipids., 64*(1–3), 275–285.
Bao-Shan, L., Chun-Hui, L., Li-Jun, F., Shu-Chun, Q., & Min, Y., (2004). Effect of TMS (nanostructured silicon dioxide) on growth of Changbai larch seedlings. *J. For. Res., 15*(2), 138–140.
Beyth, N., Houri-Haddad, Y., Domb, A., Khan, W., & Hazan, R., (2015). Alternative antimicrobial approach: Nano-antimicrobial materials. *Evid. Based Complement Alternat. Med., 2015*, 246012. doi: 10.1155/2015/246012.
Bhat, P., & Bhat, A., (2016). Silver nanoparticles for enhancement of accumulation of capsaicin in suspension culture of *Capsicum* sp. *J. Exp. Sci., 7*, 1–6.
Boxus, P. H., & Terzi, J. M., (1987). Big losses due to bacterial contaminations can be avoided in mass propagation scheme. *Acta Hortic., 212*, 91–94.
Brooker, R. J., (2011). *Genetics: Analysis and Principles* (p. 868). McGraw-Hill, New York.
Bruchez, M., Moronne, M., Gin, P., Weiss, S., & Alivisatos, A. P., (1998). Semiconductor nanocrystals as fluorescent biological labels. *Science, 281*(5385), 2013–2016.
Chamani, E., Ghalehtaki, S. K., Mohebodini, M., & Ghanbari, A., (2015b). Secondary metabolite production by *Lilium ledebourii* Boiss under *in vitro* conditions. *Iran. J. Genet. Plant Breed., 4*(2), 11–19.
Chamani, E., Karimi, G. S., Mohebodini, M., & Ghanbari, A., (2015a). The effect of zinc oxide nanoparticles and humic acid on morphological characters and secondary metabolite production in *Lilium ledebourii* Boiss. *Iran. J. Genet. Plant Breed., 4*(2), 11–19.
Chang, F. P., Kuang, L. Y., Huang, C. A., Jane, W. N., Hung, Y., Yue-Ie, C. H., & Mou, C. Y., (2013). A simple plant gene delivery system using mesoporous silica nanoparticles as carriers. *J. Mater. Chem. B, 1*(39), 5279–5287.
Cheng, X., Chen, G., & Rodriguez, W. R., (2009). Micro- and nanotechnology for viral detection. *Anal. Bioanal. Chem., 393*(2), 487–501.
Choat, B., Jansen, S., Zwieniecki, M. A., Smets, E., & Holbrook, N. M., (2004). Changes in pit membrane porosity due to deflection and stretching: The role of vestured pits. *Journal of Experimental Botany, 55*(402), 1569–1575.
Colman, B. P., Arnaout, C. L., Anciaux, S., Gunsch, C. K., Hochella, Jr. M. F., Kim, B., et al., (2013). Low concentrations of silver nanoparticles in biosolids cause adverse ecosystem responses under realistic field scenario. *PLoS One, 8*(2), 57189.
Danby, S., Berger, F., Howitt, D. J., Wilson, A. R., Dawson, S., & Leifert, C., (1994). Fungal contaminants of *Primula*, *Coffea*, *Musa* and *Iris* tissue cultures. In: Lumsden, P. J., Nicholas, J. R., & Davies, W. J., (eds.), *Physiology, Growth and Development of Plants in Culture* (pp. 397–403). Springer, Dordrecht.
Danied, M. C., & Astruc, D., (2004). Gold nanoparticles: Assembly, supramolecular chemistry, quantum-size-related properties, and applications toward biology, catalysis, and nanotechnology. *Chem Rev., 104*(1), 293–346.
Delfani, M., Baradarn, F. M., Farrokhi, N., & Makarian, H., (2014). Some physiological responses of black-eyed pea to iron and magnesium nanofertilizers. *Commun. Soil Sci. Plant Anal., 45*(4), 530–540.
DeRosa, M. C., Monreal, C., Schnitzer, M., Walsh, R., & Sultan, Y., (2010). Nanotechnology in fertilizers. *Nat. Nanotechnol., 5*(2), 91.

Desai, C. V., Desai, H. B., Suthar, K. P., Singh, D., Patel, R. M., & Taslim, A., (2015). Phytotoxicity of zinc-nanoparticles and its influence on stevioside production in *Stevia rebaudiana* Bertoni. *Appl. Biol. Res., 17*(1), 1–7.
Doshi, R., Braida, W., Christodoulatos, C., Wazne, M., & O'Connor, G., (2008). Nanoaluminum: Transport through sand columns and environmental effects on plants and soil communities. *Environ. Res., 106*(3), 296–303.
Dreaden, E. C., Alkilany, A. M., Huang, X., Murphy, C. J., & El-Sayed, M. A., (2012). The golden age: Gold nanoparticles for biomedicine. *Chem. Soc. Rev., 41*(7), 2740–2779.
Drexler, E. K., (1992). *Nanosystems: Molecular Machinery, Manufacturing, and Computation* (p. 556). John Wiley and Sons, New York.
Drexler, K. E., (1981). Molecular engineering: An approach to the development of general capabilities for molecular manipulation. *Proc. Natl. Acad. Sci. USA, 78*(9), 5275–5278.
Eric, D. K., (1986). *Engines of Creation: The Coming Era of Nanotechnology* (p. 320). Doubleday Publisher, New York.
Fakhrfeshani, M., Bagheri, A., & Sharifi, A., (2012). Disinfecting effects of nanosilver fluids in gerbera (*Gerbera jamesonii*) capitulum tissue culture. *J. Biol. Environ. Sci., 6*(17), 121–127.
Fazal, H., Abbasi, B. H., Ahmad, N., & Ali, M., (2016). Elicitation of medicinally important antioxidant secondary metabolites with silver and gold nanoparticles in callus cultures of *Prunella vulgaris* L. *Appl Biochem. Biotechnol., 180*(6), 1076–1092.
Felgner, P. L., Gadek, T. R., Holm, M., Roman, R., Chan, H. W., Wenz, M., Northrop, J. P., et al., (1987). Lipofection: A highly efficient, lipid-mediated DNA-transfection procedure. *Proc. Natl. Acad. Sci. USA, 84*(21), 7413–7417.
Feynman, R. P., (1960). The wonders that await a micro-microscope. *Saturday Rev., 43*(2), 45–47.
Galbraith, D. W., (2007). Silica breaks through in plants. *Nat. Nanotechnol., 2*(5), 272–273.
Genady, E. A., Qaid, E. A., & Fahmy, A. H., (2016). Copper sulfate nanoparticles *in vitro* applications on *Verbena bipinnatifida* Nutt. stimulating growth and total phenolic content increasments. *Int. J. Pharm. Res. Allied Sci., 5*(1), 196–202.
Ghasemi, B., Hosseini, R., & Nayeri, F. D., (2015). Effects of cobalt nanoparticles on artemisinin production and gene expression in *Artemisia annua. Turk J. Botany, 39*(5), 769–777.
Ghorbanpour, M., & Hadian, J., (2015). Multi-walled carbon nanotubes stimulate callus induction, secondary metabolites biosynthesis and antioxidant capacity in medicinal plant *Satureja khuzestanica* grown *in vitro*. *Carbon, 94*, 749–759.
Giraldo, J. P., Landry, M. P., Faltermeier, S. M., McNicholas, T. P., Iverson, N. M., Boghossian, A. A., et al., (2014). Plant nanobionics approach to augment photosynthesis and biochemical sensing. *Nat. Mater., 13*(4), 400–408.
Gopinath, K., Gowri, S., Karthika, V., & Arumugam, A., (2014). Green synthesis of gold nanoparticles from fruit extract of *Terminalia arjuna*, for the enhanced seed germination activity of *Gloriosa superba. J. Nanostructure Chem., 4*(3), 115.
Gottschalk, F., & Nowack, B., (2011). The release of engineered nanomaterials to the environment. *J. Environ. Monit., 13*(5), 1145–1155.
Gouran, A., Jirani, M., Mozafari, A. A., Saba, M. K., Ghaderi, N., & Zaheri, S., (2014). Effect of silver nanoparticles on grapevine leaf explants sterilization at *in vitro* conditions. In: *2nd National Conference of Nanotechnology: From Theory to Application* (pp. 1–6). Jami Institute, Isfahan, Iran.

Helaly, M. N., El-Metwally, M. A., El-Hoseiny, H., Omar, S. A., & El-Sheery, N. I., (2014). Effect of nanoparticles on biological contamination of *in vitro* cultures and organogenic regeneration of banana. *Aust. J. Crop Sci., 8*(4), 612–624.

Hernandez-Viezcas, J. A., Castillo-Michel, H., Servin, A. D., Peralta-Videa, J. R., & Gardea-Torresdey, J. L., (2011). Spectroscopic verification of zinc absorption and distribution in the desert plant *Prosopis juliflora-velutina* (velvet mesquite) treated with ZnO nanoparticles. *Chem. Eng. J., 170*(2, 3), 346–352.

Huang, X., Jain, P. K., El-Sayed, I. H., & El-Sayed, M. A., (2007). Gold nanoparticles and nanorods in medicine: From cancer diagnostics to photothermal therapy. *Nanomedicine, 2*(5), 681–693.

Hussain, M. S., Fareed, S., Ansari, S., Rahman, M. A., & Ahmad, I. Z., & Saeed, M., (2012). Current approaches toward production of secondary plant metabolites. *J. Pharm. Bioallied Sci., 4*(1), 10–20.

Ioannou, M., (1989). *Agricultural Applications of Plant Tissue Culture: Rapid Clonal Propagation of Horticultural Crops*. Agricultural Research Institute, Nicosia, Cyprus.

Jamshidi, M., Ghanati, F., Rezaei, A., & Bemani, E., (2016). Change of antioxidant enzymes activity of hazel (*Corylus avellana* L.) cells by Ag NPs. *Cytotechnology, 68*(3), 525–530.

Javed, R., Usman, M., Yücesan, B., Zia, M., & Gürel, E., (2017). Effect of zinc oxide (ZnO) nanoparticles on physiology and steviol glycosides production in micro propagated shoots of *Stevia rebaudiana* Bertoni. *Plant Physiol. Biochem., 110*, 94–99.

Jeong, B. R., & Sivanesan, I., (2015). Direct adventitious shoot regeneration, *in vitro* flowering, fruiting, secondary metabolite content and antioxidant activity of *Scrophularia takesimensis* Nakai. *Plant Cell Tiss. Organ Cult., 123*(3), 607–618.

Jia, G., Wang, H., Yan, L., Wang, X., Pei, R., Yan, T., et al., (2005). Cytotoxicity of carbon nanomaterials: Single-wall nanotube, multi-wall nanotube, and fullerene. *Environ. Sci. Technol., 39*(5), 1378–1383.

Kalsaitkar, P., Tanna, J., Kumbhare, A., Akre, S., Warade, C., & Gandhare, N., (2014). Silver nanoparticles induced effect on *in vitro* callus production in *Bacopa monnieri*. *Asian J. Biol. Life Sci., 3*(3), 167–172.

Khan, I., Saeed, K., & Khan, I., (2019). Nanoparticles: Properties, applications and toxicities. *Arab. J. Chem., 12*(7), 908–931.

Khodakovskaya, M. V., Silva, K. D., Biris, A. S., Dervishi, E., & Villagarcia, H., (2012). Carbon nanotubes induce growth enhancement of tobacco cells. *ACS Nano, 6*(3), 2128–2135.

Kim, D. H., Gopal, J., & Sivanesan, I., (2017). Nanomaterials in plant tissue culture: The disclosed and undisclosed. *RSC Adv., 7*(58), 36492–36505.

Kokina, I., Gerbreders, V., Sledevskis, E., & Bulanovs, A., (2013). Penetration of nanoparticles in flax (*Linumusita tissimum* L.) calli and regenerants. *J. Biotechnol., 165*(2), 127–132.

Kokina, I., Sļedevskis, E., Gerbreders, V., Grauda, D., Jermaļonoka, M., Valaine, K., & Rashal, I., (2012). Reaction of flax (*Linumusita tissimum* L.) calli culture to supplement of medium by carbon nanoparticles. *Proceedings of the Latvian Academy of Sciences. Section B. Natural, Exact, and Applied Sciences, 66*(4, 5), 200–209.

Kroneck, P. M., & Torres, M. E. S., (2015). *Sustaining Life on Planet Earth: Metalloenzymes Mastering Dioxygen and Other Chewy Gases*. Springer International Publishing, Switzerland.

Landa, P., Prerostova, S., Petrova, S., Knirsch, V., Vankova, R., & Vanek, T., (2015). The transcriptomic response of *Arabidopsis thaliana* to zinc oxide: A comparison of the impact of nanoparticle, bulk, and ionic zinc. *Environ. Sci. Technol., 49*(24), 14537–14545.

Lei, Z., Mingyu, S., Chao, L., Liang, C., Hao, H., Xiao, W., & Fashui, H., (2007). Effects of nanoanatase TiO_2 on photosynthesis of spinach chloroplasts under different light illumination. *Biol. Trace Elem. Res., 119*(1), 68–76.

Lei, Z., Mingyu, S., Xiao, W., Chao, L., Chunxiang, Q., Liang, C., et al., (2008). Antioxidant stress is promoted by nano-anatase in spinach chloroplasts under UV-B radiation. *Biol. Trace Elem. Res., 121*(1), 69–79.

Leifert, C., Morris, C. E., & Waites, W. M., (1994). Ecology of microbial saprophytes and pathogens in tissue culture and field-grown plants: Reasons for contamination problems *in vitro*. *Crit. Rev. Plant Sci., 13*(2), 139–183.

Leifert, C., Waites, W. M., Nicholas, J. R., & Keetley, J. W., (1990). Yeast contaminants of micropropagated plant cultures. *Journal of Applied Bacteriology, 69*(4), 471–476.

Leslie-Pelecky, D. L., & Rieke, R. D., (1996). Magnetic properties of nanostructured materials. *Chemistry of Materials, 8*(8), 1770–1783.

Lin, C., Fugetsu, B., Su, Y., & Watari, F., (2009). Studies on toxicity of multi-walled carbon nanotubes on *Arabidopsis* T87 suspension cells. *Journal of Hazardous Materials, 170*(2, 3), 578–583.

Lin, D. H., & Xing, B. S., (2007). Phytotoxicity of nanoparticles: Inhibition of seed germination and root elongation. *Environ. Pollut., 150*(2), 243–250.

Liu, J., Wang, F. H., Wang, L. L., Xiao, S. Y., Tong, C. Y., Tang, D. Y., & Liu, X. M., (2008). Preparation of fluorescence starch-nanoparticle and its application as plant transgenic vehicle. *J. Cent. South Univ. T., 15*(6), 768–773.

Ma, C., Chhikara, S., Xing, B., Musante, C., White, J. C., & Dhankher, O. P., (2013). Physiological and molecular response of *Arabidopsis thaliana* (L.) to nanoparticle cerium and indium oxide exposure. *ACS Sustain. Chem. Eng., 1*(7), 768–778.

Ma, C., Liu, H., Guo, H., Musante, C., Coskun, S. H., Nelson, B. C., et al., (2016). Defense mechanisms and nutrient displacement in *Arabidopsis thaliana* upon exposure to CeO_2 and In_2O_3 nanoparticles. *Environmental Science: Nano, 3*(6), 1369–1379.

Mansha, M., Khan, I., Ullah, N., & Qurashi, A., (2017). Synthesis, characterization and visible-light-driven photoelectrochemical hydrogen evolution reaction of carbazole-containing conjugated polymers. *Int. J. Hydrog. Energy, 42*(16), 10952–10961.

Marslin, G., Sheeba, C. J., & Franklin, G., (2017). Nanoparticles alter secondary metabolism in plants via ROS burst. *Front. Plant Sci., 8*, 832.

McKnight, T. E., Melechko, A. V., Griffin, G. D., Guillorn, M. A., Merkulov, V. I., Serna, F., & Simpson, M. L., (2003). Intracellular integration of synthetic nanostructures with viable cells for controlled biochemical manipulation. *Nanotechnology, 14*(5), 551.

McKnight, T. E., Melechko, A. V., Hensley, D. K., Mann, D. G., Griffin, G. D., & Simpson, M. L., (2004). Tracking gene expression after DNA delivery using spatially indexed nanofiber arrays. *Nano Letters, 4*(7), 1213–1219.

Moharrami, F., Hosseini, B., Sharafi, A., & Farjaminezhad, M., (2017). Enhanced production of hyoscyamine and scopolamine from genetically transformed root culture of *Hyoscyamus reticulatus* L. elicited by iron oxide nanoparticles. *In vitro Cellular and Developmental Biology-Plant, 53*(2), 104–111.

Moore, M. N., (2006). Do nanoparticles present ecotoxicological risks for the health of the aquatic environment? *Environ. Int., 32*(8), 967–976.

Nahar, M., Dutta, T., Murugesan, S., Asthana, A., Mishra, D., Rajkumar, V., & Jain, N. K., (2006). Functional polymeric nanoparticles: An efficient and promising tool for active delivery of bioactives. *Crit. Rev. Ther. Drug Carrier Syst., 23*(4), 259–318.

Nair, R., Varghese, S. H., Nair, B. G., Maekawa, T., Yoshida, Y., & Kumar, D. S., (2010). Nanoparticulate material delivery to plants. *Plant Science, 179*(3), 154–163.

Nalwa, H. S., (2004). *Encyclopedia of Nanoscience and Nanotechnology* (Vol. 10). American Scientific Publishers, California, USA.

Naqvi, S., Maitra, A. N., Abdin, M. Z., Akmal, M. D., Arora, I., & Samim, M. D., (2012). Calcium phosphate nanoparticle-mediated genetic transformation in plants. *J. Mater. Chem., 22*(8), 3500–3507.

Navarro, E., Baun, A., Behra, R., Hartmann, N. B., Filser, J., Miao, A. J., et al., (2008). Environmental behavior and ecotoxicity of engineered nanoparticles to algae, plants, and fungi. *Ecotoxicology, 17*(5), 372–386.

Oloumi, H., Soltaninejad, R., & Baghizadeh, A., (2015). The comparative effects of nano and bulk size particles of CuO and ZnO on glycyrrhizin and phenolic compounds contents in *Glycyrrhiza glabra* L. seedlings. *Indian J. Plant Physiol., 20*(2), 157–161.

Pasupathy, K., Lin, S., Hu, Q., Luo, H., & Ke, P. C., (2008). Direct plant gene delivery with a poly (amidoamine) dendrimer. *Biotechnology Journal: Healthcare Nutrition Technology, 3*(8), 1078–1082.

Pavel, A., Trifan, M., Bara, I. I., Creanga, D. E., & Cotae, C., (1999). Accumulation dynamics and some cytogenetical tests at *Chelidonium majus* and *Papaver somniferum* callus under the magnetic liquid effect. *J. Magn. Magn. Mater., 201*(1–3), 443–445.

Piccinno, F., Gottschalk, F., Seeger, S., & Nowack, B., (2012). Industrial production quantities and uses of ten engineered nanomaterials in Europe and the world. *J. Nanopart. Res., 14*(9), 1109.

Poborilova, Z., Opatrilova, R., & Babula, P., (2013). Toxicity of aluminium oxide nanoparticles demonstrated using a BY-2 plant cell suspension culture model. *Environ. Exper. Bot., 91*, 1–11.

Qin, Y. H., Da Silva, J. A. T., Bi, J. H., Zhang, S. L., & Hu, G. B., (2011). Response of *in vitro* strawberry to antibiotics. *Plant Growth Regul., 65*(1), 183–193.

Răcuciu, M., & Creangă, D. E., (2009). Cytogenetical changes induced by β-cyclodextrin coated nanoparticles in plant seeds. *Rom. J. Phys., 54*(1, 2), 125–131.

Raei, M., Angaji, S. A., Omidi, M., & Khodayari, M., (2014). Effect of abiotic elicitors on tissue culture of *Aloe vera*. *Int. J. Biosci., 5*(1), 74–81.

Rai, M., & Ingle, A., (2012). Role of nanotechnology in agriculture with special reference to management of insect pests. *Appl. Microbiol. Biotechnol., 94*(2), 287–293.

Rao, J. P., & Geckeler, K. E., (2011). Polymer nanoparticles: Preparation techniques and size-control parameters. *Prog. Polym. Sci., 36*(7), 887–913.

Riu, J., Maroto, A., & Rius, F. X., (2006). Nanosensors in environmental analysis. *Talanta, 69*(2), 288–301.

Rostami, A. A., & Shahsavar, A., (2009). Olive "mission" explants. *Asian J. Plant Sci., 8*(7), 505–509.

Ruttkay-Nedecky, B., Krystofova, O., Nejdl, L., & Adam, V., (2017). Nanoparticles based on essential metals and their phytotoxicity. *J. Nanobiotechnology, 15*(1), 33.

Safavi, K., Mortazaeinezahad, F., Esfahanizadeh, M., & Asgari, M. J., (2011). *In vitro* antibacterial activity of nanomaterial for use in tobacco plants tissue culture. *World Academy of Science, Engineering and Technology, 79*, 372–373.

Sarmast, M. K., & Salehi, H., (2016). Silver nanoparticles: An influential element in plant nanobiotechnology. *Mol. Biotechnol., 58*(7), 441–449.

Savithramma, N., Ankanna, S., & Bhumi, G., (2012). Effect of nanoparticles on seed germination and seedling growth of *Boswellia ovalifoliolata* an endemic and endangered medicinal tree taxon. *Nano Vision, 2*(1), 2.

Serrato-Valenti, G., Cornara, L., Modenesi, P., Piana, M., & Mariotti, M. G., (2000). Structure and histochemistry of embryo envelope tissues in the mature dry seed and early germination of *Phacelia tanacetifolia*. *Ann. Bot., 85*(5), 625–634.

Shakeran, Z., Keyhanfar, M., Asghari, G., & Ghanadian, M., (2015). Improvement of atropine production by different biotic and abiotic elicitors in hairy root cultures of *Datura metel*. *Turk. J. Biol., 39*(1), 111–118.

Sharma, P., Bhatt, D., Zaidi, M. G. H., Saradhi, P. P., Khanna, P. K., & Arora, S., (2012). Silver nanoparticle-mediated enhancement in growth and antioxidant status of *Brassica juncea*. *Appl. Biochem. Biotech., 167*(8), 2225–2233.

Shokri, S., Babaei, A., Ahmadian, M., Arab, M. M., & Hessami, S., (2013). The effects of different concentrations of nano-silver on elimination of bacterial contaminations and phenolic exudation of rose (*Rosa hybrida* L.) *in vitro* culture. *Acta Hortic., 1083*, 391–396.

Siddiqui, M. H., & Al-Whaibi, M. H., (2014). Role of nano-SiO_2 in germination of tomato (*Lycopersicum esculentum* seeds Mill.). *Saudi J. Biol. Sci., 21*(1), 13–17.

Siddiqui, M. H., Al-Whaibi, M. H., & Mohammad, F., (2015). *Nanotechnology and Plant Sciences* (p. 303). Springer International Publishing, New York.

Sivanesan, I., & Jeong, B. R., (2012). Identification of somaclonal variants in proliferating shoot cultures of *Senecio cruentus* cv. Tokyo Daruma. *Plant Cell Tiss. Organ Cult., 111*(2), 247–253.

Sivanesan, I., Saini, R. K., & Kim, D. H., (2016). Bioactive compounds in hyper hydric and normal micropropagated shoots of *Aronia melanocarpa* (Michx.) Elliott. *Ind. Crops Prod., 83*, 31–38.

Sosan, A., Svistunenko, D., Straltsova, D., Tsiurkina, K., Smolich, I., Lawson, T., et al., (2016). Engineered silver nanoparticles are sensed at the plasma membrane and dramatically modify the physiology of *Arabidopsis thaliana* plants. *Plant J., 85*(2), 245–257.

Syu, Y. Y., Hung, J. H., Chen, J. C., & Chuang, H. W., (2014). Impacts of size and shape of silver nanoparticles on *Arabidopsis* plant growth and gene expression. *Plant Physiol. Biochem., 83*, 57–64.

Taghizadeh, M., & Solgi, M., (2014). The application of essential oils and silver nanoparticles for sterilization of Bermuda grass explants in *in vitro* culture. *Int. J. Hortic. Sci. Technol., 1*(2), 131–140.

Talankova-Sereda, T. E., Liapina, K. V., Shkopinskij, E. A., Ustinov, A. I., Kovalyova, A. V., Dulnev, P. G., & Kucenko, N. I., (2016). The influence of Cu and Co nanoparticles on growth characteristics and biochemical structure of *Mentha longifolia in vitro*. In: Fesenko, O., & Yatsenko, L., (eds.), *Nanophysics, Nanophotonics, Surface Studies, and Applications* (pp. 427–436). Springer International Publishing, New York.

Tambarussi, E. V., Rogalski, M., Nogueira, F. T. S., Brondani, G. E., De Martin, V. D. F., & Carrer, H., (2015). Influence of antibiotics on indirect organogenesis of teak. *Ann. For. Res., 58*(1), 177–183.

Tarafdar, J. C., Sharma, S., & Raliya, R., (2013). Nanotechnology: Interdisciplinary science of applications. *Afr. J. Biotechnol., 12*(3), 219–226.

Thorpe, T. A., (2007). History of plant tissue culture. *Mol. Biotechnol., 37*(2), 169–180.

Torchilin, V. P., (2005). Recent advances with liposomes as pharmaceutical carriers. *Nat. Rev. Drug Discov., 4*(2), 145–160.

Torney, F., Trewyn, B. G., Lin, V. S. Y., & Wang, K., (2007). Mesoporous silica nanoparticles deliver DNA and chemicals into plants. *Nat. Nanotechnol., 2*(5), 295–300.

Tripathi, D. K., Singh, S., Singh, S., Pandey, R., Singh, V. P., Sharma, N. C., et al., (2017). An overview on manufactured nanoparticles in plants: Uptake, translocation, accumulation and phytotoxicity. *Plant Physiol. Biochem., 110*, 2–12.

Vijayakumar, P. S., Abhilash, O. U., Khan, B. M., & Prasad, B. L., (2010). Nanogold-loaded sharp-edged carbon bullets as plant-gene carriers. *Adv. Funct. Mater., 20*(15), 2416–2423.

Wang, L., Hu, C., & Shao, L., (2017). The antimicrobial activity of nanoparticles: Present situation and prospects for the future. *Int. J. Nanomed., 12*, 1227.

Wang, P., Lombi, E., Zhao, F. J., & Kopittke, P. M., (2016). Nanotechnology: A new opportunity in plant sciences. *Trends Plant Sci., 21*(8), 699–712.

Wierzbicka, M., & Obidzińska, J., (1998). The effect of lead on seed imbibition and germination in different plant species. *Plant Science, 137*(2), 155–171.

Willard, M. A., Kurihara, L. K., Carpenter, E. E., Calvin, S., & Harris, V. G., (2004). Chemically prepared magnetic nanoparticles. In: Nalwa, H. S., (ed.), *Encyclopedia of Nanoscience and Nanotechnology* (Vol. 1, pp. 815–848). American Scientific Publishers, New York.

Wu, W., He, Q., & Jiang, C., (2008). Magnetic iron oxide nanoparticles: Synthesis and surface functionalization strategies. *Nanoscale Res. Lett., 3*(11), 397.

Yang, F., Hong, F., You, W., Liu, C., Gao, F., Wu, C., & Yang, P., (2006). Influence of nano-anatase TiO_2 on the nitrogen metabolism of growing spinach. *Biol. Trace Elem. Res., 110*(2), 179–190.

Yunlong, C., & Smit, B., (1994). Sustainability in agriculture: A general review. *Agric. Ecosyst. Environ., 49*(3), 299–307.

Zafar, H., Ali, A., Ali, J. S., Haq, I. U., & Zia, M., (2016). Effect of ZnO nanoparticles on *Brassica nigra* seedlings and stem explants: growth dynamics and antioxidative response. *Front. Plant Sci., 7*, 535.

Zhang, B., Zheng, L. P., Yi, L. W., & Wen, W. J., (2013). Stimulation of artemisinin production in *Artemisia annua* hairy roots by Ag-SiO_2 core-shell nanoparticles. *Curr. Nanosci., 9*(3), 363–370.

Zhang, Y., Leu, Y. R., Aitken, R. J., & Riediker, M., (2015). Inventory of engineered nanoparticle-containing consumer products available in the Singapore retail market and likelihood of release into the aquatic environment. *Int. J. Environ. Res. Public Health, 12*(8), 8717–8743.

Zhao, J., & Castranova, V., (2011). Toxicology of nanomaterials used in nanomedicine. *Journal of Toxicology and Environmental Health, Part B, 14*(8), 593–632.

Zhao, M. X., & Zhu, B. J., (2016). The research and applications of quantum dots as nano-carriers for targeted drug delivery and cancer therapy. *Nanoscale Res. Lett., 11*(1), 1–9.

CHAPTER 10

VARIOUS APPROACHES TO TRANSFER MACROMOLECULES INTO PLANTS USING NANOPARTICLES

ZAHRA HAJIAHMADI,[1] REZA SAYYAD,[2] REZA SHIRZADIAN-KHORRAMABAD,[1] and DEVARAJAN THANGADURAI[3]

[1]*Department of Agricultural Biotechnology, Faculty of Agricultural Sciences, University of Guilan, Rasht–4199613776, Iran*

[2]*School of Metallurgy and Materials Engineering, College of Engineering, University of Tehran, Tehran, Iran*

[3]*Department of Botany, Karnatak University, Dharwad–580003, Karnataka, India*

10.1 INTRODUCTION

Nowadays, agriculture becomes one of the best important topics in developing countries. A significant portion of gross domestic production (GPD) in developing countries depends on food production rate (De Sormeaux and Pemberton, 2011). In the case of food demand, production of high quantitative and qualitative crops in association with increasing the global population is the main challenge for plant researchers (Tomlinson, 2013). Besides, the agriculture production rate is affected by weather conditions, access to adequate water as well as utilization of suitable fertilizers and pesticides. Pests, disease, and weeds can significantly reduce crops production. To overcome these problems, farmers employ excessive and catastrophic values of agrochemicals such as pesticides, fertilizers, and fungicides which can cause considerable pollution in soil and sub-surface water, deterioration of the environment as well as increment of insects and pathogens resistance against pesticides and fungicides. Hence, extensive changes are necessary for delivery methods of nutrients and pesticides to overcome the aforementioned shortcomings.

Nanotechnology means the ability to work in a dimension of 1–100 nm with the aim of achieving the materials with improved properties to apply in different industries (Khan and Rizvi, 2014). Nanotechnology emerged as a pioneer technology with high potentials to be applied in various fields such as electronic (Yu and Meyyappan, 2006), solar cell (Vennila et al., 2018), fuel cell (Abdalla et al., 2018), medicine (Owen et al., 2014), agriculture (Chang et al., 2013) and energy (Hussein, 2015). The appearance of the new features in nanoscale materials shows their ability to be involved in new applications. The nanomaterials are classified into four groups based on their dimensions: zero-dimensional, one-dimensional, two-dimensional, and three-dimensional nanomaterials. In the first group, all dimensions are nanoscales like nanoparticles, quantum dot, and fullerene. In the second group, one dimension is outside the nanoscale (rod-shaped) like carbon nanotube (CNT), nanowires, nanorods, nanofibers. In the two-dimensional group, two dimensions are outside the nanoscale (plate-shaped) like nanocoatings, nanolayers, and graphene. Finally, in the last group, all dimensions are beyond the nanoscale, but the materials are composed of nano-structure like nano-composites and graphite (Tiwari et al., 2012). In other words, the nanoparticles (NPs) are divided into three categories according to their origin: natural, incidental, and manmade. Viruses, fungi, bacteria, and algae are the natural NPs. Some other natural NPs are formed by a natural process like eroded rocks via water, wind, and volcanic eruption. The incidental NPs are by-products of an industrial process such as dams arising from burning fossil fuels, combustion of coal, vehicle exhaust or welding operations. The manmade NPs are produced by researchers in the lab and called engineering nanoparticles (ENPs). ENPs can be manufactured with precision control over their size and shape due to different chemical, physical, mechanical, and optic features of NPs in accordance with their shape and size (Buzea et al., 2007). ENPs are organized in the five groups, including carbon-based NPs, metal-based NPs, magnetic NPs, dendrimers, and composite NPs. Carbon-based NPs compose fullerene, single-wall carbon nanotube (SWCNT), multiwall-carbon nanotube (MWCNT), and carbon fibers. Metal-based NPs contain gold (Au), silver (Ag), copper (Cu), and iron (Fe) NPs or metal oxide NPs like iron (II) oxide (FeO), aluminum oxide (Al_2O_3), titanium dioxide (TiO_2), and zinc oxide (ZnO). Magnetic nanoparticles are a combination of Fe, nickel (Ni), and cobalt (Co) oxides. Dendrimer is a polymer whose structure is built of several branches causing star-shape with a central core. Composite nanoparticles are made by combining two or more nanoparticles like core-shell structures. Different physical and chemical features of nanoparticles compared with their bulk material arise from small surface/volume ratio, and subsequently they can be used as a carrier to deliver macromolecules such as nucleic acid, protein, fertilizers, pesticides, and fungicides to the target organism (Khan et al., 2017).

In recent years, extensive researches have been conducted to assay the effect of various NPs on plants, including metal, metal oxide, and polymeric NPs (Rameshaiah et al., 2015; Nuruzzaman et al., 2016). In traditional agriculture, chemical fertilizers are usually used to deliver nutrients into plants. However, only a small portion of chemical fertilizers are absorbed by plants which cause soil pollution. These chemical materials can be transmitted through soil holes and reach underground water which is hazardous to human health. Nanotechnology can greatly decrease the undesirable effects of fertilizers, pesticides, and fungicides due to the controlled release, targeted delivery, increasing solubility, and reduction of chemical compound consumption without a significant unfavorable effect on human, plant, and environment (Khot et al., 2012; McKee and Filser, 2016). Thus, in addition to the application of smart NPs leading to enhance crop yields as a response to food demand in the word, the use of NPs in agriculture could enhance human, animal, and plant healthiness. Small size, high surface area, high delivery efficiency, and easy incorporation of nanoparticles make them good carriers to transfer macromolecules into plants. The widespread use of NPs has been attracted more attention to their toxicity on human and eco-system. Therefore, in the current chapter, the application of different types of NPs such as metal, metal oxide, chitosan, mesoporous silica, etc., to deliver macromolecules into plants are reviewed. Likewise, the mechanism of NPs uptake by plants and toxicity effects of NPs on plants are discussed.

10.2 NANOPARTICLES IN GENE DELIVERY

One of the important applications of nanotechnology is bio-macromolecules delivery into plants, especially DNA molecules. Therefore, with the aid of nanotechnology, food production in the world could significantly increase. NPs-mediated gene and protein delivery methods have been developed in animals more compared with plants due to cell wall existence in plant cells (Azencott et al., 2007). Nanomaterials have been used in the delivery systems due to their inimitable features, including small size, low toxicity at low concentration, high surface to volume ratio and ability to penetrate the plant cell wall. In recent years, gene transformation systems (transient and stable transformation approaches) have utilized them to develop more resistance plants to biotic and abiotic stresses and production of recombinant protein in a short time. Current gene transformation methods are protoplast transformation, *Agrobacterium*-mediated transformation, and biolistic. These approaches have some disadvantages such as unprotected DNA transformation, damage to target cells, protoplast hard regeneration, limited host range, cost-, and time-consuming. Controlled release of cargo without species limitation might seem to be the most important feature of NPs-mediated macromolecules delivery system. On the other hand, various macromolecules can transfer into a wide range of plant species without the need for mechanical or external force. Alternative delivery methods should be used due to the aforementioned shortcomings of the current transformation methods. Different approaches including biolistic (Martin-Ortigosa et al., 2014; Fu et al., 2015), injection into the abaxial surface of leaf (Hajiahmadi et al., 2019), spray on the abaxial surface of leaf (Sun et al., 2018; Hajiahmadi et al., 2019), magnetofection (Zhao et al., 2017), ultra-sonication (Fu et al., 2012) and incubation (Burlaka et al., 2015) can be utilized to transfer NPs containing gene of interest. The important parameters for NPs selection are: (i) Size of NPs (less than cell wall and nuclear pores); (ii) Cargo trapped capacity; and (iii) Low toxicity or damage to plant cells. Commonly used NPs to deliver bio-macromolecules into plant cells are mesoporous silica nanoparticles (MSNs), carbon-based nanoparticles, gold nanoparticles (Au NPs), zinc oxide nanoparticles (ZnO NPs), and layered double hydroxide nanoparticles (LDH NPs).

10.2.1 MESOPOROUS SILICA NANOPARTICLES (MSNs)

Torney et al. (2007) reported the co-delivery of DNA and chemical materials using gold-capped MSNs (100–200 nm). They successfully transferred pGFP:MSNs into tobacco protoplast using biolistic method (Torney et al., 2007). However, one of the important disadvantages of this method is to regenerate the plants. Au-MSNs were used in other studies (Martin-Ortigosa et al., 2012; Martin-Ortigosa et al., 2014). Martin-Ortigosa et al. (2012) applied Au-MSNs to deliver fluorescent-labeled bovine serum albumin and GFP (green fluorescent protein) plasmid simultaneously into white onion and tobacco plants via the biolistic method (Martin-Ortigosa et al., 2012). Then, the authors successfully transferred Au-MSNs containing Cre recombinase protein into maize embryo via the biolistic method (Martin-Ortigosa et al., 2014). Although a low amount of DNA is needed in this method, using gene gun causes cell damages and needs expensive equipment. Therefore, alternative approaches with the least damage to plant tissues are preferable. Chang et al. (2013) investigated the potential of functionalized MSNs to transfer GFP plasmid into intact *Arabidopsis* plants under *in vitro* conditions. They revealed that the MSNs with the average particle size of 50 nm can easily penetrate the cell wall and pass through the nuclear pore (Chang et al., 2013). This method needs sterile and special condition. The pore size of the cell wall is species-dependent. For example, copper NPs (50 nm), gold NPs (10–50 nm), silver, and silica NPs (14, 50, and 200 nm) have been used in different plant species and they were able to pass through cell wall (Nair et al., 2011; Slomberg and Schoenfisch, 2012; Wang et al., 2012b). Therefore, the nuclear pore size (~ 70 nm) is more important than cell wall pore size. Hence, MSNs with particle size of less than 70 nm can be easily applied to transfer foreign DNA into the plant nuclear. Hussain et al. (2013) showed that fluorescein isothiocyanate-labeled MSNs (20 nm) can be easily uptaken by *Arabidopsis* roots and transported into vascular tissue (Hussain et al., 2013). Fu et al. (2015) demonstrated that the functionalized MSNs (500 nm) could deliver the *GUS* gene into tobacco plants via the biolistic method. They also successfully developed transgenic tobacco plants that in which transformation efficiency was 47.11% (Fu et al., 2015). However, using biolistic method need expensive equipment such as gene gun and causes damage to the cell. Nanocarriers (NCs) can protect proteins (as cargo) against internal degrading enzymes. Therefore, MSNs can carry them into the plant cell via endocytosis phenomena (Deodhar et al., 2017). In our previous study, the MSNs with the particle size of 40 nm was used to transfer plasmids (pDNA) containing *GUS* and *cryIAb* genes (Hajiahmadi et al., 2019). Three different methods have been used to transfer pDNA:MSNs into tomato plants under *in vivo* conditions, including injection into the abaxial surface of leaf, injection into the stem, and spray into the abaxial surface of leaf. Based on our findings, the best transient gene transformation via functionalized MSNs was the injection into the abaxial surface of the leaf (Hajiahmadi et al., 2019). This method is affordable, equipment-independent, and no need for sterile or special condition. Due to the benefits of this method, the authors applied pDNA:MSNs complex in order to develop stable transgenic tomato plants. The integration of foreign DNA into the plant genome can occur through homologous recombination (Primrose and Twyman, 2013). Therefore, researchers may use transposon vectors to increase homologous recombination between the foreign gene and host genome (Weeks et al., 2016).

10.2.2 CARBON NANOTUBES (CNTs)

CNTs are one of the most commonly used NPs to deliver bio-macromolecules into plants. Diameter of CNTs is varied between 0.4 nm and 100 nm. Due to their special shape (needle-like), they can penetrate into the cell wall and plasma membrane (Liu et al., 2009). The potential of single-walled carbon nanotubes (SWCNTs) as a DNA carrier to pass the plant cell wall was evaluated (Liu et al., 2009). They transferred DNA:FITC-SWCNTs into a tobacco cell suspension, and observed the fluorescence for 80% of treated cells. According to their results, temperature, and incubation time affected transformation efficiency. The highest transformation rate has been obtained at 27°C and two hours incubation of tobacco cells with FITC-CNTs. They used wortmannin as an endocytosis inhibitor as well. Therefore, a significant decrease in florescence (64%) has been observed, suggesting endocytosis phenomena as the main pathway to enter the plant cell. Burlaka et al. (2015) appraised the ability of multi-walled carbon nanotubes (MWCNTs) and SWCNTs to deliver pGreen 0029 into protoplast, callus, and leaf of *Nicotiana tabacum*. They used MWCNTs and SWCNTs at a concentration of 15 and 20 μg/L for protoplast treatment, respectively. They also applied 40 and 30 μg/L of SWCNTs and MWCNTs for callus and leaf transformation, respectively. Based on their results, both MWCNTs and SWCNTs could be applied as DNA carriers for protoplast treatment, but only SWCNTs were able to transfer pDNA into treated callus and leaf disc due to the low ability of MWCNTs to penetrate plant cell wall (Burlaka et al., 2015). Demirer et al. (2019) showed that SWCNTs and MWCNTs are able to deliver pDNA and siRNA into leaves of *Eruca sativa*. They observed that pDNA (containing GFP) has been released into plant nuclear and GFP expression has been detected in leaves mesophylls cells (Demirer et al., 2019). They also delivered pDNA:CNT into isolated protoplast of *E. sativa* (transformation efficiency, 72%). Their findings illustrated that CNTs pass through plasma membrane within seconds after exposing the samples to CNTs, and no adverse effect in their experiments was found. Meanwhile, they successfully introduced CNTs containing siRNA (21 bp, specific to GFP mRNA) into transgenic tobacco leaves. 24 hours after injection into the abaxial surface of leaves, GFP gene silencing (95%) was detected (Demirer et al., 2019). The authors examined SWCNTs (~20 nm) applicability as DNA carrier in *N. benthamiana, E. sativa*, *Triticum aestivum*, *Gossypium hirsutum* leaves, and *E. sativa* protoplast (Demirer et al., 2019). They recommend this system as a species-independent transient transformation method.

This method has other advantages including convenient, non-destructive, fast, and affordable.

10.2.3 OTHER NANOPARTICLES

A few studies related to protein and gene delivery systems into plants using other nanoparticles have been reported. Polymer nanoparticles (PNPs) have been applied to deliver siRNA into tobacco protoplast using electroporation (Silva et al., 2010). Although they were able to transfer siRNA:PNPs into tobacco protoplast, but this method is not efficient enough due to hard plant regeneration. Bao et al. (2016) used a positive charged LDH-lactate (30–60 nm) to transfer ssDNA molecules (60 bp) into *Arabidopsis* root. Their experiment showed that LDH NPs could be used as an efficient alternative method for common transient transformation approaches (Bao et al., 2016). They can easily pass through the root cell wall and enter the cells via both endocytic and non-endocytic pathways. The most important advantage of this system is the short-time incubation (15 minutes) of root cells with NPs. LDHs are biocompatible, non-toxic for plants growth and development, and degradable (Bao et al., 2017). Mitter et al. (2017) also investigated the ability of LDH NPs (20–80 nm) to deliver RNAi into tobacco leaf. They sprayed dsRNA:LDH NPs on tobacco leaves following tolerance assessment of the related plants to virus attack. Treated plants demonstrated an enhanced resistance against disease for at least 20 days (Mitter et al., 2017). Calcium phosphate NPs were used as macromolecule carriers (Naqvi et al., 2012; Rafsanjani et al., 2016). pCambia1301 plasmids were encapsulated in calcium phosphate NPs (CaP NPs, 20–50 nm) by Naqavi et al. (2012). Then, they were introduced into cabbage hypocotyl to generate transgenic cabbages. The transformation efficiency was 80.7% which was more than the *Agrobacterium*-mediated transformation method (54.4%) (Naqvi et al., 2012). Rafsanjani et al. (2016) illustrated that CaP NPs (<20 nm) containing pBinAR plasmid could easily transferred into *Cichorium intybus* leaves. The transformation frequency was 9.6% which was lower when compared with the obtained results by Naqvi et al. (2012). Calcium phosphate is naturally included in plant cell walls; thus, it is biocompatible. Fu et al. (2012) evaluated the potential of zinc sulfide (ZnS) nanoparticles (ZnS NPs, 3–5 nm) to deliver pBI121 into tobacco leaves using ultrasound. The best transformation efficiency (52.67%) was obtained under the ultrasonic condition of 60 W for 20 minutes, while the transformation efficiency of the conventional ultrasound-mediated transformation method was 30–40% that confirmed the role of ZnS NPs in protecting DNA molecules against ultrasonic damage (Fu et al., 2012). Disadvantages of this approach are the need for expensive equipment and special conditions. Magnetic Au NPs (mAu NPs, 25 nm) were applied to deliver pBI121 into *Brassica napus* cells using magnetic field (Hao et al., 2013). They successfully transferred pBI121:mAu NPs into plant cells with or without cell walls. Due to the non-toxicity feature of gold NPs, they may be used as nano-carrier to deliver foreign DNA into plants in the future. Zhao et al. (2017) introduced a new stable transformation approach using magnetic nanoparticles (MNPs). They developed transgenic cotton plants using pollen magnetofection method (Zhao et al., 2017). Plant pollens contain several pores with a diameter of 5–10 μM, which allow nanoparticles to enter into them. They mixed DNA:MNPs with pollens before pollination under magnetic field condition. This system is very effective for hard regeneration plants such as cotton. Other advantages of this system are culture-free, no need for expensive equipment, and species-independent.

Taken collectively, nanomaterials are fine alternatives for common transformation approaches. *Agrobacterium*-mediated transformation method is widely used due to lower cost compared with other conventional transformation methods. However, using NPs to deliver an interested gene into plants has more advantages, including saving energy and time when compared to *Agrobacterium*-mediated transformation method. Recently, nuclease-based genome editing systems such as CRISPR/Cas9 have been widely used in plants and animals (Fan et al., 2015). Consequently, researchers can deliver the CRISPR/Cas9 cassette by use of NPs in into target plants to develop new transgenic crops. CRISPR/Cas9 system is based on homology of the target region in the plant genome and sgRNA (single guide RNA), thus, it increases the efficiency of NPs-mediated gene transformation system.

10.3 USING NANOPARTICLES AS A VECTOR

10.3.1 NANOFERTILIZERS

Fertilizers are applied to improve plant growth and development (Behera and Panda, 2009). Artificial fertilizers are inorganic compounds which include three elements: nitrogen (N), potassium (K), and phosphorus (P) in a suitable concentration. N improves leaf growth and forms protein and chlorophyll. K promotes root and stem growth and protein synthesis. P participates in the root, flower, and fruit development (Gu et al., 2009; Corradini et al., 2010).

The world demand for nitrogen, potassium, and phosphorus was predicted to increase from 1.4 to 2.6% and is estimated to considerable increases every year (FAOSTAT and Data, 2013). On the other hand, about 70% of nitrogen, 70% of potassium and 90% phosphorus in normal fertilizers leak to the environment and cannot be absorbed by the plant that causes economic losses and ecosystem pollution (Alfaro et al., 2008).

Nanotechnology is widely used in the production of high-performance nanofertilizers. Nanofertilizers with the size less than 100 nm have been generally applied in agriculture and could be conveniently absorbed by the root, cuticles, and stomata due to their small size and high surface area (DeRosa et al., 2010; Wang et al., 2012b). In recent years, researchers focused on the use of nanofertilizers with a controlled release system (Table 10.1) (Lubkowski and Grzmil, 2007; Xiao et al., 2008). For this purpose, fertilizers are physically coated with some materials to increase their solubility rate. Application of controlled release systems reduce fertilizer loss, prevent soil, and underground water contaminant, and decrease dose of applied fertilizers compared to the conventional method (Han et al., 2009; Davidson and Gu, 2012). Chitosan is a polymer which is widely used in pharmaceutical, biomedical, and agriculture due to its interesting properties including biocompatibility, biodegradability, and non-toxicity. Therefore, it was proposed as an appropriate choice in agricultural applications (Muxika et al., 2017; Kumaraswamy et al., 2018). Ha et al. (2019) synthesized nitrogen, potassium, and phosphorus enriched chitosan nanoparticles. They investigated the effect of synthesized nanofertilizer on growth of coffee plant. The results showed that the amount of absorbed nutrients by the leaves of coffee increased (Ha et al., 2019). Furthermore, the photosynthesis rate and total content of chlorophyll were increased compared to non-treated plants (71.7% and 30.68%, respectively). In another study, Khalifa et al. (2018) evaluated the impact of different concentrations of chitosan-polymerizing methacrylic acid (PMMA)-NPK nanofertilizer complex on pea (*Pisum sativum*) plants. Interestingly, at the low concentrations (0.125 and 0.0625 serial dilutions of a stock solution containing 500 ppm of N, 400 ppm of K, and 60 ppm of P), mitotic cell division was induced (Khalifa and Hasaneen, 2018). In addition, the amount of some proteins like vicilin, convicilin, and legumin β were raised in the treated plants. Some effective parameters on the release frequency such as the concentration of urea (fertilizer), chitosan, crosslinker (a cross-link is a bond that links one polymer chain to another), and temperature have also been assessed by Hussain et al. (2012). The results revealed that a higher concentration of urea loaded in chitosan enhanced the

TABLE 10.1 Recent Studies Evaluating NPs Potential to Deliver Fertilizers, Fungicides, Insecticides, and Herbicides to Plants

Type of Nanocarrier	Type of Cargo	Plant or Pest Species	Feature	References
Fertilizer				
Hydroxyapatite	Phosphorus	*Glycine max* (Soybean)	Improved growth	Liu and Lal (2014)
Manganese hollow core-shell	Zinc	*Oryza sativa* (Rice)	Improved growth	Yuvaraj and Subramanian (2015)
Zeolite	Urea	Maize	Improved growth	Manikandan and Subramanian (2016)
Carbon nanofibers	Copper	*Cicer arietinum* (Chickpea)	Increased water uptake, germination rate, shoot, and root elongation, and chlorophyll and protein content	Ashfaq et al. (2017)
Hydroxyapatite-montmorillonite nanohybrid	Urea	*Oryza sativa* (Rice)	Increased yield	Madusanka et al. (2017)
Zeolite	Phosphorus, Potassium	*Ipomoea aquatica* (Kalmi)	Improved growth	Rajonee et al. (2017)
Hydroxyapatite	Urea	*Oryza sativa* (Rice)	Increased yield	Kottegoda et al. (2017)
Carbon nanofibers	Copper-Zinc	*Cicer arietinum* (Chickpea)	Improved growth	Kumar et al. (2018)
Fungicide				
Polymeric nanoparticles	Thiram	*Botrytis* spp.	Increased release time, prevented degradation of thiram	Kaushik et al. (2013)
Porous hollow silica nanospheres	Tebuconazole	–	Slower release in lower temperature and pH, larger particle size of porous hollow silica nanospheres	Qian et al. (2013b)

TABLE 10.1 *(Continued)*

Type of Nanocarrier	Type of Cargo	Plant or Pest Species	Feature	References
Polymeric nanoparticles	Carbendazim	*Rhizoctonia solani*	Slow-release of fungicide	Koli et al. (2015)
Polymeric nanoparticle (poly(3-hydroxybutyrate))	Tebuconazole	*Fusarium moniliforme*, *Fusarium solani*	Better performance of fungicide compared to free fungicide	Volova et al. (2016)
Polymeric nanoparticles (chitosan-pectin)	Carbendazim	*Fusarium oxysporum*, *Aspergillus parasiticus*	Higher efficacy at lower concentration of carbendazim compared to free carbendazim, low toxicity for seed germination and root growth of *Cucumis sativa*, *Zea mays* and *Lycopersicum esculantum*	Kumar et al. (2017)
Chitosan	Hexaconazole	*Rhizoctonia solani*	More effective control of fungi compared to the commercial formulation, better efficiency in alkaline soil and lower toxicity than commercial formulations	Chauhan et al. (2017)
Chitosan	Clove essential oil	*Aspergillus niger*	Impressive antifungal performance	Hasheminejad et al. (2019)
Insecticide				
Chitosan	Imidacloprid	–	Prolonged-release time	Guan et al. (2008)
Amphiphilic copolymers	Thiamethoxam	–	Lower release compared to commercial formulation	Sarkar et al. (2012)
Chitosan	Pyrifluquinazon	*Myzus persicae*	Prolonged-release time	Kang et al. (2012)
poly(ε-caprolactone)	Neem (*Azadirachta indica*)	*Plutella xylostella*	100% larval mortality, improved the stability of neem products against UV radiation, increased the neem dispersion in the aqueous phase	Forim et al. (2013)
Sodium alginate nanoparticles	Imidacloprid	Leafhoppers	Increased efficacy of insecticide and reduced toxicity	Kumar et al. (2014)

TABLE 10.1 *(Continued)*

Type of Nanocarrier	Type of Cargo	Plant or Pest Species	Feature	References
Chitosan	Methomyl	Armyworm larvae	Higher efficiency than free methomyl	Sun et al. (2014a)
Mesoporous silica	Chlorantraniliprole	*Plutella xylostella*	Preserved chlorantraniliprole against degradation under thermal conditions and UV radiation, higher morality of *Plutella xylostella* compared to the commercial formulation	Kaziem et al. (2017)
Herbicide				
Poly(epsilon-caprolactone) nanoparticles	Atrazine	*Brassica* sp.	More effective against the target organism than free herbicide, no damage to the non-target organism, increased mobility of atrazine, less genotoxic compared to the free herbicide	Pereira et al. (2014)
Solid-lipid nanoparticles	Simazine-atrazine	*Raphanus raphanistrum*	No adverse effect on *Zea mays* growth	De Oliveira et al. (2015)
Chitosan-alginate	Imazapic-imazapyr	–	Reduced toxicity, improved the herbicide performance	Maruyama et al. (2016)
Magnesium aluminum silicate-PVA	Glyphosate	*Zoysia matrella*	Decreased loss of glyphosate under simulated rainfall, improved herbicidal activity, high stability in aqueous solution for at least three months	Chi et al. (2017)
Mesoporous silica nanoparticles	2,4-dichlorophenoxy acetic acid	–	Decreased soil leaching of herbicide, good bioactivity on target plant with no impact on the non-target plants	Cao et al. (2017)
Mesoporous silica nanoparticles	Diquat dibromide	*Datura stramonium* L.	Good herbicidal activity	Shan et al. (2019)

release rate and in opposite higher concentration of chitosan and cross-linker reduced fertilizer release rate. Increase in the environment temperature can enhance the release rate as well (Hussain et al., 2012).

Copper is a micronutrient playing an important role in metabolism processes, especially in cell wall synthesis. In addition, copper can enhance photosynthesis and consequently plant growth (Bernal et al., 2007; Yruela, 2009; Ghorbanpour et al., 2016). El-Aziz et al. (2019) synthesized two types of chitosan-based fertilizers: chitosan-copper nanocomposite (CN-Cu-NCs) and chitosan/polyacrylic acid/copper hydrogel nanocomposites (CS/PAA/Cu-HNCs). They studied their impacts on onion growth (El-Aziz et al., 2019). The results demonstrated that the release rate of CS/PAA/Cu-HNCs was higher than CN-Cu-NCs. It attributed to higher swelling properties of CS/PAA/Cu-HNCs than CN-Cu-NCs. The treated plants with both nanocomposites containing copper showed an enhance in growth, bulb diameter, number of roots, and chlorophyll content. Of course, it should be noticed that plant growth and nutrient content of onion bulbs in samples exposed to CS/PAA/Cu-HNCs nanofertilizer were higher than exposing plants to CN-Cu-NCs.

Calcium (Ca) exists in both plant cell wall and membrane structures. Boron (B) is a micronutrient material which effects calcium uptake and cell wall synthesis (Marschner, 2011). On the other hand, ZnO and titan oxide as micronutrients cannot enhance the uptake of nutrients by plants but enhance plant resistance against pathogens. Therefore, Wang et al. (2018) investigated the ability of Zn/B-chitosan nanofertilizer to enhance the growth in coffee plants. Based on their results, an increase in plant growth was observed due to increased uptake of micronutrients (Zn, B, and Ca) (Wang and Nguyen, 2018). Dapkekar et al. (2018) appraised Zn-chitosan nanofertilizer effects on wheat plants compared to common Zn fertilizer ($ZnSO_4$, 400 mg/L). They showed that the zinc concentration has been increased and was in order 50% and 36% in plants exposed to $ZnSO_4$ fertilizer and Zn-chitosan nanofertilizer, respectively, while nanofertilizer concentration was 10 times less than traditional fertilizer (Dapkekar et al., 2018).

Nitrogen is known as one of the important nutrients for biomass production in agriculture. In traditional farming, approximately 50–70% of nitrogen is lost by leaching to the environment (Kottegoda et al., 2011). In recent years, nanostructure clay materials attracted more attention due to their structure. They could be applied as a nanocarrier to deliver fertilizers into plants. The layer structure of clays leads to a highly active surface area which can provide suitable sites to load fertilizers. Kaolinite is a clay with the chemical formula of $Al_2Si_2O_5(OH)_4$ which is abundant in tropical regions. Roshanravan et al. (2014) studied the nitrogen release from urea fertilizer intercalated kaolinite. The 20 wt.% of urea intercalated kaolinite granulates and coated with water-based resin epoxy (Roshanravan et al., 2014). The effect of coating thickness and granulates size on the release rate of urea have been investigated as well. The results illustrated that the release time of nitrogen from urea intercalated kaolinite was 3 times longer than urea non-intercalated. Furthermore, the increment of coating thickness and granules size prolonged the release time. The evaluation of growth parameters on rice plants exposed to urea-kaolinite nanofertilizer indicated a significantly increase in productivity compared to applying urea fertilizer (Roshanravan et al., 2014). Layer double hydroxides (LDHs) is an inorganic compound with general formula $[(M^{2+})_{1-x}(M^{3+})_x(OH)_2]^{x+}[A^{n-}]_{x/n}\cdot yH_2O$, where M^{2+} and M^{3+} are divalent and trivalent metallic cations, respectively, A^{n-} is a charge-balancing anion, and x is the molar ratio of $M^{3+}/(M^{3+}+M^{2+})$. LDHs structure consists of hydroxide layers of divalent (M^{2+}) and trivalent (M^{3+}) in which trivalent cations replace instead of divalent cations at the octahedral sites. The positive charge of the hydroxide layers neutralizes A^{n-} species between the layers. LDHs have a good anionic exchange capability due to the structural characteristic and can allocate appropriate positions in the interlayer spaces for phosphor incorporation, and subsequently, through this, reduction in direct contact of these ions with soil occur. The studies have confirmed that the release of phosphates into the water from [Mg, Al-PO_4]-LDH was 10 times more than the release from KH_2PO_4 (Bernardo et al., 2018). Unlike initial assumption, the interaction between P with Fe^{3+} and Al^{3+} cations had no impact on ionic exchange. The experiment results on wheat plant showed that [Mg, Al-PO_4]-LDH can provide availability of phosphate for plants over a longer time period compared with other conventional P supplier sources. Zeolites are aluminosilicates porous structures which are able to absorb and desorb of nutrients. Thus, they improve crop yields. Zeolites have a high cation exchange capacity and can retain a high amount of nutrients in their honeycomb structure. So, it can be added to the soil as a nanocarrier of nutrient and enhanced crop production efficiency (Rehakova et al., 2004; Ramesh et al., 2010). It is demonstrated that Zn release from the zeolite-Zn occurs in a long-period (1176 h), while Zn release from $ZnSO_4$ fertilizer stops after a short period (216 h) (Yuvaraj and Subramanian, 2018). In another study, the spinach growth was improved in the presence of zeolite as a carrier of NH_4 and potassium without changing vitamin C content (Li et al., 2013).

10.3.2 NANOPESTICIDES

The plant pests are considered as serious problems in crop production which affect crop quality and quantity (Chakraborty and Newton, 2011). Common plant pests are fungi, viruses, insects, nematodes, and parasites (Labandeira and Prevec, 2014). In conventional agriculture, a diverse range of chemical and biological pesticides has been employed to control pests, but incorrect use of them can lead to pathogens resistivity and environment pollution (Ghormade et al., 2011). So, it seems purposeful for researches to develop new material and formulation of pesticides which can reduce contamination of environment and pathogens resistivity caused by traditional pesticides. Hereby, application of nanotechnology can improve crop quality, and decrease the shortcomings. Nanopesticides are more effective than common pesticides due to the higher surface area, higher mobility, higher solubility as well as low toxicity, which result in the possibility of organic solvent elimination. Nanocages, nanoformulations, nanoemulsions, and nanocontainers are different techniques based-nanotechnology which has been developed to control pests (Table 10.1) (Merlan and Raftery, 2009; Bergeson, 2010; Memarizadeh et al., 2014b).

Dendrimers have tree-shape structure and empty space between their branches which can be suitable places for the interested cargo (Liu et al., 2015). Memarizadeh et al. (2014) illustrated the positive effect of encapsulated indoxacarb with a dendrimer polymer structure. They demonstrated that the degradation rate of encapsulated indoxacarb was lower than free indoxacarb when they exposed to UV and natural lights (Memarizadeh et al., 2014a). The amphiphilic polymers are composed of micelles which have hydrophilic and hydrophobic parts; thus, they spontaneously accumulate when exposed to organic and aqueous solvents, which can be considered as the appropriate choice for the encapsulation of nanopesticides. For the controlled release systems of various pesticides, high dissolution capacity and low critical micelle concentration of amphiphilic polymers were developed (Kumar et al., 2010; Loha et al., 2011). Imidacloprid is a new formulation of pesticides and can cease the action of sucking pests in different crops. It can be also employed as a soil and foliar treatment in order to reduce damages of the pathogen activity in plants such as rice, cotton, maize, vegetables, and potatoes. Adak et al. (2012) successfully synthesized encapsulated imidacloprid by nano-micelles through self-assembly amphiphilic polymers in chloroform solvent, then, they showed that the release of encapsulated imidacloprid is very slow compared with commercial imidacloprid pesticides

(Adak et al., 2012). The low release of pesticide can prevent their repeated application and thus decrease the leaching losses in water and unpleasant impact of the environment.

Silica-based nanoparticles have shown a high potential to use in controlled release systems due to good dispersion, uniformity of porous size, with the possibility to adjust porous size, flexibility to synthesize different structures with high capacity to load active molecules, higher mechanical strength than polymer structures, and biocompatibility (Torney et al., 2007; Lou et al., 2008). Therefore, they have been employed as a carrier to deliver molecules such as proteins (Santos et al., 1999), nucleic acids (Castillo et al., 2017), and drugs (Vivero-Escoto et al., 2010). The previous researches have shown the capability of silica-based nanoparticles, including nanoporous silica nanoparticles (Ao et al., 2012), MSNs (Chen et al., 2010; Popat et al., 2012; Wanyika et al., 2012), and hollow-core silica-shell nanoparticles (Wen et al., 2005; Liu et al., 2006; Tan et al., 2012), in slow released of various agrochemical. These nanostructures can promote the performance efficiency of pesticides and decrease their dosage and reusability. Prochloraz is an imidazole fungicide which extensively used to prevent the activity of *Botrytis cinerea* in cucumber. Since cucumber is freshly consumed, the residual pesticide can be dangerous for human health. Zhao et al. (2018) prepared prochloraz-loaded mesoporous silica nanoparticles (p-MSN) and studied on the pesticide features in cucumber. They also applied fluorescein isothiocyanate-labeled nanoparticles to pursue the track of carrier distribution in the plant. They dropwise added the aqueous solution containing p-MSN on leaves of cucumber. The results showed a quick uptake by leaves and translocation to other parts of the plant. New nanoformulation showed a better uptake compared with the plants exposed to the conventional pesticide. Moreover, the results suggested that the fastest and slowest release of pesticide occurred in leaves and fruits, respectively. It confirmed a low risk of p-MSN accumulation in the cucumber fruits (Zhao et al., 2018a). In another research, Wanyika et al. (2013) studied the sustained release of metalaxyl stored in MSNs as a nanocarrier. Metalaxyl is an acylanilide fungicide which could control fungi of the order Peronosporales through systematic activity. Wanyika (2013) synthesized mesoporous silica with an average particle diameter of 162 nm and average pore size of 3.2 nm by sol-gel method. Then, they loaded 14 wt.% metalaxyl into porous of mesoporous silica and investigated the release behavior of metalaxyl in soil and water. The results indicated that only 11.5 and 47% of metalaxyl loaded in the MSNs released in soil and water, respectively, while 76% of free metalaxyl released in soil. It exhibited a slower and more sustained release of the metalaxyl-MSNs compared to free metalaxyl within 30 days (Wanyika, 2013). Zhang et al. (2014) employed a novel functionalized microcapsule using silica cross-linked with alginate in order to accurately control the release of pesticide from MSNs. The precise control of release rate was obtained through the loading of prochloraz into cavities of MSNs, then modification of MSNs surface by functionalizing with alginate and finally, linking of micronutrient elements as Fe, Cu, Zn, and Ca to surface functional group. Prochloraz includes a wide range of imidazole fungicide which is widely applied to protect crops against diseases and also applied as a fruit and vegetable fresh-keeping agent in the process of storage and transportation. Application of prochloraz in plants is restricted due to its sensitivity to UV, alkaline, and acidic conditions. Alginate is a linear anionic polysaccharide which not only preserves the order of mesoporous channels without effect on pore size, but can reinforce the interaction between functional group and loaded molecule in MSNs. The protection of prochloraz against UV light, alkaline, and acidic media is one of the other benefits of alginate linked to MSNs. Microcapsules showed a high capacity to retain prochloraz, sustained release of prochloraz for at least 60 days, and increase the resistivity against disease due to micronutrient on the capsule surface. Also, the microcapsules can effectively prohibit degradation of prochloraz by UV light and pH changes (Zhang et al., 2014). Wibowo et al. (2014) reported a controlled release fipronil insecticide through a new biocompatible oil-core silica-shell nanocapsule. They demonstrated that the sustained release can be adjustable by thickness control of the silica shell. The study of the release system of fipronil showed the decrease release of insecticide with increasing the thickness of silica-shell. The release treatment of fipronil from silica-shell in different thickness of silica (8, 25, and 44 nm) against termites demonstrated that time interval between the use of nanocapsule and morality of termites has been prolonged with increasing the thickness of silica-shell. The dispersion treatment of nanocapsules with 44 nm shell in α-cellulose feeding bait against worker and soldier termites showed that worker termites can transmit the silica nanocapsules to recipient termites through horizontal transfer. 100% of termites were dead after treatment with silica nanocapsules which was 3 days longer than the commercial Termidor. The results suggested that application of the loaded fipronil on oil-core silica-shell nanocapsules could be an insecticide release system with having biocompatibility, high encapsulation efficiency, and/with sustained-release property which can suggest the use of silica nanocapsules as a delivery system in agriculture (Wibowo et al., 2014).

Straw is residual of crop production and is known as biomass. The biomass combustion could be utilized to generate electrical energy. Sub product of straw combustion process is straw ash which has no reuse due to technological problems of recycling resulting in environmental pollution (Liu et al., 2010; Xie et al., 2010). Straw ash is consisting of a large amount of biosilica and biochar, which have high surface area due to micro/nanostructures in their surface and can be a super absorbent of organic material. Furthermore, it is demonstrated that straw ash could improve the plant growth and development (Liang et al., 2006; Lehmann et al., 2011). The micro/nanometer porous can be an appropriate site for loading large amount of pesticide. Chlorpyrifos is extensively applied as an insecticide for controlling of rice leafhopper, rice thrips, planthoppers, gall midge, termites, etc. It is known that pesticides usually sprayed onto the leaves of plants against plant pests. However, they can conveniently eliminate through washing by rain and leaching into the environment resulting in soil pollution and decrease pesticides efficiency to protect plants against various pests. Hence, the stability of pesticides on plant leaves is important in order to maintain its performance. Cai et al. (2013) conducted a study on the new formulation of pesticide through the entering chlorpyrifos into micro/nanoporous of straw ash and investigated the controlled release of chlorpyrifos and the adhesion efficiency of the new formulation to plants leaves compared to free chlorpyrifos. The results implied that chlorpyrifos well retained on the leaf surface and illustrated good adhesion performance as well as lower loss via washing and leaching compared to chlorpyrifos alone. Also, the controlled release of chlorpyrifos increased pesticide efficiency and reduced pesticide dosage and environment pollution (Cai et al., 2013).

Kumar et al. (2015) introduced a new formulation of alginate-chitosan nanocapsules as an appropriate carrier for the controlled release of acetamiprid. Acetamiprid is a neonicotinoid insecticide and can repel the insect attacks of mainly Aphids, Lepidoptera, Hemiptera, and Thysanoptera in vegetables, fruits, and tea crops. Alginate is a polymeric material with low stability, which limits its use as an independent carrier system. Incorporation of chitosan to alginate is a method to overcome the problem of fast delivery in alginate. The results indicated that acetamiprid release is dependent on soil pH and acidic pH can enhance the release of acetamiprid pesticide. They also founded the sustained release of acetamiprid nanoformulation compared to commercial acetamiprid, which reduce the frequency of use and pesticide consumption and prevent the entry of acetamiprid to natural water resources (Kumar et al., 2015).

Paraquat is a common herbicide which broadly used in worldwide. Due to desiccating and defoliating properties, it can control the growth of weeds. The high solubility of paraquat in water and fast sorption in soil can have destructive effects on the environment. Grillo et al. (2014) suggested that the encapsulation of paraquat by chitosan nanoparticles prohibit the swift release of herbicide and could diminish problems of toxicity in non-target organisms and soil. The results of toxicity investigations on *Allium cepa* showed that encapsulated paraquat with chitosan caused less chromosome damage, compared to the free herbicide, while the herbicidal activity of paraquat preserved after encapsulation with chitosan (Grillo et al., 2014). Halloysite nanotube (HNT) is a natural aluminum silicate mineral from the kaolin group with layer by layer that empty space between layers can be an appropriate site for loading of active molecules. However, the controlled delivery systems based on HNT exhibit low ability to long-term sustained release. Zeng et al. (2019) served poly(vinyl alcohol)/starch (PVA/ST) composites as a coating agent for reduction of the release rate of active ingredients of *Eupatorium adenophorum* Spreng (AIEAS) as a botanical herbicide loaded in HNT. The results indicated slower and more sustained release of AIEAS loaded in HNT-PVA/ST compared to AIEAS with PVA/ST film. In addition, HNT significantly inhibited the erosion of PVA/ST film through reinforcement of the film skeleton which can again reduce the release rate of herbicide into the soil (Zeng et al., 2019).

10.4 MECHANISMS OF NANOPARTICLE UPTAKE BY PLANT CELL

Uptake of NPs is generally dependent on plant species, size, type, and charge of NPs (Rico et al., 2011). NPs before penetration into plant cells have to pass through the cell wall and plasma membrane. The cell wall is a natural barrier to the entry of NPs (Eggenberger et al., 2010). NPs with particle size of less than cell wall pores diameter can easily pass through them. They can uptake via exposure leaves or roots. NPs that are exposed to roots first entered the root epidermis and endodermis, and then transported to aerial parts of plants via xylem (Zhang et al., 2015a). However, those exposed NPs to leaves, needed the stomatal apparatus to enter the intercellular space and then transport to other parts of plants via the phloem (Zhang et al., 2015a). In this chapter, the effective uptake of commonly used NPs has been evaluated.

10.4.1 SILICA NANOPARTICLES (SiO_2)

Several techniques such as flow cytometry, transmission electron microscope (TEM), and confocal laser scanning microscope (CLSM) were employed to assess MSNs uptake by plants. The results have shown that two main pathways including pinocytosis and endocytosis were involved in the uptake of MSNs by plant cells (Chang et al., 2013). In endocytosis pathway, pDNA:MSN complex entered the *Arabidopsis* cells by binding to specific cell surface receptors. Uptake of fluorescent MSNs (20 nm) has been investigated in *Lupinus albus*, *T. aestivum*, *Zea mays*, and *Arabidopsis* by Sun et al. (2014). They revealed that MSNs were transported to plant roots via apoplastic and symplastic pathways and subsequently transported to the aerial parts of the plant through vascular tissues (Sun et al., 2014b). Slomberg et al. (2012) appraised the uptake of different size of MSNs (14, 50, and 200 nm) in *Arabidopsis* roots. They showed that uptaking of MSNs is size-dependent, and the smallest size (14 nm) had the highest absorption by treated plants roots (Slomberg and Schoenfisch, 2012). Hajiahmadi et al. (2019) ascertained transportation of MSNs (40 nm) into other parts of plants through vascular tissues. They injected MSNs containing *GUS* gene into the tomato stem and detected *GUS* expression in leaves due to the movement of MSNs in the phloem (Hajiahmadi et al., 2019). They also injected pDNA (containing *GUS* gene):MSNs complex into the abaxial surface of leaves and illustrated that the MSNs entered into plant cells and passed through the cell wall. According to the size of stomatal pore diameter (10–30 μM), the MSNs with the particle size of 40 nm can easily enter the intercellular space.

10.4.2 TITANIUM DIOXIDE NANOPARTICLES (TiO_2)

Root uptake of NPs TiO_2 in *T. aestivum* and *Arabidopsis* has been assessed by Larue et al. (2012) and Kurepa et al. (2010), respectively. Kurepa et al. (2010) demonstrated that TiO_2 coated with Alizarin red S and sucrose could pass through the cell wall and entered into *Arabidopsis* cells (Kurepa et al., 2010). Larue et al. (2012) evaluated the uptake of different size of TiO_2 NPs (14–655 nm) by wheat root. They showed that size-dependent translocation of NPs from plant root to shoot. Their results revealed that TiO_2 NPs with a particle size up to 36 nm transported from wheat root to shoot via vascular tissues (Larue et al., 2012). However, in another study, TiO_2 NPs were not translocated in the treated maize plants due to the smaller size of cell wall pore diameter (6.6. nm) compared to applied TiO_2 NPs (30 nm) (Asli and Neumann, 2009).

10.4.3 CARBON-BASED NPs

SWCNTs can easily pass through the plant cell wall and plasma membrane (Liu et al., 2009). As a rule of thumb, penetration of NPs into the seed coat is harder than cell wall and plasma membrane due to the thickness of seed coat (Srinivasan and Saraswathi, 2010). However, one of the important advantages of CNTs compared with other NPs is its ability to penetrate the seed coat (Ganguly et al., 2014). Lin et al. (2009) evaluated the uptake of FITC dye-labeled SWCNTs (<500 nm) by BY-2 cell line of tobacco. Based on their findings, the SWCNTs entered into the cells via endocytosis phenomenon (Liu et al., 2009). Likewise, their entry into the cells through endocytosis pathway has been confirmed in *Arabidopsis* and *Oryza sativa* (Shen et al., 2010). The ability of SWCNTs to pass through *Catharanthus roseus* plasma membrane has also been ascertained (Serag et al., 2011). Uptake of MWCNTs was evaluated in some plant species as well. For instance, in tomato plants, uptake of MWCNTs by seed was observed (Khodakovskaya et al., 2009). While, in cell suspension of rice, cell wall inhibited the entry of MWCNTs into the cytoplasm (Tan and Fugetsu, 2007). At high concentration of MWCNTs (400 mg/L), they accumulated at the root surface resulting in preventing water and nutrient uptake by plants, thus, delayed flowering in the treated rice was observed (Shukla et al., 2016).

10.4.4 OTHER NANOPARTICLES

Uptake of ZnO NPs (8 nm, 500–4,000 mg/L) by seedling of soybean was appraised and dose-dependent uptake of ZnO NPs has been reported (López-Moreno et al., 2010). Based on their results, the highest uptake efficiency of ZnO NPs was obtained at the concentration of 500 mg/L. Agglomeration of NPs at higher concentration (more than 1,000 mg/L) led to the hard passage of agglomerated NPs through cell wall pores. Uptake and translocation of CuO NPs by *T. aestivum* (NPs<50 nm), *Z. mays* (NPs: 20–40 nm), and *Phaseolus radiatus* (NPs ~30 nm) have been reported (Lee et al., 2008; Dimkpa et al., 2012; Wang et al., 2012b). Lee et al. (2008) evaluated the uptake of CuO NPs by root of *P. radiatus* under *in vitro* conditions. They showed that the NPs entered the mung bean cells (Lee et al., 2008). Transportation of CuO NPs from root to shoot via xylem has been confirmed by Wang et al. (2012b) in maize plants. Larue et al. (2014) ascertained foliar uptake of silver NPs (Ag NPs) in lettuce plants. They sprayed Ag NPs (38.6 nm) on lettuce leaves and illustrated that Ag NPs passed through stomata and translocated to other parts of plants via vascular tissues (Larue et al., 2014). Passage of Ag NPs (~20 nm) through cell wall pores (5–20 nm) in *Vigna radiata* revealed that cell wall plays a sieve role and prevents larger NPs to enter plant cells (Mazumdar, 2014).

NPs can penetrate into plant cells via stomata, cuticle, root, and wound. Both symplastic and apoplastic transportation of NPs have been reported. In the apoplastic route, size of cell wall pore is a limiting factor against penetration of NPs into plant cells (Dietz and Herth, 2011). Likewise, in symplastic route, plasmodesmata size (3–50 nm) can limit the transportation of NPs from cell to cell (Lucas and Lee, 2004). Other penetration mechanisms of NPs into plant cells are aquaporins, endocytosis pathways, ion channels, and creating new pores in the cell wall (Dev et al., 2018). Despite the limitation of cell wall pore size, TiO_2 NPs, and Ag NPs with particle size of more than cell wall pores were entered the plant cells through creating new pores (Larue et al., 2012; Wang et al., 2015). Based on some reports, NPs affect the expression of aquaporin proteins which allows NPs to pass through them (Taylor et al., 2014). In general, effective parameters in the uptake of NPs by plants are plant species, chemical compounds of NPs, plant development stage, and size and charge of NPs.

10.5 TOXICITY EFFECTS OF NANOPARTICLES ON PLANTS

Plants are the main food source for animals and humans. Therefore, the study of adverse effects of NPs on plants growth and development is necessary. However, a few studies related to the phytotoxicity of NPs on plants have been accomplished (Zuverza-Mena et al., 2016). Structural features of NPs such as shape and size are effective on their phytotoxicity properties (Rastogi et al., 2017). In the current chapter, the effect of three commonly used NPs groups including metal-based NPs, silica NPs, and carbon-based-NPs on plants have been investigated. A summary of carrying out researches has been listed in Table 10.2.

TABLE 10.2 Recent Studies Investigating Effects of NPs on Plant Growth and Development

Plant Species	Nanoparticle	Size and Concentration of NPs	Impacts on Plant Growth and Development	References
Arabidopsis thaliana	SiO_2	14, 50 and 200 nm 250 and 1000 mg/L	No toxicity effect of Si NPs (without charge) on treated plants	Slomberg and Schoenfisch (2012)
Vicia faba	SiO_2	0, 1, 2, and 3 mM	Increased antioxidant activity and plant tolerance to salinity	Qados (2015)
Solanum lycopersicum	SiO_2	0. 1, 2, and 3 mM	Increased plant tolerance to salinity	Almutairi (2016)
Medicago sativa	SiO_2	7 nm 0.3, 1.5, and 3 mg/L	Positive effect on phytoremediation under heavy metal soil contamination	Zmeeva et al. (2017)
Oryza sativa	SiO_2	<200 nm 106 g/L	Increased leaf lignin content No effect on plant biomass	Alvarez et al. (2018)
Vicia faba	Ag	60 nm 12.5, 25, 50, and 100 mg/L	Chromosome anomalies and genome instability	Patlolla et al. (2012)
Arabidopsis thaliana	Ag	2–14 nm 0.2–3 mg/L	Negative effects of AG NPs on root elongation No effect on seed germination	Qian et al. (2013a)
Oryza sativa	Ag	20, 30–60, 70–120 and 150 nm 0.1, 1, 10, 100, and 1000 mg/L	Adverse correlation between seed germination and Ag NPs size and concentration	Thuesombat et al. (2014)
Raphanus sativus	Ag	2 nm 125, 250, and 500 mg/L	Negative effects on plant growth No effect on seed germination	Zuverza-Mena et al. (2016)
Pisum sativum	Ag	22 nm 1000 and 3000 μM	Inhibition of plant growth	Tripathi et al. (2017a)
Oryza sativa	Ag	18.16 nm 10, 20, and 40 ppm	Enhanced plant growth	Gupta et al. (2018)
Physalis peruviana	Ag	121.6 nm 0.385, 0.77, 1.54, and 15.4 mg/L	Reduction in plant growth at high concentration (more than 1.54 mg/L) Increased plant growth at low concentration (0.385 mg/L)	de Oliveira Timoteo et al. (2019)

TABLE 10.2 *(Continued)*

Plant Species	Nanoparticle	Size and Concentration of NPs	Impacts on Plant Growth and Development	References
Zea mays *Vicia narbonensis*	TiO_2	<100 nm 0.2, 1, 2, and 4%	Dose-dependent genotoxicity effects and decreased mitotic index	Castiglione et al. (2011)
Brassica napus	TiO_2	20 nm 10–2000 mg/L	Enhanced seed germination and root growth at 2000 mg/L concentration	Mahmoodzadeh et al. (2013)
Vigna radiata	TiO_2	10 mg/L	Increased plant biomass and leaf total soluble protein content	Raliya et al. (2015)
Cicer arietinum	TiO_2	5 mg/L	Increased plant tolerance to cold stress	Hasanpour et al. (2015)
Oryza sativa	ZnO	10, 100, 500, and 1000 mg/L	Inhibition of root elongation	Boonyanitipong et al. (2011)
Allium sativum	ZnO	<100 nm 25, 50, 75, and 100 mg/L	Dose- and time-dependent genotoxicity effects on treated plants	Shaymurat et al. (2012)
Zea mays *Cucumis sativus*	ZnO	30 ± 10 nm 10, 100, and 1000 mg/L	No effect on germination Negative effects on root elongation	Zhang et al. (2015b)
Brassica nigra	ZnO	>100 nm 10 mg/L	Shoot emergence and increased antioxidants	Zafar et al. (2016)
Leucaena leucocephala	ZnO	2–64 nm 25 mg/L	Enhanced plant growth rate under heavy metal stress (Cd and Pb)	Venkatachalam et al. (2017a)
Lemna minor	ZnO	20–30 nm 10 mg/L	Reduced Cd toxicity on treated plants	Sun et al. (2019)
Cucumis sativus	CuO	50 nm 10, 50, 100, 500, and 1000 mg/L	Decreased seedling biomass	Kim et al. (2012)
Lactuca sativa *Medicago sativa*	CuO	10–100 nm 0.5, 10, and 20 mg/L	Decreased root elongation	Hong et al. (2015)

TABLE 10.2 *(Continued)*

Plant Species	Nanoparticle	Size and Concentration of NPs	Impacts on Plant Growth and Development	References
Oryza sativa	CuO	<50 nm 2.5, 10, 50, 100, and 1000 mg/L	Decreased seed germination and root elongation	Da Costa and Sharma (2016)
Solanum lycopersicum	CuO	20–40 nm 110, 50, 100, and 500 mg/L	Decreased chlorophyll content Concentration-dependent increase in antioxidants	Singh et al. (2017)
Arabidopsis thaliana	CuO	360–400 nm 125 and 625 μM	Inhibition on plant growth	Ke et al. (2017)
Coriandrum sativum	CuO	20 nm 200, 400, and 800 mg/L	Decreased biomass, chlorophyll content and root elongation	AlQuraidi et al. (2019)
Triticum aestivum	Oxidized MWCNTs	50–630 nm 10–160 μg/mL	Enhanced root elongation No effect on seed germination	Wang et al. (2012a)
Triticum aestivum	Carbon nano-dots (CNDs)	20–100 nm 150 mg/mL	Enhanced plant growth under light and dark condition	Tripathi and Sarkar (2014)
Vigna radiata	CNDs	4 nm 0.1–10 mg/mL	Positive effects on seed germination and root and shoot elongation	Li et al. (2016)
Ricinus communis	MWCNTs	35–300 nm 2–500 μg/mL	Enhanced plant growth and seed germination	Fathi et al. (2017)
Arabidopsis thaliana	CNDs	3 nm 62.5, 125, 250, 500, and 1000 mg/L	Decreased root elongation at concentration more than 125 mg/L No toxic effect at concentration of 62.5 mg/L	Chen et al. (2018)

10.5.1 METAL-BASED NANOPARTICLES

The effect of metal-based NPs on the expression of *Arabidopsis* genes has been evaluated. Exposure plants to Ag NPs, zinc oxide NPs (ZnO NPs), copper oxide NPs (CuO NPs) and titanium oxide NPs (TiO_2 NPs) were revealed upregulation of 286, 660, 851, and 80 genes, respectively. Whereas, they led to down-regulate 80, 826, 869, and 76 genes, respectively (Landa et al., 2012; Kaveh et al., 2013; Taylor et al., 2014). The upregulated and downregulated genes were mostly related to antioxidant and pathogen related genes, respectively. The upregulation of antioxidant genes showed that the treated plants have been activated the protective mechanisms instead of biomass production (Hawthorne et al., 2012; Nair and Chung, 2014).

10.5.1.1 SILVER NANOPARTICLES (Ag NPs)

Silver NPs (Ag NPs) have negative effects on living organisms due to the affinity of them to cysteine amino acid in cell wall proteins (Tripathi et al., 2017b). Exposure of *Arabidopsis* plant to Ag NPs was illustrated upregulation of the aquaporin genes in the first week. It might be related to the increased nutrient and water demand under stress conditions. However, downregulation of the aquaporin genes has been obtained in the second week (Qian et al., 2013a). Based on the results of Qian et al. (2013), Ag NPs led to inhibition of root elongation due to the reduction in chlorophyll content. Dimkpa et al. (2013) assessed the toxicity impact of Ag NPs on wheat plants. Their findings showed that expression of the metallothionein gene was upregulated (Dimkpa et al., 2013). It indicated that defense mechanisms in plants cells have been activated. Therefore, Ag NPs have dangerous side effects on plants mechanisms and pathways. Study on rice revealed the adverse effect of Ag NPs (18 nm and 16 μg/ml) on plant bio-macromolecules such as DNA, RNA, and protein that led to cell apoptosis (Mirzajani et al., 2014). Nair and Chung (2015) appraised the toxic effects of Ag NPs (20 nm, 20, and 50 mg/mL) on *V. radiata* plants. The increased ROS generation and cellular damage were observed in the case of exposing plants to Ag NPs (Nair and Chung, 2015a). Ag NPs may induce ROS production in plants. For instance, exposure *Spirodela polyrhiza* plants to Ag NPs (6 nm) revealed Ag NPs concentration-dependent ROS production (Jiang et al., 2014). Toxic impacts of Ag NPs were reported in *Lemna gibba* as well (Dewez et al., 2018). They exposed plants to Ag NPs (310.5 nm, and 1 mg/L) and found decreased biomass and chlorophyll content in the treated plants (Dewez et al., 2018). A recent study on the cytotoxicity effect of Ag NPs (10, 20, 51, and 73 nm) at the concentration of 100 mg/L illustrated changes in germination time, root elongation and nuclear abnormality in *A. cepa* (Scherer et al., 2019). Cytotoxicity and genotoxicity are depended on Ag NPs sizes. A contrary relation between particle size and cytotoxicity has been observed. Due to the larger surface area of smaller NPs, they might seem more reactive and toxic (Scherer et al., 2019). More studies about the effect of Ag NPs on plant system is shown in Table 10.2.

Positive effects of Ag NPs on plant growth, development, and protection are reported as well. For instance, treatment of *P. vulgaris* and *Z. mays* with Ag NPs (20 nm) at different concentrations of 20 to 100 ppm revealed an incremental effect on plant growth at low concentration (up to 60 ppm), while high concentration (80 and 100 ppm) led to growth inhibition (Salama, 2012). The positive effect of Ag NPs (25, 50, 100, 200, and 400 ppm) on cabbage has been ascertained as well (Sharma et al., 2012). The increased ROS scavenging enzymes, decreased ROS levels and improvement of photosynthesis efficacy have been observed in exposing plants to Ag NPs (Sharma et al., 2012). Jasim et al. (2016) appraised the effect of Ag NPs (8–21 nm) on *Trigonella foenum-graecum* plants. They reported positive effects of Ag NPs on secondary metabolite generation (Jasim et al., 2017). It can be a turning point in the production of secondary metabolites in medical plants. Dose-dependent effect of Ag NPs has been demonstrated in a study of Mujeeb et al. (2018). They evaluated the effects of various concentrations of Ag NPs (405 nm) on *Cassia occidentalis* plants (Mujeeb et al., 2018). Based on their findings, increased seed germination and seedling biomass have been observed at Ag NPs concentration up to 80 mg/L, while inhibition of plant growth occurred at 100 and 1,000 mg/L Ag NPs (Mujeeb et al., 2018). Similar to other NPs, the effect of Ag NPs on plant physiological parameters might be related to their size, shape, and concentration (Mujeeb et al., 2018; Scherer et al., 2019).

10.5.1.2 TITANIUM DIOXIDE NANOPARTICLES (TIO_2)

Plant response to TiO_2 NPs depends on the dose of NPs and plant species (Table 10.2). In general, exposing plants to TiO_2 NPs produce ROS scavenging enzymes (Yang et al., 2017). Negative effects of TiO_2 NPs such as inhibition of seed germination and root elongation have been reported in both monocots and dicots (Castiglione et al., 2011). Silva et al. (2017) investigated the toxic impact of TiO_2 NPs on lettuce and basil. Although no negative effects on seed germination found, reduction in biomass and root length was observed (Silva et al., 2017b). Further experiments revealed that exposure time could affect the amount of damage (Silva et al., 2017a). TiO_2 NPs cause damage to the root cell plasma membrane, thus, water and mineral nutrient absorption can be affected. Therefore, long-term exposure to TiO_2 NPs increased the amount of injury in plants (Silva et al., 2017a). Based on these findings, several variables' elements such as size, concentration, plant species, and exposure time could determine TiO_2 NPs toxicity in plants.

In contrast, positive effects of TiO_2 NPs on plants were also reported (Abdel Latef et al., 2018; Amini et al., 2018). Abdel Latef et al. (2018) investigated the effect of three concentrations of TiO_2 NPs (0.01%, 0.02%, and 0.03%) on bean growth and also in response to salinity stress. They revealed that 0.01% of TiO_2 NPs improved plant tolerance under salinity stress condition, while 0.02% and 0.03% of TiO_2 NPs led to intermediate response to salinity stress in association with no specific effect on the plant growth (Abdel Latef et al., 2018). Therefore, TiO_2 NPs can lead to positive effects depending on NPs concentration and plant species. The effect of TiO_2 NPs (5 mg/L) on chickpea under cold stress conditions was found in another assessment (Amini et al., 2018). No negative effect was detected, and TiO_2 NPs-treated plants showed an enhanced tolerance against cold stress due to a decrease in plasma membrane electrolyte leakage and induction of *receptor-like kinase*, *ethylene response transcription factor* and *vascular sorting receptor 6* genes (Amini et al., 2018). Mahmoodzadeh et al. (2014) studied the effect of TiO_2 NPs on wheat germination and growth. Their results indicated that 1200 ppm of TiO_2 NPs increased germination rate and biomass, while TiO_2 in low concentrations (10, 100, and 1000 ppm) decreased germination rate (Mahmoodzadeh and Aghili, 2014). The results founded by Mahmoodzadeh et al. (2014) are in agreement with those of Boonyanitipong et al. (2011). Based on their results, TiO_2 NPs have no negative effect on root growth and seed germination of rice (Boonyanitipong et al., 2011). Taken together, the use of proper concentration of TiO_2 NPs may increase plant growth, germination, and tolerance to abiotic stress.

10.5.1.3 ZINC OXIDE NANOPARTICLES (ZnO NPs)

Zinc oxide nanoparticles (ZnO NPs) are widely used in agriculture as fertilizer, pesticide, and fungicide (Rastogi, 2019). Therefore, the investigation of their toxicity effects on plants is necessary. Lee et al. (2010) reported that the use of ZnO NPs with particle size of less than 44 nm led to negative effects on *Arabidopsis* growth and germination (Lee et al., 2010). Consistent with that, Wang et al. (2016) revealed that ZnO NPs with the concentration of 200 and 300 mg/L reduced *Arabidopsis* growth (20% and 80%, respectively) due to inhibition of expression of photosystem genes (Wang et al., 2016). Wan et al. assessed the effect of ZnO NPs on *Arabidopsis* growth as well. Their findings illustrated that ZnO NPs had negative effects on primary root growth due to the disturbance in the root cell wall structure (Wan et al., 2019). Ghosh et al. (2016) appraised the ZnO NPs (less than 100 nm with 0.2, 0.4, and 0.8 mg/L concentration) toxicity on *Vicia faba*, *N. tabacum*, and *A. cepa* plants. They observed DNA damage, lipid peroxidation, increased ROS production and cell death in plants expose to ZnO NPs (Ghosh et al., 2016). However, their findings showed a decrease in toxicity effects at the highest concentration of ZnO NPs, while the lowest concentration gives a higher toxicity effect. The phytotoxicity impact of ZnO NPs has been investigated in other plant species such as rice, river tamarind, cucumber, corn, and cabbage (Table 10.2) (Lin and Xing, 2008; López-Moreno et al., 2010; Boonyanitipong et al., 2011; Kouhi et al., 2015; Yang et al., 2015; Venkatachalam et al., 2017a).

However, there are several studies about the positive effects of ZnO NPs (Raliya and Tarafdar, 2013; Tarafdar et al., 2014; Venkatachalam et al., 2017b). For instance, Latef et al. (2017) investigated different concentrations of ZnO NPs (21.3 nm) including 20, 40, and 60 mg/L on *Lupinus termis* plants under salinity stress. The results revealed that pre-treated plants by ZnO NPs showed an enhanced tolerance against salinity condition due to increase induction of antioxidant enzymes activities such as superoxide dismutase, ascorbate peroxidase and catalase (Latef et al., 2017). Tarafdar et al. (2014) evaluated ZnO NPs (15–25 nm) as nanofertilizer on pearl millet plant (*Pennisetum americanum*) growth. Their results ascertained increase in shoot and root length, chlorophyll content, and dry biomass in the targeted plants (Tarafdar et al., 2014). Moreover, the impact of ZnO NPs under abiotic stresses was investigated on plants. For example, *O. sativa*, *Glycine max*, and *L. termis* plants have been treated by 1000 mg/L, 0.5–1 mg/L, and 20–60 mg/L, respectively (Sedghi et al., 2013; Latef et al., 2017; Pavithra et al., 2017) following their tolerance to drought stress. The pre-treated plants by ZnO NPs illustrated an increase in growth and development and enhanced their tolerance against salinity stress. Based on the aforementioned findings, the effect of ZnO NPs on plants strongly depends on their size, concentration, and plant species. Therefore, the toxic effects of ZnO NPs can be avoided by considering these parameters.

10.5.1.4 COPPER OXIDE NPs (CuO NPs)

CuO NPs have been used due to their biocide features and can be used to protect plants against phytopathogenic fungi (Giannousi et al., 2013; Giorgetti, 2019). Copper is an essential micronutrient for plant as well. However, at high concentration, Cu has negative effects on seed germination, root, and shoot elongation and damage to DNA (Table 10.2) (Rajput et al., 2018b). Lee et al. (2013) investigated CuO NPs (<50 nm) toxicity on buckwheat (*Fagopyrum esculentum*) plants. Dose-dependent negative effects such as DNA damage and reduction of root growth at a concentration of 2,000 and 4,000 mg/L was observed (Lee et al., 2013). *Brassica juncea* plants were exposed to CuO NPs (20–500 mg/L) and concentration-dependent adverse effects of CuO NPs on root and shoot elongation have been observed (Nair and Chung, 2015b). Deng et al. (2016) also appraised various concentration of CuO NPs (5, 10, 20, 40, and 80 mg/L) on *A. cepa* root elongation. A decreased in root growth was observed at a concentration of more than 5 mg/L due to inhibition of apical meristem division (Deng et al., 2016). Shaw et al. (2013) evaluated CuO NPs (<50 nm, 0.5–1.5 mM) effects on rice plants. They found that at the concentration of more than 1 mM CuO NPs, ROS accumulation, reduction in growth and DNA damage occurs. Based on findings of Wang et al. (2015), CuO NPs (5 mg/L) led to ROS overproduction in association with inhibition of plant growth in rice plants. The toxicity effect of CuO NPs (>10 mg/kg) on wheat plants was also appraised by Adams et al. (2017). Deceased root elongation and cell death were observed in the exposing plants to CuO NPs (Adams et al., 2017). In a recent study, CuO NPs (30–50 nm, 10 g/L) toxicity effects have been investigated on plant growth in *Hordeum sativum*. Their results indicated a decreased plant growth in response to CuO NPs (Rajput et al., 2018a).

However, in a few studies, no adverse effects have been observed for CuO NPs. For instance, Juárez-Maldonado et al. (2016) investigated the effect of Cu NPs in a chitosan hydrogel (100 mg/kg) on tomato growth. They demonstrated that combination of Cu NPs/chitosan hydrogel caused an increase in growth and lycopene content of the treated plants, whereas the use of Cu NPs alone had no significant effect on tested parameters (Juarez-Maldonado et al., 2016). These results are in agreement with those of Choudhary et al. (2017). They showed that a mixture of Cu NPs-Chitosan increase biomass and antioxidants activities in maize, causing an enhanced resistance against *Curvularia* leaf spot disease (Choudhary et al., 2017). Rawat et al. (2018) revealed that the effect of CuO NPs (20–100 nm) at different concentrations

(125, 250, and 500 mg/kg) on growth and development of *Capsicum annum* was not significant (Rawat et al., 2018). Taken together, the effect of CuO NPs strongly depends on plant species, thus, more researches are needed to determine their toxicity on different plant species.

10.5.2 SILICA NANOPARTICLES (SiO_2 NPs)

SiO_2 NPs have been used in many fields such as medicine and agriculture. In agriculture, they can be used to deliver DNA and protein (Napierska et al., 2010; Hajiahmadi et al., 2019). Si is naturally absorbed by plants roots and usually added as fertilizer to plants. However, nanomaterials have different physicochemical features, thus, assessment of their toxicity on plants is essential. The effect of SiO_2 NPs on seed germination and root elongation has been appraised. Their negative effects depend on the size and concentration of NPs and plant species. Adhikari et al. (2013) assessed the effects of SiO_2 NPs on rice and found no toxic effect of SiO_2 NPs. The Si NPs positive effects have been ascertained by other researchers as well (Table 10.2). In this subgroup, MSNs have been widely used due to their unique physicochemical properties such as biodegradable and biocompatibility (Anjum et al., 2015). The MSNs can be used to deliver macromolecules such as DNA and phytohormones (Chang et al., 2013; Hussain et al., 2013; Sun et al., 2018; Hajiahmadi et al., 2019). One of the most important parameters that determine the toxic impact of MSNs is the size. The mesoporous silica size should be 20–2500 nm. The smaller or larger MSNs can be more toxic to human cells (Napierska et al., 2009). Therefore, the size of MSNs should be considered to minimize the researchers' health risk is associated with applying MSNs. The MSNs have positive effects on plants as well. For example, they can increase plant tolerance to abiotic and abiotic stresses (Breitler et al., 2004; Ma and Yamaji, 2008).

A few studies reported adverse effects of high concentration SiO_2 NPs on plant growth and development. Slomberg and Schoenfisch (2012) confirmed that no negative effect related to size and concentration of SiO_2 NPs was detected, while a decrease in plant growth and chlorosis was the results of adsorption of micronutrients and macronutrients to SiO_2 NPs surface. In contrast, concentration-dependent negative effects of SiO_2 NPs have also been reported in *T. aestivum* L. (Karimi and Mohsenzadeh, 2016). They illustrated that high concentration of SiO_2 NPs (more than 200 mg/L) can have negative effects on the development of tested plants. Accordingly, toxicity effects of SiO_2 NPs depend on environmental factors, plant species, and their size and concentration. In conclusion, the advantages of Si NPs are more than their disadvantages.

10.5.3 CARBON-BASED NPs

Carbon-based nanoparticles are divided into two main groups: (1) single-walled carbon nanotube and (2) MWCNTs. They have been applied in many studies due to their strange features (Table 10.2). Srinivasan and Saraswathi (2010) investigated the effects of carbon nanotubes (CNTs: 10, 20, and 40 mg/L) on tomato seed germination and seedling growth. Enhanced plant growth and water absorption were found in their research (Srinivasan and Saraswathi, 2010). The toxicity effect of MWCNTs (500, 1000, and 2000 mg/L) on *Amaranthus tricolor*, *Amaranthus lividus*, *B. oleracea*, *Lactuca sativa*, and *Lycopersicon esculentum* have been evaluated (Begum et al., 2011). Species- and concentration-dependent negative effects of CNTs were reported by Begum et al. (2012). ROS overproduction was detected when red spinach, tomato, and cabbage were exposed to MWCNTs, while no significant toxic effect was found in lettuce. In another study, the effect of MWCNTs has been appraised on wheat biomass (Wang et al., 2012a). The use of MWCNTs enhanced plant growth and root elongation in wheat (1.4-fold compared to control). The negative effect of graphene oxide (GO) in various concentrations of 100, 200, and 1600 mg/L on *V. faba* has been appraised (Anjum et al., 2013). Decreased in antioxidant activities were dose-dependent and the lowest enzyme activity has been reported at the concentration of 1600 mg/L. The negative effect of MWCNTs (5–15 nm) in different concentrations of 125, 250, 500, and 1000 μg/mL was reported on seed germination of *Cucurbita pepo* under both normal and drought conditions as well (Hatami, 2017). In summary, the negative effect of carbon-based NPs on plants depends on their size, concentration, and plant species.

In some other studies, positive effects of carbon-based NPs in several plants were reported. MWCNTs effects on *Cicer arietinum* plants has been ascertained (Tripathi et al., 2011). Based on their findings, enhanced plant growth and water absorption have been obtained in plants treated by MWCNTs (400 nm, up to 6 μg/mL). Khodakovskaya et al. (2012) displayed positive effect of MWCNTs (5–500 mg/L) on tobacco plants. They demonstrated that the growth rate increased by 55–64% when plants are treated by NPs (Khodakovskaya et al., 2012). MWCNTs and SWCNTs can easily penetrate into seed coat leading to an increase in water absorption, thus, enhanced seed germination and plant growth can be obtained in plants treated by carbon-based nanoparticles (Husen and Siddiqi, 2014). Khodakovskaya et al. (2013) revealed that SWCNTs (10–25 nm, 50–200 μg/mL) can enhance tomato production. Positive effects of MWCNTs on seed germination and plant growth has been confirmed in maize as well (Tiwari et al., 2014). Tiwari et al. (2014) tested the impact of MWCNTs on seed germination in maize plants. Their results emphasized concentration-dependent positive effects of MWCNTs on maize seed germination, and reported that the best germination rate was 20 mg/L (low dose) of NPs, while a higher dose could prevent germination (Tiwari et al., 2014). Carbon-based nanoparticles can be applied as plant growth regulators in many plant species. Lahiani et al. (2015) evaluated the effect of SWCNTs (50–100 nm, 25, 50, and 100 μg/ mL) on barley, rice, tomato, and tobacco, soybean, and corn plants resulting in considerable enhancement in seed germination and growth parameters (Lahiani et al., 2015). Martínez-Ballesta et al. (2016) investigated the effect of MWCNTs (100–500 nm, 10–60 mg/L) on *B. oleracea.* They displayed that an enhanced tolerance under salinity stress in plants occurred due to the increased water uptake (Martínez-Ballesta et al., 2016). Assessment of MWCNTs (50–500 nm, 50 mg/L) toxicity on *Arabidopsis* also illustrated no toxic effects on plant growth (Fan et al., 2018). In general, the reported positive effects of carbon-based NPs are more than their toxic and negative effects.

10.6 OUTLOOK

In this chapter, different applied NPs-mediated delivery methods have been reviewed and discussed. NCs in agriculture have been widely used to deliver pesticides, fungicides, and fertilizers. However, a few studies exist in the field of NPs-mediated gene transformation (transient or stable transformation) in plants, and therefore further researches are necessary. Nowadays, one of the important genome editing tools is CRISPR/Cas9 (clustered regularly interspaced short palindromic repeat-associated protein 9) system which has several benefits such as cost-efficient and high transformation efficiency. Considering this system as uses homology of sgRNA with the target sequence in plant genome, NCs containing CRISPR/Cas9 expression cassette can be applied to develop transgenic plants with no limitation of host range. Taken together, by selecting the best NPs with the least toxicity on target plant species, unconditional NPS-mediated macromolecule delivery system can be used in a wide range of plant species.

KEYWORDS

- chitosan
- gene delivery
- layered double hydroxide nanoparticles
- macromolecules delivery
- mesoporous silica nanoparticles
- nanocarriers
- nanocomposites
- nanofertilizers
- nanoparticles
- nanopesticides
- nanotechnology
- pesticides delivery

REFERENCES

Abdalla, A. M., Hossain, S., Azad, A. T., Petra, P. M. I., Begum, F., Eriksson, S. G., & Azad, A. K., (2018). Nanomaterials for solid oxide fuel cells: A review. *Renewable and Sustainable Energy Reviews, 82,* 353–368.

Abdel, L. A. A. H., Srivastava, A. K., El-sadek, M. S. A., Kordrostami, M., & Tran, L. S. P., (2018). Titanium dioxide nanoparticles improve growth and enhance tolerance of broad bean plants under saline soil conditions. *Land Degradation and Development, 29*(4), 065–1073.

Adak, T., Kumar, J., Shakil, N., & Walia, S., (2012). Development of controlled release formulations of imidacloprid employing novel nano-ranged amphiphilic polymers. *Journal of Environmental Science and Health, Part B, 47*(3), 217–225.

Adams, J., Wright, M., Wagner, H., Valiente, J., Britt, D., & Anderson, A., (2017). Cu from dissolution of CuO nanoparticles signals changes in root morphology. *Plant Physiology and Biochemistry, 110,* 108–117.

Alfaro, M., Salazar, F., Iraira, S., Teuber, N., Villarroel, D., & Ramírez, L., (2008). Nitrogen, phosphorus and potassium losses in a grazing system with different stocking rates in a volcanic soil. *Chilean Journal of Agricultural Research, 68*(2), 146–155.

Almutairi, Z. M., (2016). Effect of nano-silicon application on the expression of salt tolerance genes in germinating tomato (*Solanum lycopersicum* L.) seedlings under salt stress. *Plant Omics,* 1, 106.

Alquraidi, A. O., Mosa, K. A., & Ramamoorthy, K., (2019). Phytotoxic and genotoxic effects of copper nanoparticles in coriander (*Coriandrum sativum*—Apiaceae). *Plants,8*(1), 19.

Alvarez, R. D. C. F., De Mello, P. R., Felisberto, G., Deus, A. C. F., & De Oliveira, R. L. L., (2018). Effects of soluble silicate and nanosilica application on rice nutrition in an oxisol. *Pedosphere, 28*(4), 597–606.

Amini, S., Moali, A. R., & Mohammadi, V., (2018). Effect of TiO_2 nanoparticles on relative expression of some genes involved in response to cold stress in chickpea. *Iranian Journal of Field Crop Science, 49*(1), 155–165.

Anjum, N. A., Adam, V., Kizek, R., Duarte, A. C., Pereira, E., Iqbal, M., Lukatkin, A. S., & Ahmad, I., (2015). Nanoscale copper in the soil-plant system-toxicity and underlying potential mechanisms. *Environmental Research, 13,* 306–8325.

Anjum, N. A., Singh, N., Singh, M. K., Shah, Z. A., Duarte, A. C., Pereira, E., & Ahmad, I., (2013). Single-bilayer graphene oxide sheet tolerance and glutathione redox system significance assessment in faba bean (*Vicia faba* L.). *Journal of Nanoparticle Research, 15*(7), 1770.

Ao, M., Zhu, Y., He, S., Li, D., Li, P., Li, J., & Cao, Y., (2012). Preparation and characterization of 1-naphthylacetic acid-silica conjugated nanospheres for enhancement of controlled-release performance. *Nanotechnology, 24*(3), 035601.

Ashfaq, M., Verma, N., & Khan, S., (2017). Carbon nanofibers as a micronutrient carrier in plants: Efficient translocation and controlled release of Cu nanoparticles. *Environmental Science: Nano, 4*(1), 138–148.

Asli, S., & Neumann, P. M., (2009). Colloidal suspensions of clay or titanium dioxide nanoparticles can inhibit leaf growth and transpiration via physical effects on root water transport. *Plant, Cell and Environment, 32*(5), 577–584.

Azencott, H. R., Peter, G. F., & Prausnitz, M. R., (2007). Influence of the cell wall on intracellular delivery to algal cells by electroporation and sonication. *Ultrasound in Medicine and Biology, 33*(11), 805–1817.

Bao, W., Wan, Y., & Baluška, F., (2017). Nanosheets for delivery of biomolecules into plant cells. *Trends in Plant Science, 22*(6), 445–447.

Bao, W., Wang, J., Wang, Q., O'Hare, D., & Wan, Y., (2016). Layered double hydroxide nanotransporter for molecule delivery to intact plant cells. *Scientific Reports,* 6, 26738.

Begum, P., Ikhtiari, R., & Fugetsu, B., (2011). Graphene phytotoxicity in the seedling stage of cabbage, tomato, red spinach, and lettuce. *Carbon, 49*(12), 907–3919.

Behera, S., & Panda, R., (2009). Integrated management of irrigation water and fertilizers for wheat crop using field experiments and simulation modeling. *Agricultural Water Management, 96*(11), 532–1540.

Bergeson, L. L., (2010). Nanosilver pesticide products: What does the future hold? *Environmental Quality Management, 19,* 73–82.

Bernal, M., Cases, R., Picorel, R., & Yruela, I., (2007). Foliar and root Cu supply affect differently Fe-and Zn-uptake and photosynthetic activity in soybean plants. *Environmental and Experimental Botany, 60*(2), 145–150.

Bernardo, M. P., Guimarães, G. G., Majaron, V. F., & Ribeiro, C., (2018). Controlled release of phosphate from layered double hydroxide structures: Dynamics in soil and application as smart fertilizer. *ACS Sustainable Chemistry and Engineering, 6*(4), 152–5161.

Boonyanitipong, P., Kositsup, B., Kumar, P., Baruah, S., & Dutta, J., (2011). Toxicity of ZnO and TiO_2 nanoparticles on germinating rice seed *Oryza sativa* L. *International Journal of Bioscience, Biochemistry and Bioinformatics, 4,* 282.

Breitler, J. C., Vassal, J. M., Del, M. C. M., Meynard, D., Marfà, V., Melé, E., Royer, M., et al., (2004). Bt rice harboring cry genes controlled by a constitutive or wound-inducible promoter: Protection and transgene expression under Mediterranean field conditions. *Plant Biotechnology Journal, 2*(5), 417–430.

Burlaka, O., Pirko, Y. V., Yemets, A., & Blume, Y. B., (2015). Plant genetic transformation using carbon nanotubes for DNA delivery. *Cytology and Genetics, 49*(6), 349–357.

Buzea, C., Pacheco, I. I., & Robbie, K., (2007). Nanomaterials and nanoparticles: Sources and toxicity. *Biointerphases, 2*(4), R17–MR71.

Cai, D., Wang, L., Zhang, G., Zhang, X., & Wu, Z., (2013). Controlling pesticide loss by natural porous micro/nano composites: Straw ash-based biochar and biosilica. *ACS Applied Materials and Interfaces, 5*(18), 212–9216.

Cao, L., Zhou, Z., Niu, S., Cao, C., Li, X., Shan, Y., & Huang, Q., (2017). Positive-charge functionalized mesoporous silica nanoparticles as nanocarriers for controlled 2,4-dichlorophenoxy acetic acid sodium salt release. *Journal of Agricultural and Food Chemistry, 66*(26), 594–6603.

Castiglione, M. R., Giorgetti, L., Geri, C., & Cremonini, R., (2011). The effects of nano-TiO_2 on seed germination, development and mitosis of root tip cells of *Vicia narbonensis* L. and *Zea mays* L. *Journal of Nanoparticle Research, 13*(6), 443–2449.

Castillo, R. R., Baeza, A., & Vallet-Regí, M., (2017). Recent applications of the combination of mesoporous silica nanoparticles with nucleic acids: Development of bio-responsive devices, carriers and sensors. *Biomaterials Science, 5*(3), 353–377.

Chakraborty, S., & Newton, A. C., (2011). Climate change, plant diseases and food security: An overview. *Plant Pathology, 60*(1), 2–14.

Chang, F. P., Kuang, L. Y., Huang, C. A., Jane, W. N., Hung, Y., Yue-ie, C. H., & Mou, C. Y., (2013). A simple plant gene delivery system using mesoporous silica nanoparticles as carriers. *Journal of Materials Chemistry B, 1*(39), 279–5287.

Chauhan, N., Dilbaghi, N., Gopal, M., Kumar, R., Kim, K. H., & Kumar, S., (2017). Development of chitosan nanocapsules for the controlled release of hexaconazole. *International Journal of Biological Macromolecules, 97*, 616–624.

Chen, J., Liu, B., Yang, Z., Qu, J., Xun, H., Dou, R., Gao, X., & Wang, L., (2018). Phenotypic, transcriptional, physiological and metabolic responses to carbon nanodot exposure in *Arabidopsis thaliana* (L.). *Environmental Science: Nano, 5*(11), 672–2685.

Chen, J., Wang, W., Xu, Y., & Zhang, X., (2010). Slow-release formulation of a new biological pesticide, pyoluteorin, with mesoporous silica. *Journal of Agricultural and Food Chemistry, 59*(1), 307–311.

Chi, Y., Zhang, G., Xiang, Y., Cai, D., & Wu, Z., (2017). Fabrication of a temperature-controlled-release herbicide using a nanocomposite. *ACS Sustainable Chemistry and Engineering, 5*(6), 969–4975.

Choudhary, R. C., Kumaraswamy, R., Kumari, S., Sharma, S., Pal, A., Raliya, R., Biswas, P., & Saharan, V., (2017). Cu-chitosan nanoparticle boost defense responses and plant growth in maize (*Zea mays* L.). *Scientific Reports, 7*(1), 9754.

Corradini, E., De Moura, M., & Mattoso, L., (2010). A preliminary study of the incorporation of NPK fertilizer into chitosan nanoparticles. *Express Polymer Letters, 4*(8), 509–515.

Da Costa, M., & Sharma, P., (2016). Effect of copper oxide nanoparticles on growth, morphology, photosynthesis, and antioxidant response in *Oryza sativa*. *Photosynthetica, 54*(1), 110–119.

Dapkekar, A., Deshpande, P., Oak, M. D., Paknikar, K. M., & Rajwade, J. M., (2018). Zinc use efficiency is enhanced in wheat through nanofertilization. *Scientific Reports, 8*(1), 6832.

Davidson, D., & Gu, F. X., (2012). Materials for sustained and controlled release of nutrients and molecules to support plant growth. *Journal of Agricultural and Food Chemistry, 60*(4), 870–876.

De Oliveira, J. L., Campos, E. N. V. R., Gonçalves Da, S. C. M., Pasquoto, T., Lima, R., & Fraceto, L. F., (2015). Solid lipid nanoparticles co-loaded with simazine and atrazine:

Preparation, characterization, and evaluation of herbicidal activity. *Journal of Agricultural and Food Chemistry, 63*(2), 422–432.

De Oliveira, T. C., Paiva, R., Dos, R. M. V., Claro, P. I. C., Ferraz, L. M., Marconcini, J. M., De Oliveira, J. E., (2019). *In vitro* growth of *Physalis peruviana* L. affected by silver nanoparticles. *3 Biotech, 4,* 145.

De Sormeaux, A., & Pemberton, C., (2011). *Factors Influencing Agriculture's Contribution to GDP: Latin America and the Caribbean.* Department of Agricultural Economics and Extension, The University of the West Indies, St Augustine Campus, Trinidad and Tobago.

Demirer, G. S., Zhang, H., Matos, J. L., Goh, N. S., Cunningham, F. J., Sung, Y., Chang, R., et al., (2019). *High Aspect Ratio Nanomaterials Enable Delivery of Functional Genetic Material without DNA Integration in Mature Plants* (p. 1–32). BioRxiv.

Deng, F., Wang, S., & Xin, H., (2016). Toxicity of CuO nanoparticles to structure and metabolic activity of *Allium cepa* root tips. *Bulletin of Environmental Contamination and Toxicology, 97*(5), 702–708.

Deodhar, G. V., Adams, M. L., & Trewyn, B. G., (2017). Controlled release and intracellular protein delivery from mesoporous silica nanoparticles. *Biotechnology Journal, 12*(1), 1600408.

DeRosa, M. C., Monreal, C., Schnitzer, M., Walsh, R., & Sultan, Y., (2010). Nanotechnology in fertilizers. *Nature Nanotechnology, 5*(2), 91.

Dev, A., Srivastava, A. K., & Karmakar, S., (2018). Nanomaterial toxicity for plants. *Environmental Chemistry Letters, 16*(1), 85–100.

Dewez, D., Goltsev, V., Kalaji, H. M., & Oukarroum, A., (2018). Inhibitory effects of silver nanoparticles on photosystem II performance in *Lemna gibba* probed by chlorophyll fluorescence. *Current Plant Biology, 1*, 15–21.

Dietz, K. J., & Herth, S., (2011). Plant nanotoxicology. *Trends in Plant Science, 16*(11) 582–589.

Dimkpa, C. O., McLean, J. E., Latta, D. E., Manangón, E., Britt, D. W., Johnson, W. P., Boyanov, M. I., & Anderson, A. J., (2012). CuO and ZnO nanoparticles: Phytotoxicity, metal speciation, and induction of oxidative stress in sand-grown wheat. *Journal of Nanoparticle Research, 14*(9), 1125.

Dimkpa, C. O., McLean, J. E., Martineau, N., Britt, D. W., Haverkamp, R., & Anderson, A. J., (2013). Silver nanoparticles disrupt wheat (*Triticum aestivum* L.) growth in a sand matrix. *Environmental Science and Technology, 47*(2), 1082–1090.

Eggenberger, K., Frey, N., Zienicke, B., Siebenbrock, J., Schunck, T., Fischer, R., Bräse, S., et al., (2010). Use of nanoparticles to study and manipulate plant cells. *Advanced Engineering Materials, 12*(9), 406–B412.

El-Aziz, M. A., Morsi, S., Salama, D. M., Abdel-Aziz, M., Elwahed, M. S. A., Shaaban, E., & Youssef, A., (2019). Preparation and characterization of chitosan/polyacrylic acid/copper nanocomposites and their impact on onion production. *International Journal of Biological Macromolecules, 12,* 856–3865.

Fan, D., Liu, T., Li, C., Jiao, B., Li, S., Hou, Y., & Luo, K., (2015). Efficient CRISPR/Cas9-mediated targeted mutagenesis in *Populus* in the first generation. *Scientific Reports, 5,* 12217.

Fan, X., Xu, J., Lavoie, M., Peijnenburg, W., Zhu, Y., Lu, T., Fu, Z., Zhu, T., & Qian, H., (2018). Multiwall carbon nanotubes modulate paraquat toxicity in *Arabidopsis thaliana. Environmental Pollution, 233,* 633–641.

Fathi, Z., Nejad, R. A. K., Mahmoodzadeh, H., & Satari, T. N., (2017). Investigating of a wide range of concentrations of multi-walled carbon nanotubes on germination and growth of castor seeds (*Ricinus communis* L.). *Journal of Plant Protection Research, 57*(3), 228–236.

Forim, M. R., Costa, E. S., Da Silva, M. F. G. F., Fernandes, J. B., Mondego, J. M., Boiça, Jr. A. L., (2013). Development of a new method to prepare nano-/microparticles loaded with extracts of *Azadirachta indica*, their characterization and use in controlling *Plutella xylostella*. *Journal of Agricultural and Food Chemistry, 61*(38), 131–9139.

Fu, Y. Q., Li, L. H., Wang, P. W., Qu, J., Fu, Y. P., Wang, H., & Sun, J. R., (2012). Delivering DNA into plant cell by gene carriers of ZnS nanoparticles. *Chemical Research in Chinese Universites, 28*(4), 672–676.

Fu, Y., Li, L., Wang, H., Jiang, Y., Liu, H., Cui, X., Wang, P., & Lü, C., (2015). Silica nanoparticles-mediated stable genetic transformation in *Nicotiana tabacum*. *Chemical Research in Chinese Universities, 31*(6), 976–981.

Ganguly, S., Das, S., & Dastidar, S., (2014). Effect of zinc sulphide nano particles on germination of seeds of *Vigna radiata* and their subsequent acceleration of growth in presence of the nanoparticles. *Euro J. Biomed. Pharma. Sci., 1*(2), 273–280.

Ghorbanpour, M., Asgari, L. H., & Hadian, J., (2016). Influence of copper and zinc on growth, metal accumulation and chemical composition of essential oils in sweet basil *(Ocimum basilicum* L.). *Journal of Medicinal Plants, 3*(59), 132–144.

Ghormade, V., Deshpande, M. V., & Paknikar, K. M., (2011). Perspectives for nano-biotechnology enabled protection and nutrition of plants. *Biotechnology Advances, 29*(6), 792–803.

Ghosh, M., Jana, A., Sinha, S., Jothiramajayam, M., Nag, A., Chakraborty, A., Mukherjee, A., & Mukherjee, A., (2016). Effects of ZnO nanoparticles in plants: Cytotoxicity, genotoxicity, deregulation of antioxidant defenses, and cell-cycle arrest. *Mutation Research/Genetic Toxicology and Environmental Mutagenesis, 80,* 725–732.

Giannousi, K., Avramidis, I., & Dendrinou-Samara, C., (2013). Synthesis, characterization and evaluation of copper based nanoparticles as agrochemicals against *Phytophthora infestans*. *RSC Adv., 3*(44), 2143–21752.

Giorgetti, L., (2019). Effects of nanoparticles in plants: Phytotoxicity and genotoxicity assessment. In: Tripathi, D. K., Ahmad, P., Sharma, S., Chauhan, D. K., & Dubey, N. K., (eds.), *Nanomaterials in Plants, Algae and Microorganisms* (pp. 65–87). Elsevier, Neherlands.

Grillo, R., Pereira, A. E., Nishisaka, C. S., De Lima, R., Oehlke, K., Greiner, R., & Fraceto, L. F., (2014). Chitosan/tripolyphosphate nanoparticles loaded with paraquat herbicide: An environmentally safer alternative for weed control. *Journal of Hazardous Materials, 278,* 163–171.

Gu, Y., Zhang, X., Tu, S., & Lindström, K., (2009). Soil microbial biomass, crop yields, and bacterial community structure as affected by long-term fertilizer treatments under wheat-rice cropping. *European Journal of Soil Biology, 45*(3), 239–246.

Guan, H., Chi, D., Yu, J., & Li, X., (2008). A novel photodegradable insecticide: Preparation, characterization and properties evaluation of nano-Imidacloprid. *Pesticide Biochemistry and Physiology, 92*, 83–91.

Gupta, S. D., Agarwal, A., & Pradhan, S., (2018). Phytostimulatory effect of silver nanoparticles (Ag NPs) on rice seedling growth: An insight from antioxidative enzyme activities and gene expression patterns. *Ecotoxicology and Environmental Safety, 161,* 624–633.

Ha, N. M. C., Nguyen, T. H., Wang, S. L., & Nguyen, A. D., (2019). Preparation of NPK nanofertilizer based on chitosan nanoparticles and its effect on biophysical characteristics and growth of coffee in green house. *Research on Chemical Intermediates, 45*, 51–63.

Hajiahmadi, Z., Shirzadian-Khorramabad, R., Kazemzad, M., & Sohani, M. M., (2019). Enhancement of tomato resistance to *Tuta absoluta* using a new efficient mesoporous silica nanoparticle-mediated plant transient gene expression approach. *Scientia Horticulturae, 243*, 367–375.

Han, X., Chen, S., & Hu, X., (2009). Controlled-release fertilizer encapsulated by starch/polyvinyl alcohol coating. *Desalination, 240*(1), 21–26.

Hao, Y., Yang, X., Shi, Y., Song, S., Xing, J., Marowitch, J., Chen, J., & Chen, J., (2013). Magnetic gold nanoparticles as a vehicle for fluorescein isothiocyanate and DNA delivery into plant cells. *Botany, 91*(7), 457–466.

Hasanpour, H., Maali-Amir, R., & Zeinali, H., (2015). Effect of TiO_2 nanoparticles on metabolic limitations to photosynthesis under cold in chickpea. *Russian Journal of Plant Physiology, 62*(6), 779–787.

Hasheminejad, N., Khodaiyan, F., & Safari, M., (2019). Improving the antifungal activity of clove essential oil encapsulated by chitosan nanoparticles. *Food Chemistry, 275*, 113–122.

Hatami, M., (2017). Toxicity assessment of multi-walled carbon nanotubes on *Cucurbita pepo* L. under well-watered and water-stressed conditions. *Ecotoxicology and Environmental Safety, 142*, 274–283.

Hawthorne, J., Musante, C., Sinha, S. K., & White, J. C., (2012). Accumulation and phytotoxicity of engineered nanoparticles to *Cucurbita pepo*. *International Journal of Phytoremediation, 14*(4), 429–442.

Hong, J., Rico, C. M., Zhao, L., Adeleye, A. S., Keller, A. A., Peralta-Videa, J. R., & Gardea-Torresdey, J. L., (2015). Toxic effects of copper-based nanoparticles or compounds to lettuce (*Lactuca sativa*) and alfalfa (*Medicago sativa*). *Environmental Science: Processes and Impacts, 17*(1), 177–185.

Husen, A., & Siddiqi, K. S., (2014). Carbon and fullerene nanomaterials in plant system. *Journal of Nanobiotechnology, 2*(1), 16.

Hussain, H. I., Yi, Z., Rookes, J. E., Kong, L. X., & Cahill, D. M., (2013). Mesoporous silica nanoparticles as a biomolecule delivery vehicle in plants. *Journal of Nanoparticle Research, 15*(6), 1676.

Hussain, M. R., Devi, R. R., & Maji, T. K., (2012). Controlled release of urea from chitosan microspheres prepared by emulsification and cross-linking method. *Iranian Polymer Journal, 21*(8), 473–479.

Hussein, A. K., (2015). Applications of nanotechnology in renewable energies: A comprehensive overview and understanding. *Renewable and Sustainable Energy Reviews, 42*, 460–476.

Jasim, B., Thomas, R., Mathew, J., & Radhakrishnan, E., (2017). Plant growth and diosgenin enhancement effect of silver nanoparticles in Fenugreek (*Trigonella foenum-graecum* L.). *Saudi Pharmaceutical Journal, 25*(3), 443–447.

Jiang, H. S., Qiu, X. N., Li, G. B., Li, W., & Yin, L. Y., (2014). Silver nanoparticles induced accumulation of reactive oxygen species and alteration of antioxidant systems in the aquatic plant *Spirodela polyrhiza*. *Environmental Toxicology and Chemistry, 33*(6), 398–1405.

Juarez-Maldonado, A., Ortega-Ortíz, H., Pérez-Labrada, F., Cadenas-Pliego, G., & Benavides-Mendoza, A., (2016). Cu nanoparticles absorbed on chitosan hydrogels positively alter morphological, production, and quality characteristics of tomato. *Journal of Applied Botany and Food Quality, 89*, 183–189.

Kang, M. A., Seo, M. J., Hwang, I. C., Jang, C., Park, H. J., Yu, Y. M., & Youn, Y. N., (2012). Insecticidal activity and feeding behavior of the green peach aphid, *Myzus persicae*, after treatment with nano types of pyrifluquinazon. *Journal of Asia-Pacific Entomology, 15*(4), 533–541.

Karimi, J., & Mohsenzadeh, S., (2016). Effects of silicon oxide nanoparticles on growth and physiology of wheat seedlings. *Russian Journal of Plant Physiology, 63*(1), 119–123.

Kaushik, P., Shakil, N. A., Kumar, J., Singh, M. K., Singh, M. K., & Yadav, S. K., (2013). Development of controlled release formulations of thiram employing amphiphilic polymers and their bio-efficacy evaluation in seed quality enhancement studies. *Journal of Environmental Science and Health, Part B, 48*(8), 677–685.

Kaveh, R., Li, Y. S., Ranjbar, S., Tehrani, R., Brueck, C. L., & Van, A. B., (2013). Changes in *Arabidopsis thaliana* gene expression in response to silver nanoparticles and silver ions. *Environmental Science and Technology, 47*(18), 1037–10644.

Kaziem, A. E., Gao, Y., He, S., & Li, J., (2017). Synthesis and insecticidal activity of enzyme-triggered functionalized hollow mesoporous silica for controlled release. *Journal of Agricultural and Food Chemistry, 65*(36), 854–7864.

Ke, M., Zhu, Y., Zhang, M., Gumai, H., Zhang, Z., Xu, J., & Qian, H., (2017). Physiological and molecular response of *Arabidopsis thaliana* to CuO nanoparticle (nCuO) exposure. *Bulletin of Environmental Contamination and Toxicology, 99*(6), 713–718.

Khalifa, N. S., & Hasaneen, M. N., (2018). The effect of chitosan-PMAA–NPK nanofertilizer on *Pisum sativum* plants. *3 Biotech,* 4, 193.

Khan, M. N., Mobin, M., Abbas, Z. K., AlMutairi, K. A., & Siddiqui, Z. H., (2017). Role of nanomaterials in plants under challenging environments. *Plant Physiology and Biochemistry, 110,* 194–209.

Khan, M. R., & Rizvi, T. F., (2014). Nanotechnology: Scope and application in plant disease management. *Plant Pathol. J., 13*(3), 214–231.

Khodakovskaya, M. V., De Silva, K., Biris, A. S., Dervishi, E., & Villagarcia, H., (2012). Carbon nanotubes induce growth enhancement of tobacco cells. *ACS Nano, 6*(3), 128–2135.

Khodakovskaya, M., Dervishi, E., Mahmood, M., Xu, Y., Li, Z., Watanabe, F., & Biris, A. S., (2009). Carbon nanotubes are able to penetrate plant seed coat and dramatically affect seed germination and plant growth. *ACS Nano, 3*(10), 221–3227.

Khot, L. R., Sankaran, S., Maja, J. M., Ehsani, R., & Schuster, E. W., (2012). Applications of nanomaterials in agricultural production and crop protection: A review. *Crop Protection, 5,* 64–70.

Kim, S., Lee, S., & Lee, I., (2012). Alteration of phytotoxicity and oxidant stress potential by metal oxide nanoparticles in *Cucumis sativus*. *Water, Air, and Soil Pollution, 223*(5), 799–2806.

Koli, P., Singh, B. B., Shakil, N. A., Kumar, J., & Kamil, D., (2015). Development of controlled release nanoformulations of carbendazim employing amphiphilic polymers and their bioefficacy evaluation against *Rhizoctonia solani*. *Journal of Environmental Science and Health, Part B, 50*(9), 674–681.

Kottegoda, N., Munaweera, I., Madusanka, N., & Karunaratne, V., (2011). A green slow-release fertilizer composition based on urea-modified hydroxyapatite nanoparticles encapsulated wood. *Current Science, 101*, 73–78.

Kottegoda, N., Sandaruwan, C., Priyadarshana, G., Siriwardhana, A., Rathnayake, U. A., Berugoda, A. D. M., Kumarasinghe, A. R., et al., (2017). Urea-hydroxyapatite nanohybrids for slow release of nitrogen. *ACS Nano, 11*(2), 214–1221.

Kouhi, S. M. M., Lahouti, M., Ganjeali, A., & Entezari, M. H., (2015). Comparative effects of ZnO nanoparticles, ZnO bulk particles, and Zn^{2+} on *Brassica napus* after long-term exposure: Changes in growth, biochemical compounds, antioxidant enzyme activities, and Zn bioaccumulation. *Water, Air, and Soil Pollution, 226*(11), 364.

Kumar, J., Shakil, N. A., Singh, M. K., Pankaj, Singh, M. K., Pandey, A., & Pandey, R. P., (2010). Development of controlled release formulations of azadirachtin-A employing poly (ethylene glycol) based amphiphilic copolymers. *Journal of Environmental Science and Health Part B, 45*(4), 310–314.

Kumar, R., Ashfaq, M., & Verma, N., (2018). Synthesis of novel PVA–starch formulation-supported Cu-Zn nanoparticle carrying carbon nanofibers as a nanofertilizer: Controlled release of micronutrients. *Journal of Materials Science, 53*(10), 150–7164.

Kumar, S., Bhanjana, G., Sharma, A., Sidhu, M., & Dilbaghi, N., (2014). Synthesis, characterization and on field evaluation of pesticide loaded sodium alginate nanoparticles. *Carbohydrate Polymers, 101,* 061–1067.

Kumar, S., Chauhan, N., Gopal, M., Kumar, R., & Dilbaghi, N., (2015). Development and evaluation of alginate-chitosan nanocapsules for controlled release of acetamiprid. *International Journal of Biological Macromolecules, 81,* 631–637.

Kumar, S., Kumar, D., & Dilbaghi, N., (2017). Preparation, characterization, and bio-efficacy evaluation of controlled release carbendazim-loaded polymeric nanoparticles. *Environmental Science and Pollution Research, 24*(1), 926–937.

Kumaraswamy, R., Kumari, S., Choudhary, R. C., Pal, A., Raliya, R., Biswas, P., & Saharan, V., (2018). Engineered chitosan based nanomaterials: Bioactivities, mechanisms and perspectives in plant protection and growth. *International Journal of Biological Macromolecules, 113,* 494–506.

Kurepa, J., Paunesku, T., Vogt, S., Arora, H., Rabatic, B. M., Lu, J., Wanzer, M. B., Woloschak, G. E., & Smalle, J. A., (2010). Uptake and distribution of ultrasmall anatase TiO_2 Alizarin red S nanoconjugates in *Arabidopsis thaliana*. *Nano Letters, 10*(7), 296–2302.

Labandeira, C. C., & Prevec, R., (2014). Plant paleopathology and the roles of pathogens and insects. *International Journal of Paleopathology 4,* 1–16.

Lahiani, M. H., Chen, J., Irin, F., Puretzky, A. A., Green, M. J., & Khodakovskaya, M. V., (2015). Interaction of carbon nanohorns with plants: Uptake and biological effects. *Carbon, 81,* 607–619.

Landa, P., Vankova, R., Andrlova, J., Hodek, J., Marsik, P., Storchova, H., White, J. C., & Vanek, T., (2012). Nanoparticle-specific changes in *Arabidopsis thaliana* gene expression after exposure to ZnO, TiO_2, and fullerene soot. *Journal of Hazardous Materials, 21,* 55–62.

Larue, C., Castillo-Michel, H., Sobanska, S., Cécillon, L., Bureau, S., Barthès, V., Ouerdane, L., et al., (2014). Foliar exposure of the crop *Lactuca sativa* to silver nanoparticles: Evidence for internalization and changes in Ag speciation. *Journal of Hazardous Materials, 26,* 98–106.

Larue, C., Laurette, J., Herlin-Boime, N., Khodja, H., Fayard, B., Flank, A. M., Brisset, F., & Carriere, M., (2012). Accumulation, translocation and impact of TiO_2 nanoparticles in wheat (*Triticum aestivum* spp.): Influence of diameter and crystal phase. *Science of the Total Environment, 431,* 197–208.

Latef, A. A. H. A., Alhmad, M. F. A., & Abdelfattah, K. E., (2017). The possible roles of priming with ZnO nanoparticles in mitigation of salinity stress in lupine (*Lupinus termis*) plants. *Journal of Plant Growth Regulation, 36*, 60–70.

Lee, C. W., Mahendra, S., Zodrow, K., Li, D., Tsai, Y. C., Braam, J., & Alvarez, P. J., (2010). Developmental phytotoxicity of metal oxide nanoparticles to *Arabidopsis thaliana*. *Environmental Toxicology and Chemistry: An International Journal, 29*(3), 669–675.

Lee, S., Chung, H., Kim, S., & Lee, I., (2013). The genotoxic effect of ZnO and CuO nanoparticles on early growth of buckwheat, *Fagopyrum esculentum*. *Water, Air, and Soil Pollution, 224*(9), 1668.

Lee, W. M., An, Y. J., Yoon, H., & Kweon, H. S., (2008). Toxicity and bioavailability of copper nanoparticles to the terrestrial plants mung bean (*Phaseolus radiatus*) and wheat (*Triticum aestivum*): Plant agar test for water-insoluble nanoparticles. *Environmental Toxicology and Chemistry: An International Journal, 27*(9), 915–1921.

Lehmann, J., Rillig, M. C., Thies, J., Masiello, C. A., Hockaday, W. C., & Crowley, D., (2011). Biochar effects on soil biota: A review. *Soil Biology and Biochemistry, 43*(9), 812–1836.

Li, W., Zheng, Y., Zhang, H., Liu, Z., Su, W., Chen, S., Liu, Y., et al., (2016). Phytotoxicity, uptake, and translocation of fluorescent carbon dots in mung bean plants. *ACS Applied Materials and Interfaces, 8*(31), 1939–19945.

Li, Z., Zhang, Y., & Li, Y., (2013). Zeolite as slow release fertilizer on spinach yields and quality in a greenhouse test. *Journal of Plant Nutrition, 36*(10), 496–1505.

Liang, B., Lehmann, J., Solomon, D., Kinyangi, J., Grossman, J., O'neill, B., Skjemstad, J., et al., (2006). Black carbon increases cation exchange capacity in soils. *Soil Science Society of America Journal, 70*(5), 719–1730.

Lin, D., & Xing, B., (2008). Root uptake and phytotoxicity of ZnO nanoparticles. *Environmental Science and Technology, 42*(15), 580–5585.

Liu, F., Wen, L. X., Li, Z. Z., Yu, W., Sun, H. Y., & Chen, J. F., (2006). Porous hollow silica nanoparticles as controlled delivery system for water-soluble pesticide. *Materials Research Bulletin, 41*(12), 268–2275.

Liu, H., Polenske, K. R., Xi, Y., & Guo, J. E., (2010). Comprehensive evaluation of effects of straw-based electricity generation: A Chinese case. *Energy Policy, 38*(10), 153–6160.

Liu, Q., Chen, B., Wang, Q., Shi, X., Xiao, Z., Lin, J., & Fang, X., (2009). Carbon nanotubes as molecular transporters for walled plant cells. *Nano Letters, 9*(3), 1007–1010.

Liu, R., & Lal, R., (2014). Synthetic apatite nanoparticles as a phosphorus fertilizer for soybean (*Glycine max*). *Scientific Reports 4,* 5686.

Liu, X., He, B., Xu, Z., Yin, M., Yang, W., Zhang, H., Cao, J., & Shen, J., (2015). A functionalized fluorescent dendrimer as a pesticide nanocarrier: Application in pest control. *Nanoscale, 7*(2), 445–449.

Loha, K. M., Shakil, N. A., Kumar, J., Singh, M. K., Adak, T., & Jain, S., (2011). Release kinetics of β-Cyfluthrin from its encapsulated formulations in water. *Journal of Environmental Science and Health Part B, 46*(3), 201–206.

López-Moreno, M. L., De La Rosa, G., Hernández-Viezcas, J. Á., Castillo-Michel, H., Botez, C. E., Peralta-Videa, J. R., & Gardea-Torresdey, J. L., (2010). Evidence of the differential biotransformation and genotoxicity of ZnO and CeO_2 nanoparticles on soybean (*Glycine max*) plants. *Environmental Science and Technology, 44*(19), 315–7320.

Lou, X. W., Archer, L. A., & Yang, Z., (2008). Hollow micro-/nanostructures: Synthesis and applications. *Advanced Materials, 20*(21), 987–4019.

Lubkowski, K., & Grzmil, B., (2007). Controlled release fertilizers. *Polish Journal of Chemical Technology, 9,* 83–84.

Lucas, W. J., & Lee, J. Y., (2004). Plasmodesmata as a supracellular control network in plants. *Nature Reviews Molecular Cell Biology, 9,* 712.

Ma, J., & Yamaji, N., (2008). Functions and transport of silicon in plants. *Cellular and Molecular Life Sciences, 65*(19), 049–3057.

Madusanka, N., Sandaruwan, C., Kottegoda, N., Sirisena, D., Munaweera, I., De Alwis, A., Karunaratne, V., & Amaratunga, G. A., (2017). Urea-hydroxyapatite-montmorillonite nanohybrid composites as slow release nitrogen compositions. *Applied Clay Science, 150,* 303–308.

Mahmoodzadeh, H., & Aghili, R., (2014). Effect on germination and early growth characteristics in wheat plants (*Triticum aestivum* L.) seeds exposed to TiO_2 nanoparticles. *Journal of Chemical Health Risks, 4,* 29–36.

Mahmoodzadeh, H., Nabavi, M., & Kashefi, H., (2013). Effect of nanoscale titanium dioxide particles on the germination and growth of canola (*Brassica napus*). *Journal of Ornamental Plants, 3,* 25–32.

Manikandan, A., & Subramanian, K., (2016). Evaluation of zeolite based nitrogen nano-fertilizers on maize growth, yield and quality on inceptisols and alfisols. *Int. J. Plant Soil Sci., 4,* 1–9.

Marschner, H., (2011). *Marschner's Mineral Nutrition of Higher Plants* (p. 684). Academic Press, USA.

Martínez-Ballesta, M. C., Zapata, L., Chalbi, N., & Carvajal, M., (2016). Multiwalled carbon nanotubes enter broccoli cells enhancing growth and water uptake of plants exposed to salinity. *Journal of Nanobiotechnology, 4*(1), 42.

Martin-Ortigosa, S., Peterson, D. J., Valenstein, J. S., Lin, V. S. Y., Trewyn, B. G., Lyznik, L. A., & Wang, K., (2014). Mesoporous silica nanoparticle-mediated intracellular Cre protein delivery for maize genome editing via loxP site excision. *Plant Physiology, 164*(2), 537–547.

Martin-Ortigosa, S., Valenstein, J. S., Lin, V. S. Y., Trewyn, B. G., & Wang, K., (2012). Gold functionalized mesoporous silica nanoparticle mediated protein and DNA codelivery to plant cells via the biolistic method. *Advanced Functional Materials, 22*(17), 576–3582.

Maruyama, C. R., Guilger, M., Pascoli, M., Bileshy-José, N., Abhilash, P., Fraceto, L. F., & De Lima, R., (2016). Nanoparticles based on chitosan as carriers for the combined herbicides imazapic and imazapyr. *Scientific Reports, 6,* 19768.

Mazumdar, H., (2014). The impact of silver nanoparticles on plant biomass and chlorophyll content. *Int. J. Eng. Sci., 4,* 12–20.

McKee, M. S., & Filser, J., (2016). Impacts of metal-based engineered nanomaterials on soil communities. *Environmental Science: Nano, 3*(3), 506–533.

Memarizadeh, N., Ghadamyari, M., Adeli, M., & Talebi, K., (2014a). Linear-dendritic copolymers/indoxacarb supramolecular systems: Biodegradable and efficient nano-pesticides. *Environmental Science: Processes and Impacts, 16*(10), 380–2389.

Memarizadeh, N., Ghadamyari, M., Adeli, M., & Talebi, K., (2014b). Preparation, characterization and efficiency of nanoencapsulated imidacloprid under laboratory conditions. *Ecotoxicology and Environmental Safety, 17,* 77–83.

Merlan, F., & Raftery, D., (2009). *Tracking Rural Change: Community, Policy and Technology in Australia* (p. 194). New Zealand and Europe, ANU Press, Australia.

Mirzajani, F., Askari, H., Hamzelou, S., Schober, Y., Römpp, A., Ghassempour, A., & Spengler, B., (2014). Proteomics study of silver nanoparticles toxicity on *Oryza sativa* L. *Ecotoxicology and Environmental Safety, 108,* 335–339.

Mitter, N., Worrall, E. A., Robinson, K. E., Li, P., Jain, R. G., Taochy, C., Fletcher, S. J., et al., (2017). Clay nanosheets for topical delivery of RNAi for sustained protection against plant viruses. *Nature Plants, 3,* 16207.

Mujeeb, M., Ismail, M., Aqil, M., Khan, A., Ikram, S., & Ahmad, A., (2018). Evaluation of phytotoxic impact of plant mediated silver nanoparticles on seed germination and growth of seedling of *Cassia occidentalis*. *Journal of Pharmaceutical Innovation 7*, 8–12.

Muxika, A., Etxabide, A., Uranga, J., Guerrero, P., & De La Caba, K., (2017). Chitosan as a bioactive polymer: Processing, properties and applications. *International Journal of Biological Macromolecules, 105,* 358–1368.

Nair, P. M. G., & Chung, I. M., (2014). Impact of copper oxide nanoparticles exposure on *Arabidopsis thaliana* growth, root system development, root lignificaion, and molecular level changes. *Environmental Science and Pollution Research, 21*(22), 1209–12722.

Nair, P. M. G., & Chung, I. M., (2015a). Physiological and molecular level studies on the toxicity of silver nanoparticles in germinating seedlings of mung bean (*Vigna radiata* L.). *Acta Physiologiae Plantarum, 37*(1), 1719.

Nair, P. M. G., & Chung, I. M., (2015b). Study on the correlation between copper oxide nanoparticles induced growth suppression and enhanced lignification in Indian mustard (*Brassica juncea* L.). *Ecotoxicology and Environmental Safety, 113,* 302–313.

Nair, R., Poulose, A. C., Nagaoka, Y., Yoshida, Y., Maekawa, T., & Kumar, D. S., (2011). Uptake of FITC labeled silica nanoparticles and quantum dots by rice seedlings: Effects on seed germination and their potential as biolabels for plants. *Journal of Fluorescence, 21*(6), 2057.

Napierska, D., Thomassen, L. C., Lison, D., Martens, J. A., & Hoet, P. H., (2010). The nanosilica hazard: Another variable entity. *Particle and Fibre Toxicology,7*(1), 39.

Napierska, D., Thomassen, L. C., Rabolli, V., Lison, D., Gonzalez, L., Kirsch-Volders, M., Martens, J. A., & Hoet, P. H., (2009). Size-dependent cytotoxicity of monodisperse silica nanoparticles in human endothelial cells. *Small, 5*(7), 846–853.

Naqvi, S., Maitra, A., Abdin, M., Akmal, M., Arora, I., & Samim, M., (2012). Calcium phosphate nanoparticle mediated genetic transformation in plants. *Journal of Materials Chemistry, 22*(8), 500–3507.

Nuruzzaman, M., Rahman, M. M., Liu, Y., & Naidu, R., (2016). Nanoencapsulation, nano-guard for pesticides: A new window for safe application. *Journal of Agricultural and Food Chemistry, 64*(7), 447–1483.

Owen, A., Dufès, C., Moscatelli, D., Mayes, E., Lovell, J. F., Katti, K. V., Sokolov, K., et al., (2014). The application of nanotechnology in medicine: Treatment and diagnostics. *Nanomedicine, 9*(9), 291–1294.

Patlolla, A. K., Berry, A., May, L., & Tchounwou, P. B., (2012). Genotoxicity of silver nanoparticles in *Vicia faba*: A pilot study on the environmental monitoring of nanoparticles. *International Journal of Environmental Research and Public Health, 9*(5), 649–1662.

Pavithra, G., Reddy, B. R., Salimath, M., Geetha, K., & Shankar, A., (2017). Zinc oxide nano particles increases Zn uptake, translocation in rice with positive effect on growth, yield and moisture stress tolerance. *Indian Journal of Plant Physiology, 22*(3), 287–294.

Pereira, A. E., Grillo, R., Mello, N. F., Rosa, A. H., & Fraceto, L. F., (2014). Application of poly (epsilon-caprolactone) nanoparticles containing atrazine herbicide as an alternative technique to control weeds and reduce damage to the environment. *Journal of Hazardous Materials, 268,* 207–215.

Popat, A., Liu, J., Hu, Q., Kennedy, M., Peters, B., Lu, G. Q. M., & Qiao, S. Z., (2012). Adsorption and release of biocides with mesoporous silica nanoparticles. *Nanoscale, 4*(3), 970–975.

Primrose, S. B., & Twyman, R., (2013). *Principles of Gene Manipulation and Genomics* (p. 672). Wiley-Blackell, USA.

Qados, A. M. A., (2015). Mechanism of nanosilicon-mediated alleviation of salinity stress in faba bean (*Vicia faba* L.) plants. *American Journal of Experimental Agriculture, 7*, 78–95.

Qian, H., Peng, X., Han, X., Ren, J., Sun, L., & Fu, Z., (2013a). Comparison of the toxicity of silver nanoparticles and silver ions on the growth of terrestrial plant model *Arabidopsis thaliana*. *Journal of Environmental Sciences, 25*(9), 947–1956.

Qian, K., Shi, T., He, S., Luo, L., & Cao, Y., (2013b). Release kinetics of tebuconazole from porous hollow silica nanospheres prepared by miniemulsion method. *Microporous and Mesoporous Materials, 169,* 1–6.

Rafsanjani, M., Kiran, U., Ali, A., & Abdin, M., (2016). Transformation efficiency of calcium phosphate nanoparticles for genetic manipulation of *Cichorium intybus* L. *Indian Journal of Biotechnology, 15*(2), 145–152.

Rajonee, A. A., Zaman, S., & Huq, S. M. I., (2017). Preparation, characterization and evaluation of efficacy of phosphorus and potassium incorporated nano fertilizer. *Advances in Nanoparticles, 6*(2), 62.

Rajput, V., Minkina, T., Fedorenko, A., Sushkova, S., Mandzhieva, S., Lysenko, V., Duplii, N., et al., (2018a). Toxicity of copper oxide nanoparticles on spring barley (*Hordeum sativum distichum*). *Science of the Total Environment, 645,* 103–1113.

Rajput, V., Minkina, T., Suskova, S., Mandzhieva, S., Tsitsuashvili, V., Chapligin, V., & Fedorenko, A., (2018b). Effects of copper nanoparticles (CuO NPs) on crop plants: A mini review. *Bionanoscience, 8*, 36–42.

Raliya, R., & Tarafdar, J. C., (2013). ZnO nanoparticle biosynthesis and its effect on phosphorous-mobilizing enzyme secretion and gum contents in cluster bean (*Cyamopsis tetragonoloba* L.). *Agricultural Research, 2*, 48–57.

Raliya, R., Biswas, P., & Tarafdar, J., (2015). TiO_2 nanoparticle biosynthesis and its physiological effect on mung bean (*Vigna radiata* L.). *Biotechnology Reports, 5,* 22–26.

Ramesh, K., Biswas, A. K., Somasundaram, J., & Rao, A. S., (2010). Nanoporous zeolites in farming: Current status and issues ahead. *Current Science, 99*(6), 760–764.

Rameshaiah, G., Pallavi, J., & Shabnam, S., (2015). Nano fertilizers and nano sensors: An attempt for developing smart agriculture. *International Journal of Engineering Research and General Science, 3*(1), 314–320.

Rastogi, A., (2019). Industrial nanoparticles and their influence on gene expression in plants. In: Tripathi, D. K., Ahmad, P., Sharma, S., Chauhan, D. K., & Dubey, N. K., (eds.), *Nanomaterials in Plants, Algae and Microorganisms*, (pp. 89–101). Elsevier, Neherlands.

Rastogi, A., Zivcak, M., Sytar, O., Kalaji, H. M., He, X., Mbarki, S., & Brestic, M., (2017). Impact of metal and metal oxide nanoparticles on plant: A critical review. *Frontiers in Chemistry, 5,* 78.

Rawat, S., Pullagurala, V. L., Hernandez-Molina, M., Sun, Y., Niu, G., Hernandez-Viezcas, J. A., Peralta-Videa, J. R., & Gardea-Torresdey, J. L., (2018). Impacts of copper oxide nanoparticles on bell pepper (*Capsicum annum* L.) plants: A full life cycle study. *Environmental Science: Nano, 5*, 83–95.

Rehakova, M., Čuvanová, S., Dzivak, M., Rimár, J., & Gavalova, Z., (2004). Agricultural and agrochemical uses of natural zeolite of the clinoptilolite type. *Current Opinion in Solid State and Materials Science, 8*(6), 397–404.

Rico, C. M., Majumdar, S., Duarte-Gardea, M., Peralta-Videa, J. R., & Gardea-Torresdey, J. L., (2011). Interaction of nanoparticles with edible plants and their possible implications in the food chain. *Journal of Agricultural and Food Chemistry, 59*(8), 485–3498.

Roshanravan, B., Mahmoud, S. S., Mahdavi, F., Abdul, R. S., & Khanif, Y. M., (2014). Preparation of encapsulated urea-kaolinite controlled release fertilizer and their effect on rice productivity. *Chemical Speciation and Bioavailability, 26*(4), 249–256.

Salama, H. M., (2012). Effects of silver nanoparticles in some crop plants, common bean (*Phaseolus vulgaris* L.) and corn (*Zea mays* L.). *Int. Res. J. Biotechnol., 3*(10), 190–197.

Santos, E. M., Radin, S., & Ducheyne, P., (1999). Sol-gel derived carrier for the controlled release of proteins. *Biomaterials, 20*(18), 695–1700.

Sarkar, D. J., Kumar, J., Shakil, N., & Walia, S., (2012). Release kinetics of controlled release formulations of thiamethoxam employing nano-ranged amphiphilic PEG and diacid based block polymers in soil. *Journal of Environmental Science and Health, Part A, 47*(11), 701–1712.

Scherer, M. D., Sposito, J. C., Falco, W. F., Grisolia, A. B., Andrade, L. H., Lima, S. M., Machado, G., et al., (2019). Cytotoxic and genotoxic effects of silver nanoparticles on meristematic cells of *Allium cepa* roots: A close analysis of particle size dependence. *Science of the Total Environment, 660,* 459–467.

Sedghi, M., Hadi, M., & Toluie, S. G., (2013). Effect of nano zinc oxide on the germination parameters of soybean seeds under drought stress. *Annales of West University of Timisoara Series of Biology, 6*(2), 73.

Serag, M. F., Kaji, N., Venturelli, E., Okamoto, Y., Terasaka, K., Tokeshi, M., Mizukami, H., et al., (2011). Functional platform for controlled subcellular distribution of carbon nanotubes. *ACS Nano, 5*(11), 264–9270.

Shan, Y., Cao, L., Xu, C., Zhao, P., Cao, C., Li, F., Xu, B., & Huang, Q., (2019). Sulfonate-functionalized mesoporous silica nanoparticles as carriers for controlled herbicide diquat dibromide release through electrostatic interaction. *International Journal of Molecular Sciences, 20*(6), 1330.

Sharma, P., Bhatt, D., Zaidi, M., Saradhi, P. P., Khanna, P., & Arora, S., (2012). Silver nanoparticle-mediated enhancement in growth and antioxidant status of *Brassica juncea. Applied Biochemistry and Biotechnology, 167*(8), 225–2233.

Shaymurat, T., Gu, J., Xu, C., Yang, Z., Zhao, Q., Liu, Y., & Liu, Y., (2012). Phytotoxic and genotoxic effects of ZnO nanoparticles on garlic (*Allium sativum* L.): A morphological study. *Nanotoxicology, 6*(3), 241–248.

Shen, C. X., Zhang, Q. F., Li, J., Bi, F. C., & Yao, N., (2010). Induction of programmed cell death in *Arabidopsis* and rice by single-wall carbon nanotubes. *American Journal of Botany, 97*(10), 602–1609.

Shukla, P. K., Misra, P., & Kole, C., (2016). Uptake, translocation, accumulation, transformation, and generational transmission of nanoparticles in plants. In: Kole, C., Kumar, D. S., & Khodakovskaya, M. V., (eds.), *Plant Nanotechnology*, (pp. 183–218). Springer International Publishing, Swtzerland.

Silva, A. T., Nguyen, A., Ye, C., Verchot, J., & Moon, J. H., (2010). Conjugated polymer nanoparticles for effective siRNA delivery to tobacco BY-2 protoplasts. *BMC Plant Biology, 1*(1), 291.

Silva, S., Craveiro, S. C., Oliveira, H., Calado, A. J., Pinto, R. J., Silva, A. M., & Santos, C., (2017a). Wheat chronic exposure to TiO_2-nanoparticles: Cyto- and genotoxic approach. *Plant Physiology and Biochemistry, 11,* 89–98.

Silva, S., Oliveira, H., Silva, A., & Santos, C., (2017b). The cytotoxic targets of anatase or rutile+anatase nanoparticles depend on the plant species. *Biologia Plantarum, 61*(4), 717–725.

Singh, A., Singh, N., Hussain, I., & Singh, H., (2017). Effect of biologically synthesized copper oxide nanoparticles on metabolism and antioxidant activity to the crop plants *Solanum lycopersicum* and *Brassica oleracea* var. *botrytis*. *Journal of Biotechnology, 22,* 11–27.

Slomberg, D. L., & Schoenfisch, M. H., (2012). Silica nanoparticle phytotoxicity to *Arabidopsis thaliana*. *Environmental Science and Technology, 46*(18), 1047–10254.

Srinivasan, C., & Saraswathi, R., (2010). Nano-agriculture-carbon nanotubes enhance tomato seed germination and plant growth. *Current Science, 99*(3), 274–275.

Sun, C., Shu, K., Wang, W., Ye, Z., Liu, T., Gao, Y., Zheng, H., et al., (2014a). Encapsulation and controlled release of hydrophilic pesticide in shell cross-linked nanocapsules containing aqueous core. *International Journal of Pharmaceutics, 463*(1), 108–114.

Sun, D., Hussain, H. I., Yi, Z., Rookes, J. E., Kong, L., & Cahill, D. M., (2018). Delivery of abscisic acid to plants using glutathione responsive mesoporous silica nanoparticles. *Journal of Nanoscience and Nanotechnology, 18*(3), 615–1625.

Sun, D., Hussain, H. I., Yi, Z., Siegele, R., Cresswell, T., Kong, L., & Cahill, D. M., (2014b). Uptake and cellular distribution, in four plant species, of fluorescently labeled mesoporous silica nanoparticles. *Plant Cell Reports, 33*(8), 389–1402.

Sun, S., Li, X., Sun, C., Cao, W., Hu, C., Zhao, Y., & Yang, A., (2019). Effects of ZnO nanoparticles on the toxicity of cadmium to duckweed *Lemna minor*. *Science of the Total Environment, 662,* 697–702.

Tan, W. M., Hou, N., Pang, S., Zhu, X. F., Li, Z. H., Wen, L. X., & Duan, L. S., (2012). Improved biological effects of uniconazole using porous hollow silica nanoparticles as carriers. *Pest Management Science, 68*(3), 437–443.

Tan, X. M., & Fugetsu, B., (2007). Multi-walled carbon nanotubes interact with cultured rice cells: Evidence of a self-defense response. *Journal of Biomedical Nanotechnology, 3*(3), 285–288.

Tarafdar, J., Raliya, R., Mahawar, H., & Rathore, I., (2014). Development of zinc nanofertilizer to enhance crop production in pearl millet (*Pennisetum americanum*). *Agricultural Research, 3*(3), 257–262.

Taylor, A. F., Rylott, E. L., Anderson, C. W., & Bruce, N. C., (2014). Investigating the toxicity, uptake, nanoparticle formation and genetic response of plants to gold. *PLoS One, 9*(4), e93793.

Thuesombat, P., Hannongbua, S., Akasit, S., & Chadchawan, S., (2014). Effect of silver nanoparticles on rice (*Oryza sativa* L. cv. KDML 105) seed germination and seedling growth. *Ecotoxicology and Environmental Safety, 104,* 302–309.

Tiwari, D., Dasgupta-Schubert, N., Cendejas, L. V., Villegas, J., Montoya, L. C., & García, S. B., (2014). Interfacing carbon nanotubes (CNT) with plants: Enhancement of growth, water and ionic nutrient uptake in maize (*Zea mays*) and implications for nanoagriculture. *Applied Nanoscience, 4*(5), 577–591.

Tiwari, J. N., Tiwari, R. N., & Kim, K. S., (2012). Zero-dimensional, one-dimensional, two-dimensional and three-dimensional nanostructured materials for advanced electrochemical energy devices. *Progress in Materials Science, 57*(4), 724–803.

Tomlinson, I., (2013). Doubling food production to feed the 9 billion: A critical perspective on a key discourse of food security in the UK. *Journal of Rural Studies, 9,* 81–90.

Torney, F., Trewyn, B. G., Lin, V. S. Y., & Wang, K., (2007). Mesoporous silica nanoparticles deliver DNA and chemicals into plants. *Nature Nanotechnology, 5*, 295.

Tripathi, D. K., Singh, S., Singh, S., Srivastava, P. K., Singh, V. P., Singh, S., Prasad, S. M., et al., (2017a). Nitric oxide alleviates silver nanoparticles (AgNps)-induced phytotoxicity in *Pisum sativum* seedlings. *Plant Physiology and Biochemistry, 110,* 167–177.

Tripathi, D. K., Tripathi, A., Singh, S., Singh, Y., Vishwakarma, K., Yadav, G., Sharma, S., et al., (2017b). Uptake, accumulation and toxicity of silver nanoparticle in autotrophic plants, and heterotrophic microbes: A concentric review. *Frontiers in Microbiology, 8,* 7.

Tripathi, S., & Sarkar, S., (2014). Influence of water soluble carbon dots on the growth of wheat plant. *Applied Nanoscience, 5*(5), 609–616.

Tripathi, S., Sonkar, S. K., & Sarkar, S., (2011). Growth stimulation of gram (*Cicer arietinum*) plant by water soluble carbon nanotubes. *Nanoscale, 3*(3), 176–1181.

Venkatachalam, P., Jayaraj, M., Manikandan, R., Geetha, N., Rene, E. R., Sharma, N., & Sahi, S., (2017a). Zinc oxide nanoparticles (ZnONPs) alleviate heavy metal-induced toxicity in *Leucaena leucocephala* seedlings: A physiochemical analysis. *Plant Physiology and Biochemistry, 10,* 59–69.

Venkatachalam, P., Priyanka, N., Manikandan, K., Ganeshbabu, I., Indiraarulselvi, P., Geetha, N., Muralikrishna, K., et al., (2017b). Enhanced plant growth promoting role of phycomolecules coated zinc oxide nanoparticles with P supplementation in cotton (*Gossypium hirsutum* L.). *Plant Physiology and Biochemistry, 110,* 118–127.

Vennila, R., Kamaraj, P., Arthanareeswari, M., Sridharan, M., Sudha, G., Devikala, S., Arockiaselvi, J., et al., (2018). Biosynthesis of ZrO nanoparticles and its natural dye sensitized solar cell studies. *Materials Today: Proceedings, 5*(2), 691–8698.

Vivero-Escoto, J. L., Slowing, I. I., Trewyn, B. G., & Lin, V. S. Y., (2010). Mesoporous silica nanoparticles for intracellular controlled drug delivery. *Small, 6*(18), 952–1967.

Volova, T., Zhila, N., Vinogradova, O., Shumilova, A., Prudnikova, S., & Shishatskaya, E., (2016). Characterization of biodegradable poly-3-hydroxybutyrate films and pellets loaded with the fungicide tebuconazole. *Environmental Science and Pollution Research, 23*(6), 243–5254.

Wan, J., Wang, R., Wang, R., Ju, Q., Wang, Y., & Xu, J., (2019). Comparative physiological and transcriptomic analyses reveal the toxic effects of ZnO nanoparticles on plant growth. *Environ. Sci. Technol., 53*(8), 235–4244.

Wang, P., Menzies, N. W., Lombi, E., Sekine, R., Blamey, F. P. C., Hernandez-Soriano, M. C., Cheng, M., et al., (2015). Silver sulfide nanoparticles (Ag_2S-NPs) are taken up by plants and are phytotoxic. *Nanotoxicology, 9*(8), 041–1049.

Wang, S. L., & Nguyen, A. D., (2018). Effects of Zn/B nanofertilizer on biophysical characteristics and growth of coffee seedlings in a greenhouse. *Research on Chemical Intermediates, 44*(8), 889–4901.

Wang, X., Han, H., Liu, X., Gu, X., Chen, K., & Lu, D., (2012a). Multi-walled carbon nanotubes can enhance root elongation of wheat (*Triticum aestivum*) plants. *Journal of Nanoparticle Research, 1*(6), 841.

Wang, X., Yang, X., Chen, S., Li, Q., Wang, W., Hou, C., Gao, X., Wang, L., & Wang, S., (2016). Zinc oxide nanoparticles affect biomass accumulation and photosynthesis in *Arabidopsis*. *Frontiers in Plant Science 6,* 1243.

Wang, Z., Xie, X., Zhao, J., Liu, X., Feng, W., White, J. C., & Xing, B., (2012b). Xylem- and phloem-based transport of CuO nanoparticles in maize (*Zea mays* L.). *Environmental Science and Technology, 46*(8), 434–4441.

Wanyika, H., (2013). Sustained release of fungicide metalaxyl by mesoporous silica nanospheres. In: Diallo, M. S., Fromer, N. A., & Jhon, M. S., (eds.), *Nanotechnology for Sustainable Development* (pp. 321–329). Springer International Publishing, Switzerland.
Wanyika, H., Gatebe, E., Kioni, P., Tang, Z., & Gao, Y., (2012). Mesoporous silica nanoparticles carrier for urea: Potential applications in agrochemical delivery systems. *Journal of Nanoscience and Nanotechnology, 12*(3), 221–2228.
Weeks, D. P., Spalding, M. H., & Yang, B., (2016). Use of designer nucleases for targeted gene and genome editing in plants. *Plant Biotechnology Journal, 14*(2), 483–495.
Wen, L. X., Li, Z. Z., Zou, H. K., Liu, A. Q., & Chen, J. F., (2005). Controlled release of avermectin from porous hollow silica nanoparticles. *Pest Management Science: Formerly Pesticide Science, 61*(6), 583–590.
Wibowo, D., Zhao, C. X., Peters, B. C., & Middelberg, A. P., (2014). Sustained release of fipronil insecticide *in vitro* and *in vivo* from biocompatible silica nanocapsules. *Journal of Agricultural and Food Chemistry, 62*(52), 1204–12511.
Xiao, Q., Zhang, F., Wang, Y., Zhang, J., & Zhang, S., (2008). Effects of slow/controlled release fertilizers felted and coated by nano-materials on crop yield and quality. *Plant Nutrition and Fertilizer Science, 14*(5), 951–955.
Xie, J., Chen, T., Qing, C., & Song, H., (2010). Adsorption of organic by using rice husk ash of power generation residue. *Transactions of the Chinese Society of Agricultural Engineering, 26*(5), 283–287.
Yang, J., Cao, W., & Rui, Y., (2017). Interactions between nanoparticles and plants: Phytotoxicity and defense mechanisms. *Journal of Plant Interactions, 12*(1), 158–169.
Yang, Z., Chen, J., Dou, R., Gao, X., Mao, C., & Wang, L., (2015). Assessment of the phytotoxicity of metal oxide nanoparticles on two crop plants, maize (*Zea mays* L.) and rice (*Oryza sativa* L.). *International Journal of Environmental Research and Public Health, 12*(12), 15000–15109.
Yruela, I., (2009). Copper in plants: Acquisition, transport and interactions. *Functional Plant Biology, 36*(5), 409–430.
Yu, B., & Meyyappan, M., (2006). Nanotechnology: Role in emerging nanoelectronics. *Solid-State Electronics, 50*(4), 536–544.
Yuvaraj, M., & Subramanian, K., (2015). Controlled-release fertilizer of zinc encapsulated by a manganese hollow core shell. *Soil Science and Plant Nutrition, 61*(2), 319–326.
Yuvaraj, M., & Subramanian, K., (2018). Development of slow release Zn fertilizer using nano-zeolite as carrier. *Journal of Plant Nutrition, 41*(3), 311–320.
Zafar, H., Ali, A., Ali, J. S., Haq, I. U., & Zia, M., (2016). Effect of ZnO nanoparticles on *Brassica nigra* seedlings and stem explants: Growth dynamics and antioxidative response. *Frontiers in Plant Science, 7,* 535.
Zeng, X., Zhong, B., Jia, Z., Zhang, Q., Chen, Y., & Jia, D., (2019). Halloysite nanotubes as nanocarriers for plant herbicide and its controlled release in biodegradable polymers composite film. *Applied Clay Science, 11,* 20–28.
Zhang, P., Ma, Y., & Zhang, Z., (2015a). Interactions between engineered nanomaterials and plants: Phytotoxicity, uptake, translocation, and biotransformation. In: Siddiqui, M. H., Al-Whaibi, M. H., & Mohammad, F., (eds.), *Nanotechnology and Plant Sciences* (pp. 77–99). Springer International Publishing, Swtzerland.
Zhang, R., Zhang, H., Tu, C., Hu, X., Li, L., Luo, Y., & Christie, P., (2015b). Phytotoxicity of ZnO nanoparticles and the released Zn (II) ion to corn (*Zea mays* L.) and cucumber (*Cucumis sativus* L.) during germination. *Environmental Science and Pollution Research, 22*(14), 1109–1117.

Zhang, W., He, S., Liu, Y., Geng, Q., Ding, G., Guo, M., Deng, Y., et al., (2014). Preparation and characterization of novel functionalized prochloraz microcapsules using silica-alginate-elements as controlled release carrier materials. *ACS Applied Materials and Interfaces, 6*(14), 1183–11790.

Zhao, P., Cao, L., Ma, D., Zhou, Z., Huang, Q., & Pan, C., (2018a). Translocation, distribution and degradation of prochloraz-loaded mesoporous silica nanoparticles in cucumber plants. *Nanoscale, 10*(4), 798–1806.

Zhao, X., Meng, Z., Wang, Y., Chen, W., Sun, C., Cui, B., Cui, J., Yu, M., Zeng, Z., & Guo, S., (2017). Pollen magnetofection for genetic modification with magnetic nanoparticles as gene carriers. *Nature Plants, 3*(12), 956.

Zmeeva, O. N., Daibova, E., Proskurina, L., Petrova, L., Kolomiets, N., Svetlichny, V., Lapin, I., & Karakchieva, N., (2017). Effects of silicon dioxide nanoparticles on biological and physiological characteristics of *Medicago sativa* L. nothosubsp. *varia* (Martyn) in natural agroclimatic conditions of the subtaiga zone in Western Siberia. *Bionanoscience, 7*(4), 672–679.

Zuverza-Mena, N., Armendariz, R., Peralta-Videa, J. R., & Gardea-Torresdey, J. L., (2016). Effects of silver nanoparticles on radish sprouts: Root growth reduction and modifications in the nutritional value. *Frontiers in Plant Science 7,* 90.

CHAPTER 11

APPLICATIONS OF NANOMATERIALS IN AGRICULTURE AND THEIR SAFETY ASPECT

LEO BEY FEN,[1,2] AHMAD HAZRI ABD. RASHID,[3] NURUL IZZA NORDIN,[3] M. A. MOTALIB HOSSAIN,[2] SYED MUHAMMAD KAMAL UDDIN,[2] MOHD. RAFIE JOHAN,[2] and DEVARAJAN THANGADURAI[4]

[1]*Faculty of Medicine, University of Malaya, Kuala Lumpur–50603,Malaysia*

[2]*Nanotechnology and Catalysis Research Center, Institute for Advanced Studies, University of Malaya, Kuala Lumpur–50603,Malaysia*

[3]*Industrial Biotechnology Research Center (IBRC), SIRIM Berhad, Shah Alam, Malaysia*

[4]*Department of Botany, Karnatak University, Dharwad–580003, Karnatka, India*

11.1 INTRODUCTION

Agriculture plays a vital role to meet the growing demand for food worldwide. With the increasing trend, the global population is estimated to reach 9 billion by 2050. Hence, international organizations and agencies (e.g., UN Food and Agriculture Organization (FAO) and World Bank) are demanding for advanced researches to introduce innovative approaches to face the upcoming challenges of food shortage resulting from increased population. Nanotechnology, the emerging sector of the 21st century, is an emerging scientific approach with sophisticated devices capable of enhancing food productivity in a sustainable manner through efficient control of harmful pests and weeds (Marchiol, 2018).

Agriculture and food science have recently started their journey with nanoscience and technology with a view to bringing a revolutionary change (Kanjana, 2015). Given their small size, large surface to volume ratio, chemical reactivity, increased solubility, and unique magnetic and optical properties, nanoparticles (NPs) can exert sustainable positive effects on agricultural productivity through enhancing germination rate, improving gene delivery system, controlling agrochemicals delivery process as well as monitoring pest and disease occurrence (Kanjana, 2015). Due to their unique physicochemical properties, nanomaterials (NMs) have the potential to flourish agriculture and food sectors (Kaphle et al., 2018; Navya et al., 2018). The European Commission has categorized nanotechnology as one of its six "Key Enabling Technologies" in recognition of its contribution for sustainable growth in a large number of sectors of industrial application (Parisi et al., 2015).

Various types of ENMs are available worldwide, including metal, metal oxide, polymeric NPs and silicates, quantum dots, nanoemulsions, nanotubes, nanoliposomes, nanosensors, and nanofibers to be used as building blocks for making new structures and introducing novel properties in the nanoscale level (Kanjana, 2015). Researchers in the field of biology and chemistry are continuously trying to synthesize organic, inorganic, hybrid, and metal ENMs as well as various types of NPs possessing unique optical, physical, and biological properties. Given these exceptional properties, NPs have versatile applications in a number of fields (Rai and Ingle, 2012). The remarkable success in the application of NPs has been documented through achieving increased percentage in seed germination (Nair et al., 2010; Gopinath et al., 2014; Gowri et al., 2014), enhanced root and shoot length (Liu et al., 2005; Hafeez et al., 2015; Razzaq et al., 2015), increased fruit production (Kole et al., 2013), increased content of phytomedicine (Kole et al., 2013), and a remarkable increase in vegetative biomass of seedlings (Misra et al., 2016; Shukla et al., 2016) (Figure 11.1).

With the advancement in nanotechnology, a variety of modern techniques are introduced to improve precision farming practices which may help precise control at nanometer scale (Ditta, 2012). Nanotechnology plays a vital role in increasing the life span, freshness, and overall quality of food through resisting gas penetration, enhancing the tensile strength, and increasing micronutrient and antioxidant absorption (Kanjana, 2015). Due to insufficiency of fuel like natural gas and petroleum, the production cost of chemical fertilizers and pesticides tends to increase day by day. Thus, the problem can be effectively solved through precision farming using nanotechnology, resulting in reduced production costs with maximm output.

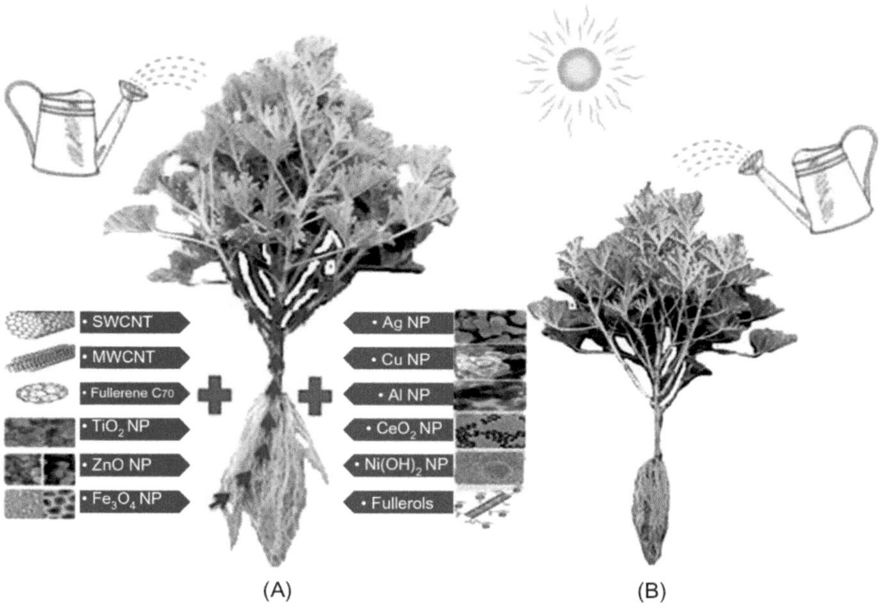

FIGURE 11.1 Effect of different NPs in plant growth. (A) Representing enhanced growth of plant treated with NPs in addition to other basic requirements in comparison to (B) plant representing growth in the absence of NPs.

Source: Reproduced with permission from: Mishra et al. (2014); Copyright © Elsevier (2014).

Besides, nanotechnology has also been applied in the agriculture sector including treatment of plant diseases, quick identification of pathogens utilizing nano-based kits, and enhancing the plants' capability to absorb nutrients and thus might bring a radical change in the existing agricultural and food industry (Srivastava et al., 2018). Nanotechnology may also help in revolutionizing agriculture and the food industry through the innovation of novel techniques to use inputs more efficiently, withstand environmental pressures, and improve systems for processing (Singh et al., 2015). It is expected that nano-structured catalysts would enhance the efficacy of commercial pesticides and insecticides as well as decrease the level of doses required for plants in the near future (Joseph and Morrisn, 2006).

11.2 APPLICATION OF ENGINEERED NANOMATERIALS (ENMS) IN AGRICULTURE RELATED AREAS

11.2.1 NANOMATERIAL IN AGRICULTURE APPLICATION

In the agricultural and food production sector, the need for new innovations is urgently needed due to strong concerns in food security for the rapidly increasing world population and challenges and issues in climate change and environmental degradation. Recently, extensive research has been made in nanotechnology for the improvement in agricultural yield. Nanotechnology has successfully been used in diverse fields of science and the promising results in many other fields encouraged the researchers to continuously look out the potential scopes in the agriculture areas.

11.2.1.1 EFFECTS OF NANOPARTICLE IN PLANT GROWTH

The effect of different NPs on the growth and metabolic functions of plants is being studied extensively nowadays. Plants' cell walls, by serving as a barrier, inhibit the entry of any external agent like NPs into plant cells. The majority of cell wall components is composed of carbohydrates and proteins (Heredia et al., 1993; Knox, 1995), and being semi-permeable, this allows the entry of only small molecules while the larger ones are blocked. The sieving properties are determined by the pore diameter of the cell walls having a thickness between 5 nm and 20 nm (Fujino and Itoh, 1998; Zemke-White et al., 2000). Thus, NPs, smaller than the size of the largest pore can enter the cell wall very easily and reach the plasma membrane. Upon interaction with NPs, the pores may be enlarged, or new cell wall pores may be induced, thus NPs uptake through the cell wall is increased. For example, permeability was found to be increased with the use of ZnO NPs and even "holes" were created in bacterial cell walls with pore size similar to plant cell walls (Carpita et al., 1979; Brayner et al., 2006; Ahmed et al., 2013). After passing through the cell wall, the NPs reach the plasma membrane. Consequently, a cavity-like structure is formed surrounding the NPs and it is pulled by the plasma membrane into the cell during the endocytic process. With the use of embedded transport carrier proteins or ion channels, the NPs able to cross the cell membranes. After entering the cell, the NPs may bind with various types of organelles (e.g., Golgi, endoplasmic reticulum, and endolysosomal system), interfering the metabolic activities due to the formation of reactive oxygen species (ROS) (Jia et al., 2005; Ahmed et al., 2013).

Furthermore, plants are exposed to ENMs available in atmosphere as well as those in terrestrial environments. Nanomaterials available in the air interact with leaves and other aerial parts of plants. On the other hand, the soil-material-associated ENMs are attached to the roots. Hence, it is assumed that the plants with higher leaf area indexes also have a higher interception possibility for the ENMs available in air, and these ultimately enhance their entrance into trophic webs. After being applied on the leaf surface, the ENMs may penetrate the plants *via* the bases of trichomes or through the openings of stomata and are translocated to various tissues. Due to the obstruction in stomata, the accumulation of ENMs in photosynthetic surfaces may cause provocation of foliar heating resulting in a change of the gas exchange. This heating may occur some alterations in certain physiological and cellular functions of plants (Da Silva et al., 2006; Ahmed et a., 2013).

11.2.1.2 APPLICATIONS OF ENMs IN CROP IMPROVEMENT

11.2.1.2.1 Application in Seed Germination

Researchers have studied the influence of various types of ENMs on seed germination and plant growth in the production of crops. The influence of metal and metal oxide NPs such as zinc (Zn), gold (Au), titanium (Ti), aluminum (Al), silver (Si) and copper (Cu) is very high on the seed germination and plant growth (Monica and Cremonini, 2009). Khodakovskaya et al. (2009) documented that multi-walled carbon nanotubes (MWCNTs) could penetrate tomato seeds and through the increase in water uptake, they could increase the germination rate (Khodakovskaya et al., 2009). The seed germination was increased by 90% by the MWCNTs in comparison with the control (71%) and plant biomass in 20 days. The seed germination process can be accelerated, and germination time can also be shortened remarkably by the addition of carbon nanotubes (CNTs) to agar medium (Figure 11.2). For 20 days, seed germination was 71% in the case of regular medium; on the other hand, after supplementation with MWCNTs, it was increased to 90%. Additionally, it was noticed that CNTs are capable of penetrating both the seeds and the root systems of plants. This reveals that seeds able to uptake CNTs and their biological activities are significantly affected by CNTs, increasing the amount of water penetrating inside the seed during the period of germination (Khodakovskaya et al., 2009). Mondal et al. (2011) also obtained a similar result for mustard plant, but the exact mechanism of uptake of water is not fully established yet (Mondal et al., 2011). The effect of TiO_2 nanoparticles (NPs) on the growth of naturally-aged spinach seeds was studied. The growth rate of spinach seeds was found to be inversely correlated with the size, reflecting that the smaller the ENMs, the better the seed germination (Zheng et al., 2005; Khot et al., 2012). The authors suggested that the TiO_2 NPs might be responsible for stress resistance of the seed and help in fast germination by promoting capsule penetration for intake of water and oxygen. On the other hand, Lin and Xing (2007) studied the phytotoxicity of ENMs (MWCNTs, ZnO, Al_2O_3, Zn, and Al) on the germination rate of radish, canola, rape, corn, ryegrass, cucumber, and lettuce. They observed that the use of increased concentrations (2000 mg/L) of nano-sized ZnO (~20 nm) and Zn (35 nm) caused the inhibition of the germination in corn and ryegrass, respectively. Magnetic NPs are of special interest because these particles can be specifically localized to release their load, which is significantly important for the nanoparticulate delivery to be applied for plants. Some researchers have studied magnetic NPs in connection with the uptake, translocation, and specific localization in pumpkin plants. No negative effect was observed on plant growth indicating the safety of such NPs for nanoparticulate delivery in plants (González-Melendi et al., 2007; Zhu et al., 2008; Corredor et al., 2009). The nanobiologists have been increasingly interested in the genetic effect of ferrofluids because it causes chromosomal aberrations in young plants (Pavel and Creanga, 2005; Racuciu and Creanga, 2007). The effect of magnetic NPs coated with tetramethylammonium hydroxide has been studied by Racuciu et al. (2007) on the growth of *Zea mays* plant during early ontogenetic stages (Racuciu and Creanga, 2007). They observed that these NPs show the magnetic effect on the enzymatic structures during several phases of photosynthesis. Low concentration of ferrofluid enhanced the level of 'chlorophyll,' nevertheless, at higher concentrations of ferrofluid, it was inhibited. Therefore, the knowledge of the appropriate ranges of ferrofluid concentration is particularly important in order to achieve a better yield with improved photosynthetic pigment levels. Zhu et al. (2008) revealed that during the use of magnetite NPs in an aqueous medium to grow pumpkin plants (*Cucurbita maxima*), the absorption, movement, and accumulation of particles occur in the plant tissues. However, lima bean plants (*Phaseolus limensis*) are unable to absorb the magnetite NPs because it was observed that no magnetic signal was produced in any parts of those plants. Therefore, there are different responses of different plants to the same NP (Zhu et al., 2008; Ahmed et al., 2013).

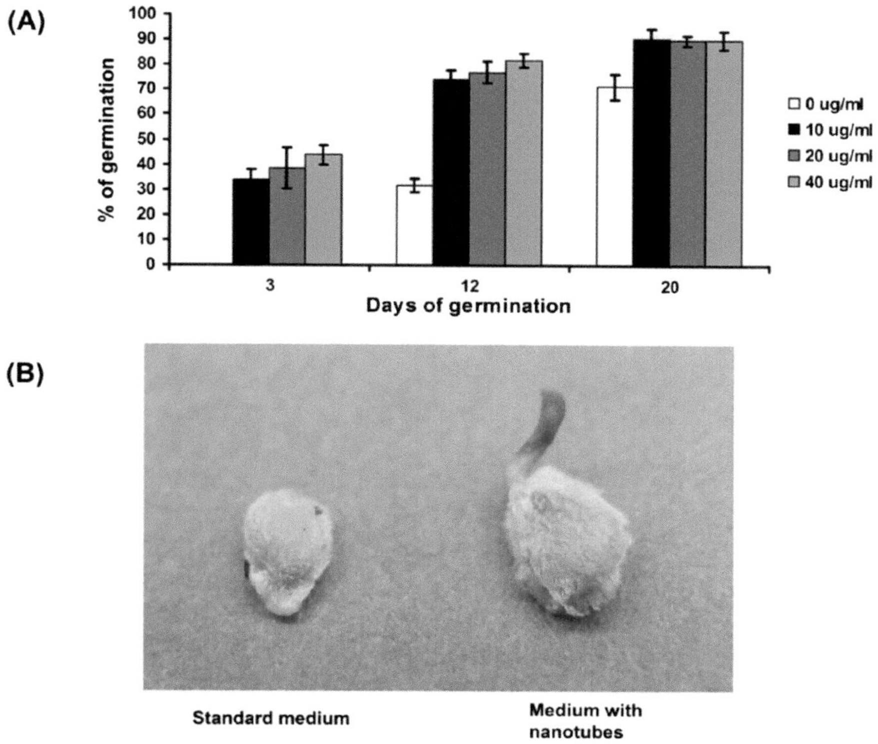

FIGURE 11.2 Effect of CNTs on tomato seed germination: (A) Time of germination and germination percentages of seeds incubated with and without CNTs during 20 days. Seedlings with developed cotyledons and root systems were recognized as fully germinated in this experiment. (B) The phenotype of tomato seeds incubated during three days without (left) or with (right) CNTs on MS (Murashige and Skoog) medium.

Source: Adapted with permission from: Khodakovskaya et al. (2009); Copyright © American Chemical Society (2009).

11.2.1.2.2 Application in Genetic Modification

Biotechnological research has been emphasizing on the improvement of plant resilience against a variety of environmental stresses, including diseases, drought, salinity, and others. Nanotechnology may contribute to the advancement of agricultural productivity through genetically improving plants, modification of the genetic constitution of crop plants, distribution of drug molecules and genes to the specific sites at cells, and gene technologies based on nano-array for gene expressions in plants during conditions of stress. Nanobiotechnology studies have been successfully utilizing DNA molecule to develop pathogen, pest, and stress-resistant strains of crop plants through gene delivery system (Kanjana, 2015). Plant genetic transformation through viral gene delivery vector, *Agrobacterium*-mediated transformation, microinjection, and microprojectile bombardment faces some constraints including decreased utilization efficiency of transportation or applicability only for dicotyledons (Gelvin, 2003). Therefore, these impediments could be overcome by using functionalized ENMs as a vehicle that can carry various genes and substances capable of triggering gene expression or controlling the continuous release of genetic material in plants, especially in both monocotyledonous and dicotyledonous (Sivamani et al., 2009; Nair et al., 2010; Gruère et al., 2011). DNA-coated NPs have easy access into the plant cells given their smaller size and increased transformation efficiency. Moreover, there is the likelihood of minimum cell damage in NP applied plants with enhanced plant regeneration. Several NPs have been applied to deliver DNA genetic material into plant cells. For example, gold nanoparticles (Au NPs, 5–25 nm) embedded carbon matrices (Vijayakumar et al., 2010), poly-*L*-lysine coated starch NPs (50–100 nm) conjugated with fluorescent material $Ru(bpy)_3^{2+}\cdot 6H_2O$ (Liu et al., 2008). With the use of honeycomb mesoporous silica NPs (MSN) system having 3-nm pores, DNA, and chemicals can be transported into isolated plant cells and intact leaves. In an experiment, MSN was loaded with the gene and its chemical inducer and the ends were capped with Au NPs to avoid leaching out of the molecules. By uncapping the Au NPs, the chemicals were released, and gene expression was triggered in the plants in a controlled release manner (Torney et a., 2007).

11.2.1.3 APPLICATIONS OF ENMs IN CROP PRODUCTION

Nutrient management planning is considered as the best agriculture management practice to improve crop yield and quality through ensuring added nutrient availability to crops, minimizing fertilizer input costs, and protecting soil. Fertilizers play a crucial role to enhance the agricultural production, but there is still low nutrient use efficiency due to several factors, including denitrification, leaching, microbial immobilization, fixation, and runoff. Generally, a large portion of the applied conventional fertilizers remains unused or without being absorbed by plants (about 80–90% of phosphorus, 40–70% of nitrogen, and 50–70% of potassium) resulting in huge economic and resource losses as well as severe environmental pollution. Therefore, nanotechnology tries to introduce a revolution in fertilizer application practices by converting conventional fertilizer to 'smart fertilizer' through the development of efficient and smart delivery systems. This may allow slow or controlled release of fertilizers by utilizing nano-encapsulation technique. For instance, diverse approaches and concepts have been practiced in various aspects, including slow release of fertilizers, transformation of insoluble fertilizers into soluble fertilizers by using nanoscience and smart sensing techniques (Wu and Liu, 2008; Kanjana, 2015).

Since the size of NPs markedly determines the behavior, the reactivity, and the toxicity of NPs, it is not unusual to observe both positive and negative effects of NPs on higher plants (Monica and Cremonini, 2009). Ma et al. (2010) evaluated the effects of four oxide NPs (CeO_2, Gadolinium (III) oxide-Gd_2O_3, Lanthanum (III) oxide-La_2O_3, Ytterbium oxide-Yb_2O_3) on the rape, radish, tomato, wheat, lettuce, cucumber, and cabbage plants (Ma et al., 2010). The authors suggested that the nano-CeO_2 affected root elongation only in lettuce at 2,000 mg/L concentration but not in other plant species. Nevertheless, root growth was significantly affected by the other three kinds of NPs (Gd_2O_3, La_2O_3, Yb_2O_3) at the same concentration. In addition, these NPs also showed inhibitory effects during different phases of root growth. Besides, Lin and Xing (2007) also evaluated the inhibitory effects of five types of NPs (alumina, Al, Zn, ZnO, and MWCNT) on root growth of six higher plant species (cucumber, ryegrass, corn, radish, lettuce, and rape) and observed that only Zn and ZnO particles among the NPs strongly inhibited the root growth of plants (Lin and Xing, 2007). Thus, the phytotoxic nature of the ENMs needs to be clearly understood prior to their utlization.

11.2.1.4 APPLICATIONS OF ENM IN WEED CONTROL

Weed management is a matter of increasing concern in agriculture since herbicides are not efficient enough in removing multi-weed species, and herbicide-resistant weeds are increasingly emerging due to repeated exposure of single herbicide. This ultimately causes the overall loss of crop of about 34% in consideration with the loss caused by other environmental factors including diseases, pests, factors related to soil and crops (Oerke, 2006). Herbicides are usually applied through foliar spray which is unable to destroy the weeds completely. However, these herbicides eventually lead to destruction of the structure and function of plant-specific chloroplast, inhibition of biosynthesis of lipids, interference with cell division through disruption of the mitotic sequence or inhibition of the plants (Wakabayashi and Böger, 2004). Therefore, following an easy process, encapsulated herbicide molecule can be more successfully applied *via* root absorption in comparison with foliar absorption because roots do not contain cuticles as do leaves. By encapsulating with NPs, a target-specific herbicide molecule was developed, which particularly aims for specific receptor in the roots of target weeds. After entering into the root system of the weeds, it is transferred to parts where there is inhibition of glycolysis of food reserve, and due to food shortage, the specific weed plant dies of starvation. Weed residues should be detoxified because extensive use of herbicides for a long time may leave the residues in the soil that may be harmful for succeeding crops (Alharby et al., 2019).

Herbicides attached with ENMs helps in facilitating the higher bioavailability, thereby ensuring effective eradication of weeds. There is adsorption, attachment, encapsulation, or entrapment of the active ingredient unto or into the nano-matrix. In achieving controlled release of the active ingredient, the factors that play an important role are slow-release properties of ENMs, attachment of the ingredients to the material and the environmental conditions. ENMs-based formulations contribute to improved efficacy due to their increased surface area, increased mobility, increased solubility, and decreased toxicity due to elimination of organic solvents (Alharby et al., 2019).

A nanoherbicide is composed of either minute particles of active components of herbicide or other minutely modified structures containing herbicidal properties. Nanomaterials and bio-composites contain salient properties including thermal stability, solubility, permeability, stiffness, crystallinity, and biodegradability, which are essential for the formulation of nanoherbicides. Nanoherbicides provide increased specific surface area, thus allowing enhanced affinity to the target. Nanoencapsules, nanoemulsions, nanocages, and nanocontainers are the examples of nanoherbicide delivery techniques to control or delay the delivery (Roy et al., 2014; Khatem et al., 2016). To overcome the present constraints regarding droplet size, NM-encapsulated or nanosized herbicides might be applied in order to have efficient spraying and reduce spray drift and splash losses.

Overall, the basic criteria of nanoformulations should be the capability to degrade faster within the soil while at a slower rate within the plants. For instance, the use of sodium dodecyl sulfate (SDS) enhances the photodegradation of NPs in the soil. To develop an SDS-modified Ag/TiO_2 imidacloprid nanoformulation, a microencapsulation technique was applied using chitosan and alginate (Baker et al., 2017). A number of other NMs may also be tried for successful development of nanoherbicides; for example, polymer stabilizers such as polyvinylpyrrolidne (PVP), poly(acrylic acid)-β-poly(butyl acrylate) (PAA-β-PBA) and polyvinyl alcohol (PVOH) have been used to synthesize bifenthrin, an efficient nanopesticide (Maarouf et al., 2016; Alharby et al., 2019).

11.2.1.5 APPLICATIONS OF ENM IN CROP PROTECTION

11.2.1.5.1 Application in Pest Control

Nowadays, the agriculture industry is facing challenges on the pest management program because of improper diagnosis of pest occurrence, lack of efficacy of pesticides, development of resistance to pesticide of the existing group, spray drift, and emergence of new pests. To overcome the existing impediments, nanotechnology might be useful to achieve enhanced efficacy of conventional agrochemicals due to their increased surface area, increased solubility, increased mobility, induction of systemic activity, and decreased toxicity (Sasson et al., 2007). The application of pesticides through normal spray involves high volume and low-value preparation of agrochemicals, while nanotechnology supported preparations are associated with low volume and high-value chemicals (Ghormade et al., 2011). Therefore, it is crucial to adopt a controlled or smart delivery system for applying targeted pesticide with an aim to release required and sufficient quantities of agrochemicals over a longer period of time, to achieve the maximum biological efficacy and to minimize the detrimental effects (Tsuji, 2001).

The basic criteria of the nano-formulations are faster degradation in the soil and at a slower rate in plants, while the residue levels should be below the level of regulatory criteria in foodstuffs. Photocatalytic TiO_2/AgNMs modified with SDS and conjugated with dimethomorph (DMM) has been successfully utilized in agriculture as a nanopesticide (Yan et al., 2005). The modified formulation with 96 nm average granularity, increased dispersivity and its decomposition in soil enhanced the efficacy of the pesticide in vegetable seedling (of cabbage and cucumber) growth. Through the modification of the ENMs using SDS, the absorption of the DMM was significantly increased. Guan et al. (2010) formulated encapsulated nano-imidacloprid having the above properties which might be applied for controlling pests in vegetable production (Guan et al., 2010). For the development of SDS modified Ag/TiO_2 imidacloprid nanoformulation, microencapsulation technique was applied with the use of chitosan and alginate. It was validated on soybean plants after transplantation to soil having 3.1% dry matter content and pH of 6.2. The degradation of formulation residues was faster in soil and the plants during the first 8 days. However, after 20 days, it was minimal to undetectable (Khot et al., 2012).

Nano-formulated pesticides and other agrochemicals have been employed to improve production efficacy in comparison with conventional formulations (Joseph and Morrison, 2006; Kah et al., 2013; Sarkar et al., 2015). To achieve optimum efficiency, the size of active ingredients and other compounds are reduced to the nano-range, mixed into nano-emulsions or nano-dispersions, or incorporated in solid lipid or polymeric nano-capsules (Frederiksen et al., 2003). The influence of nano-encapsulates based on chitosan, a polysaccharide derived from chitin, has been observed in a number of agrochemicals (Kashyap et al., 2015). Solid lipid and polymeric nano-capsules attached with carbendazim and tebuconazole have been reported as a strong fungicide formulation. The study revealed that the nano-capsules showed a high loading capacity with a slow release of the fungicides having minimum adverse effect on plant growth in comparison with the available commercial products (Campos et al., 2015).

Several polymeric NPs have been designed for the efficient release of agrochemicals. Encapsulation of bifenthrin was tried by Liu et al. (2008) with the use of polymers such as poly(acrylic acid)-β-poly(butyl acrylate), polyvinyl alcohol (PVOH) (Liu et al., 2008). Mesoporous silica NPs have been successfully used for encapsulating the agrochemicals and delivering the active ingredient in a controlled manner. To serve as an insecticide, surface-functionalized hydrophobic silica gets adsorbed into the cuticular lipids of insects causes the killing of insects through desiccation (Mewis and Ulrichs, 2001). Debnath et al. (2011) also showed a similar effect against rice weevil (*Sitophilus oryzae*) (Debnath et al., 2011). For delivery of water-soluble pesticide validamycin, it was loaded with porous hollow silica NPs (PHSNs). A high loading capacity (36% w/w) of validamycin was achieved with much higher efficiency than that of simple immersing method. Thus, PHSNS can be used as a promising carrier, particularly for controlled delivery of pesticides where immediate as well as prolonged-release are required (Liu et al., 2006).

Moreover, the use of CNTs has been reported to control plant bacterial diseases, thereby increasing crop productivity (Wang et al., 2013). CNTs may also be applicable as sensors for detecting pathogens in plants and the environment. The suitable and specific sensor has been developed to detect and quantify *Botrytis cinerea* in apparently healthy apple plants before the manifestation of disease symptoms. The highly specific and sensitive sensors could provide early signal about invading pathogens in plants (Fernández-Baldo et al., 2009; Kaphle et al., 2018). For the control of fruit pests, nano-gels containing the pheromone methyl eugenol are considered as a cost-effective approach, allowing lower application doses with full efficacy (Bhagat et al., 2013). Naturally found nanosized ashes and inorganic metal NPs are reported to show antimicrobial, insecticidal or antifungal characteristics (Stadler et al., 2010; Yildiz and Pala, 2012). NPs coated with poly-ethylene-glycol and loaded with garlic essential oils are proved to be effective in controlling stored-product pests (Yang et al., 2009). Nanotechnology not only protects crops and food products from harmful effects of pests but also fosters crop yield and quality, reducing the required quantity of fertilizers to apply (Peters et al., 2016).

11.2.1.5.2 Applications in Disease Control

Plant diseases are one of the main causes of the reduction of crop productivity in the agriculture sector. In most cases, the application of pesticides and fungicides are for precautionary measurement which ultimately results in residual toxicity and environmental hazards and may lead to crop yield losses to some extent (Chinnamuthu and Boopathi, 2009). Recently, nanotechnology has been proved as an effective and alternative tool in plant disease management through control of plant pathogens. Some of the NPs (e.g., carbon, silica, Ag, and alumina silicates) might be more effective in controlling plant diseases as compared to systemic fungicides (Prasad et al., 2014).

Silver usually exerts a few adverse effects on microorganisms, interfering with some biochemical activities in the microorganisms and affects plasma membrane (Pal et al., 2007; Jo et al., 2009; Panáček et al., 2009). It also inhibits the expression of ATP production associated proteins (Yamanaka et al., 2005). Silver NPs (Ag NPs) have been considered effective to control various plant diseases for possessing bactericidal and broad-spectrum antimicrobial effect. In recent years, the efficacy of Ag NPs in controlling many plant pathogens and bacterial strains has been documented (Gajbhiye et al., 2009; Mohammed Fayaz et al., 2009). Ag NPs having increased surface area and increased fraction of surface atoms show higher antimicrobial activity as compared to the bulk ones (Suman et al., 2010). The antimicrobial activity is particularly determined by the release of Ag^+ ions that attach with electron donor groups in molecules having oxygen, nitrogen, or sulfur. Bactericidal properties specifically depend on size; the increased activities are observed with decreasing size and increasing surface area of NPs (Morones et al., 2005). Furthermore, Ag NPs are also found potentially effective against common plant pathogenic fungi including *Phoma herbarum*, *Phoma glomerata* and *Fusarium semi-tectum*; combination of Ag NPs with antifungal drug fluconazole produced higher impact. Silver antimicrobials could easily be integrated within various materials like plastics, wares, and textiles, making it more advantageous. This offers an extra advantage, and they can be utilized as disinfectant for a long time when traditional antimicrobials would not be successful (Kaphle et al., 2018).

Silicon (Si) has been found to be absorbed by plants, inducing the resistance to disease and stress, and improving the physiological activity and growth of plants. Aqueous silicate solution is known to provide strong prevention against pathogenic microorganisms causing powdery mildew or downy mildew in plants. By combining nanosized silica and Ag, plant diseases might be controlled effectively (Sharon et al., 2010; Agrawal and Rathore, 2014). Likewise, Zinc oxide (ZnO) and magnesium oxide (MgO) NPs exhibit effective antibacterial and anti-odor effects (Wani and Shah, 2012). ZnO NPs have inhibitory effects on the fungal growth of *B. cinerea* and *Penicillium expansum* through affecting cellular functions and thereby causing deformation of the structure of fungal hyphae (He et al., 2011). Similarly, the antibacterial activities of MgO NPs were investigated against *Escherichia coli* and *Salmonella* by Jin and He (2011). It was observed that the MgO NP showed significant bactericidal effect by distorting and damaging the cell membrane, which resulted in the leakage of intracellular contents and thus ultimately caused the death of bacterial cells (Jin and He, 2011).

Abd-Elsalam (2013) documented that ZnO NPs inhibited the growth of conidiophores and conidia of *Penicillium expansum* and that ultimately caused the death of fungal mats (Abd-Elsalam, 2013). $TiO_{2,}$ in addition to the growth-enhancing effect, also shows strong properties of antifungal and antibacterial activity. TiO_2 was found efficient to control *Curvularia* leaf spot and bacteria leaf blight disease in rice plant and maize. The incidence of rice blast was also markedly decreased with 20% increase in grain weight because of the growth-promoting effect of TiO_2 NPs (Mahmoodzadeh et al., 2013).

A milestone in disease management program is the introduction of an alternative strategy through the innovation of nano-based diagnostic kit which offers detection of molecular abnormalities both at the genomic and biochemical level and identification of exact viral strains. Generally, disease diagnosis faces difficulties mainly due to the extremely low concentrations of biochemical analytes as well as very minute amounts of detectable virus, fungi, or bacteria (Misra et al., 2013). Moreover, the conventional techniques often take a long time to detect. The current nano-based diagnostic kits are designed to increase the detection speed and sensitivity. Biosystems are equipped with functional nanometric devices with the use of specific enzymes, nucleic acids and proteins which efficiently detect vital parameters in plants (Chinnamuthu and Boopathi, 2009; Kanjana, 2015).

11.2.1.6 APPLICATIONS OF ENM IN FOOD SCIENCE

11.2.1.6.1 Application in Food Additives and Food Supplements

In the food sector, engineered nanomaterials (ENMs) are used as food additives to improve food stability during processing and storage, enhance characteristics of product, or increase the efficacy and bioavailability of nutrients in the food. Among ENMs-based food additives, synthetic amorphous silica (SAS) is the most common type. SAS is generally used as clarifying agent for beverages, while as an anti-caking and free-flow agent in several food products in powdered form (E551) (Dekkers et al., 2011). Besides SAS, several formulated anti-caking agents have been utilized in food items, including calcium silicate, dicalcium phosphate, sodium ferrocyanide, sodium aluminosilicate, and microcrystalline cellulose. However, there is a lack of concrete evidence whether such materials are (partly) available at the nanoscale (Peters et al., 2016).

Titanium dioxide (TiO_2) (E171) is another useful food additive used mainly as a pigment to intensify the white color of some food items, including dairy products and candy (Weir et al., 2012). The use of TiO_2 as a food additive and flavor enhancer is also popular in several non-white food items including nuts, dried vegetables, soups, seeds, mustard, wine, and beer. Dietary intake of TiO_2 in foods is increasing with time and recent research revealed that 5–36% of the TiO_2 in available food items was found to be present in the range of nano-size (Weir et al., 2012). The nano form of TiO_2 has been used as an antimicrobial agent. Moreover, by combining with other compounds or elements like cobalt and nickel oxide, TiO_2 is used to inactivate foodborne pathogens (Amna et al., 2013). However, there was a re-evaluation by the

European Food Safety Authority (EFSA) regarding the safety of E171 for use as food additives and in feed (EFSA, 2016).

In recent years, nanotechnology has been applied in food processing, e.g., nanocarrier system to deliver nutrients and supplements, organic additives, supplements, and animal feed. Natural NPs such as casein presents in milk at a nanoscale and protein filaments of meat (Prasad et al., 2017). Many biomolecules, including carbohydrates, lipids, vitamins, and proteins are sensitive to high acidic environment and enzymatic activities of the stomach and duodenum. Being encapsulated, these biomolecules become enabled to resist such unfavorable conditions and at the same time can be assimilated instantly in food products, but in the case of non-capsulated structure, it is difficult to achieve because of lower water-solubility of these compounds. Nanoparticles-based mini edible capsules may provide remarkable health benefits through improving the delivery of vitamins, medicines, or fragile micronutrients in the daily foods (Singh et al., 2017).

Nowadays, several nutrients mostly vitamins are encapsulated and then delivered into the bloodstream through the digestion system. Fortification of many foods and drinks is successfully done with these NPs to maintain the original taste or appearance. The ice cream industry also uses NP emulsions to improve the texture and uniformity of ice creams. For instance, encapsulated omega-3 fatty acids are produced by KD Pharma Bexbach GmbH (Germany) in two distinct forms-powder and suspension, whereas the derived particles are used by the capsulation technology in nanoscale and microscale (Berekaa, 2015; Prasad et al., 2017).

11.2.1.6.2 Application in Food Packaging

Applications such as food packaging, storing boxes, crockery, cooking equipment, coatings of machines and surfaces, and nano-sieves/membranes can be improved using nanotechnology. One of the most remarkable nanotechnological applications in the food sector is to incorporate NMs as packaging materials or storage containers so that the storage time could be increased while keeping the products fresh. Nanocomposites also assist in reducing the permeability of food packaging materials to atmospheric oxygen, carbon dioxide or moisture (Peters et al., 2016). Thermoplastic polymers are usually used having nanoscale inclusions, most of which are composed of 2–8% (w/w) nanoclays, mixed with polypropylene-based nanocomposite or polyamide plastic films (e.g., Durethan® for meat-packaging) (Brody et al., 2008).

Some other fillers are also used, including CNTs as well as nanostructures of metal and metal oxide, polymeric resins or cellulose and fibers (Ramachandraiah et al., 2015; Shatkin and Kim, 2015). Besides, nanoclays have an extra advantage of improving the tensile strength and thermal properties, whereas polymer-clay nanocomposites have the potential to be used as novel food packaging material.

Thermal buffering is another important property of materials composed of polyamide layered with silicate barriers (Johnston et al., 2008). Nanocomposite films improved with silicate NPs or nanocrystals exhibit enhanced barrier properties suitable for use in plastic beer bottles (e.g., Nanocor®) and the use started in the US, but yet to start in the EU (Peters et al., 2016). Some examples of natural biopolymer-based nanocomposite packaging materials having bio-functional properties are cellulose, chitosan, and carrageenan (Sanchez-Garcia et al., 2010; Shatkin and Kim, 2015; Yang et al., 2015). Food packaging industries use CNTs for the improvement of mechanical properties and development of antimicrobial effects (Rungraeng et al., 2012). One example of ENMs incorporation in biodegradable food packaging is Ag-TiO_2 NPs in biodegradable polylactide (PLA) composite films. They were found to demonstrate not only the improvement in mechanical, thermal, and barrier properties but also the increase in the photodegradability of the packaging material due to their ultraviolet radiation absorption (Pillai et al., 2013). Moreover, phase change materials incorporated into nano- and micro-sized polycaprolactone, polystyrene or poly-lactic beads help to increase the thermal buffer capacity of the food packaging system thereby allowing a better maintenance of the cold chain (Pérez-Masiá et al., 2014; Peters et al., 2016).

Some examples of ENMs incorporated into food packaging to provide antimicrobial properties are chitosan (Cruz-Romero et al., 2013), ZnO (Tankhiwale and Bajpai, 2012; Cruz-Romero et al., 2013), Ag (Popov et al., 2015; Carbone et al., 2016), Au (Thirumurugan et al., 2013), TiO_2 (Chawengkijwanich and Hayata, 2008; Yemmireddy et al., 2015) and nisin (Cutter et al., 2001; Mauriello et al., 2005). Chitosan is an excellent ENMs suitable for antimicrobial food packaging, and it has a huge potentiality for versatile applications due to its biodegradable, biocompatible, non-toxic, and antimicrobial properties. Nanostructured TiO_2 has strong photocatalytic activity under UV irradiation and has been reported to be efficient in eradicating pathogenic microorganisms in food contact surfaces, thus increasing food safety. ZnO is also effective in protecting products from foodborne pathogens (Peters et al., 2016). ZnO-loaded polyethylene films, coated with starch as supporting matrix have been studied against *E. coli*, and they were found potentially applicable in food-packaging materials to control bacterial growth (Tankhiwale and Bajpai, 2012). A combination of ZnO and copper oxide (CuO) NM shows specific toxicity against bacteria with minimal health effects on humans and therefore is being considered for effective applications in the food sector (Ravishankar Rai and Jamuna Bai, 2011).

There are several options by which Ag nano-particles could be added to food packaging materials, for example, in the form of finely dispersed nanosilver particles embedded in containers and coatings (Popov et al., 2015; Carbone et al., 2016), as gelatin-silver nano-composites (Halder et al., 2011) or as a silver-based zeolite in polylactic acid (PLA) biocomposites (Matsumura et al., 2003; Busolo et al., 2010). The use of Ag NPs as food contact materials is also common in preventing food fouling and increasing the shelf life of food products (Kaphle et al., 2018). Some products that contain silver-modified inorganic carriers are available in the market for food contact materials, e.g., Agion®, Bactiblock®, Novaron®, Zeomic®. However, the use of (nano) Ag is not permitted in the EU for plastic materials that are supposed to come into contact with food (Peters et al., 2016). Functionalized Au NPs may be applied in packaging materials for the prevention of microbial contamination of food because of specific surface chemistry. They may be restructured into composites with polymers and may act as barrier for avoiding moisture exchange and food fouling (Thirumurugan et al., 2013; Kaphle et al., 2018).

11.2.2 AGRICULTURE SUSTAINABILITY AND SECURITY

The limitation of natural resources, e.g., cultivable land, water, soil, etc., as well as population growth in the world necessitates the agricultural development to be economically sustainable, viable, and environmentally friendly. Agricultural sustainability could be achieved through innovation, utilization, and implementation of novel techniques of farming which can increase overall crop production to meet increasing demands of a growing population while at the same time they would play a vital role in conserving and protecting the environment and its natural resources. The sustainable agricultural development is closely related to climate changes, socio-economic condition, public health, nutrition, and ecosystem strategies, energy supply, natural resources, etc., and all of these must be addressed with proper coordination to reach target-oriented goals. Therefore, sustainable agriculture is necessary to strengthen the available opportunities to get rid of poverty and hunger. Although global food production has been satisfactory for the last 50 years, existing approaches and strategies for agricultural advancement are not up to the mark to achieve food security and ensure agricultural sustainability (Arora, 2018).

The nanotechnology can play an important role in productivity through control of nutrients as well as through monitoring water quality and pesticides. In the agri-food sectors, target-oriented applications of nanotechnology have been proved useful in maintaining soil fertility. In recent years, nano-sensors have been widely applied in the field of agriculture because of their efficiency in quick environmental monitoring of contamination in soil and water. Nanotechnology-based sensors like biosensors, optical sensors, electrochemical sensors, and devices are applicable to detect the heavy metals in trace range (Prasad et al., 2017). Nanomaterials are able to help in improving the efficacy of microorganisms to degrade waste and toxic materials. Bioremediation uses living organisms to break down or remove toxins and harmful substances from agricultural soil and water (Dixit et al., 2015). Thus, bioremediation in agriculture, with the help of nanotechnology contributes to sustainable remediation for resolving and restoring the natural characters of soil (Dixit et al., 2015; Prasad et al., 2017).

ENMs may also be synthesized from plant systems and could be considered as green nanotechnology (Prasad et al., 2014). Green nanotechnology is energy efficient and safe approachable to reduce waste and lessen greenhouse gas emissions. Due to use of renewable materials in producing such products, this process has minimal effect on the environment (Prasad et al., 2016). Since these ENMs are eco-environment friendly, a significant advancement is being visible in the area of green nanotechnology. However, there is still a long way to go with green nanotechnology to make it sustainable in the true sense (Kandasamy and Prema, 2015).

The application of every agrochemical suffers from unwanted issues like contamination of water or residues with food products that might pose a potential threat to human and environmental health. Therefore, precise management and control of inputs should be done to minimize these limitations (Kah, 2015). The evolution of the engineered smart nanotools might be a fruitful strategy to introduce a radical change in agricultural practices, enhancing both the quality and quantity of yields, thereby reducing and/or eliminating the adverse effects on the environment (Liu and Lal, 2015; Prasad et al., 2017).

11.2.3 BENEFITS/NEW OPPORTUNITIES OF NANOMATERIALS IN AGRICULTURE

The widespread application of present-day nanotechnology covers diverse fields including different aspects of agriculture, food processing and food preservation, dairy industry, packaging, transportation, and quality control of agricultural products. It has immense prospective to make agriculture more accomplished and resourceful by using NPs to improve the precision in delivering the nutrients and others to the targeted part in due time. The application of nano-based agrochemicals, ceramic devices, lamination techniques, filters have huge opportunities to make agriculture more systematic and productive through transformation of the conventional agro-practices (Pandey, 2018). It is anticipated that the agri-nanotechnology, by applying the concept of precision farming, will allow the judicial use of natural resources in agricultural practices. The green synthesis of NPs with the use of plant extracts and microorganisms is a boon for promoting safe and advance research in the field of agri-nanotechnology (Pandey, 2018).

The current challenges of sustainability, food security and climate change led the researchers to explore the discipline of nanotechnology as a promising tool to make a revolutionary change in the agricultural sector (Parisi et al., 2015). In comparison with other areas of nanotechnological application, including medicine, energy, and materials, engineering, agriculture is still a marginal sector. Given the unique physicochemical properties of ENMs, there are huge opportunities for opening new paradigms and introducing new strategies in agriculture areas. Agriculture might be benefitted through the use of nanotechnology in different ways to improve crop yields, such as: (i) producing temperature tolerant crops; (ii) developing effective and specialized pesticides for specific insects; (iii) overcoming the problems of global warming; (iv) developing nanotubes to store rainwater in the soil so that plants can use it during drought (Chen and Yada, 2011).

Despite huge industrial interests for large scale application of nanotechnology in the agriculture field, most applications are either in research, experimental, and development stages or at a bench-top exploration stage; however, it is anticipated that the agriculture and food industry will experience lots of large-scale nanotechnological applications in near future as shown below:

1. **To Increase Agricultural Production by Minimizing Production Inputs:** Despite the increasing demand for crop yield, the agriculture and natural resources like land, water, and soil fertility are

finite and other production inputs including synthetic fertilizers and pesticides are also being expensive day by day. Precision farming is an important concept to minimize production inputs and maximize agricultural yield outputs to meet ever-increasing needs. Given that nanotechnology may allow precise control at nano-scale, there is huge scope to improve precision farming practices (Chen and Yada, 2011).

2. **For Controlled Delivery of Agricultural Chemicals:** Several nanoscale carriers, such as encapsulation and entrapment, polymers, and dendrimers as well as surface ionic and weak bond attachments and other related techniques may be applied for storage, protection, delivery, and controlled release of intended payloads in agriculture process. Nanoscale delivery vehicles provide improved stability of the payloads against environmental degradation with increased efficacy while reducing the amount applied. Through controlled release mechanisms, the active ingredients would be taken up slowly, thus decreasing the required quantity of chemicals with minimal waste (Chen and Yada, 2011).
3. **Nano-based Biosensors for Real-Time Monitoring of Crop and Field Condition:** Nanomaterials play a vital role in agriculture in developing nano-based biosensors. Nano-based biosensors allow the rapid identification (direct and indirect identification) of the food-borne pathogenic microorganisms, drug residues, pesticide, toxic contaminants, and heavy metal ions in food (Muniandy, 2019a, b). Moreover, nanobiosensors may also be applied in monitoring antibiotic resistance, crop stress, soil condition, moisture level, temperature, plant growth and nutrient contents and food quality. Some examples of nano-based biosensors that are used in agriculture are magnetic, Au NPs, DNA-aptamer, Quantum Dots, and CNTs. These NPs can be used to develop nanobiosensor devices capable of rapid detection with high sensitivity and selectivity (Mufamadi and Sekhejane, 2017). Wireless nanosensors placed across crop fields produce necessary data leading to agronomic intelligence processes with a view to minimizing resource inputs and maximizing output and yields. Examples of some useful information and signals may include the optimal time to plant and harvest crops as well as amount and time for water, pesticides, fertilizers, herbicides, and other necessary components that need to be delivered based on specific plant physiology, pathology, and environmental conditions (Chen and Yada, 2011).

4. **To Study Mechanisms of Plant Pathogens:** Through the development of nanofabrication and characterization tools, it is possible to study plant pathology, i.e., biological, physical, and chemical interactions between plant cell organelles and multiple disease-causing pathogens. Comprehensive knowledge of plant pathogenic mechanisms (e.g., flagella motility and biofilm formation) may lead to improved treatment strategies for controlling diseases and increasing productivity. The use of nano-featured micro-fabricated xylem vessels could be effective in understanding the mechanisms and kinetics of bacterial colonization in xylem vessels and thus help in developing novel disease control strategies (Cursino et al., 2009; Chen and Yada, 2011).
5. **To Improve Plant Resilience Against Different Environmental Stresses:** There are ongoing biotechnological researches that focus on the improvement of plant resilience against different environmental stresses like drought, diseases, salinity, and so on. Currently, there have been extensive studies with genomes of crop cultivars. It is expected that with the advancement in nanotechnology-enabled gene sequencing, rapid, and efficient identification and utilization of plant gene trait resources could be achieved within a few years (Chen and Yada, 2011).
6. **Use of Lignocellulosic Nanomaterials:** Recently it has been shown that cellulosic materials at the nanoscale could be obtained from crops and trees. This might create a new market for novel nanobiomaterials and products of crops and trees. For instance, cellulosic nanocrystals can be used as nanocomposite for lightweight reinforcement in polymeric matrices. The area of applications may include food and other packaging industry, construction, and transportation vehicle body structures (Laborie, 2009; Chen and Yada, 2011).

11.2.4 LIFE CYCLE OF NANOMATERIALS

Despite the apparently clear-cut benefits of nanotechnology, there might be unexpected health and environmental risks associated with the widespread use of ENMs, which is yet to be fully understood. Therefore, the study of a life-cycle based approach is crucial to have in-depth knowledge about their potential risks (Dhingra et al., 2010). The life cycle of manufactured ENMs is product-dependent because every product is unique in its own manufacturing process, use, and waste treatment, and therefore, each product has its own hotspot regarding ENMs release and associated risks (González-Gálvez et al., 2017). It is apparent that relevant patterns of exposure, as well as aging and transformations of ENMs, depend on the life cycles of the nano-enhanced products. When the product manufacturing part of the life cycle is performed in an industrial setting in a controlled way, the application and disposal of nano-products in the consumer level will obviously be less predictable and there will be more variables (Mitrano et al., 2015). There are several routes for the exposure of the ENMs to environment which are briefly discussed below:

1. **Release During Synthesis of Engineered Nanomaterials:** The step of ENMs synthesis within the life cycle of a nano-enabled product has been given the highest importance in the literature with respect to potential risks for human health.
2. **Release During the Use Stage:** This is the most common route of ENMs exposure to the environment and the level of exposure depends on the use of ENMs.
3. **Release During Disposal Stage:** Estimation of the release of ENMs during disposal was produced based on findings from various studies (Kiser et al., 2009), the United Nations waste generation and disposal data and 2011 world population data (Keller et al., 2013).

To date, the life cycle of ENMs applied (directly or indirectly) in agriculture sectors has not been investigated extensively. So far, the role of NP transformations and transport has been considered by few studies. Most of the publications are concerned with the effects on the plants, and few studied the possible transformations and transfer. The necessity of lifecycle study in connection with application efficacy of such materials as pesticides and fertilizers are ultimately important (Bottero et al., 2017).

After successful utilization, products containing ENMs ultimately finish their lives either by being recycled or discarded. Furthermore, waste materials containing ENMs are being produced during the manufacturing of various nanotechnology-based products. Such waste streams that are produced throughout the life cycle of products containing ENMs contribute significantly to the potential sources of ENMs into the environment. The resulting environmental exposure to ENMs is determined by the handling, treatment, and disposal of these wastes. Therefore, it is critical to develop a suitable end-of-life management strategy for waste streams containing ENMs (González-Gálvez et al., 2017).

11.3 AGRICULTURE ASSOCIATED ENMs: IMPLICATIONS FOR OCCUPATIONAL HEALTH AND ENVIRONMENT

The recent advances and the advantages of the ENMs have drawn huge attention toward their toxicology and the relationship between nanoparticles (NPs) exposure with hazardous incidences. The fact that NPs are associated with toxic side effects and hampers human health is not new. NPs present in the aerosols and air pollution have been studied for decades, and it has been established that it leads to the onset of several cardiac and respiratory diseases (Kingsley et al., 2013; Dasgupta et al., 2015). Nanotoxicology is very complex and multifaceted as it depends on a variety of physicochemical and surface properties like their size, shape, charge, area, and reactivity. Furthermore, the interaction of ENMs with the environment may affect their bioavailability and stability (Leo et al., 2013; Leo et al., 2019).

11.3.1 CHARACTERIZATION TECHNIQUES OF ENMs IN AGRICULTURAL APPLICATION

As part of its efforts to assist industry to determine if a particular product falls under the category of nanomaterials under the food law safety assessment, a comprehensive guideline on physicochemical characterization of nanomaterials in the agricultural/food sector has been published by the EFSA (Hardy et al., 2018). The first step in the safety assessment process is to determine if the product falls under the nanomaterials classification and to consider all factors which may affect this classification. This is to ensure careful planning of the characterization process performed to establish its physicochemical identity both as a pure material or when it exists in food or feed product. Changes occurring during storage or after ingestion need to be considered as well.

Under the EU FP7, the EU funded a project entitled "The NanoDefine Project" called the development of an integrated approach based on validated and standardized methods to support the implementation of the EC recommendation for a definition of nanomaterial. Based on this project, the EU has published a guidance document to select appropriate techniques of measurement of nanomaterials as well as interpretation of results. It is made up of a series of publications, technical reports, and protocols (http://www.nanodefine.eu/index.php/nanodefine-publications) and it addresses the recommended European Commission definition of nanomaterials, i.e., size,

size distribution and volume-specific surface area (VSSA). It also provides a 2-tier decision flow scheme in which the first tier is based on screening methods (VSSA by the Brunauer Emmet Teller (BET) method) while the second tier consists of more sophisticated particle size analysis methods such as electron microscopy.

The guideline also addresses the techniques and methods to be used for characterization of nanomaterials and differentiating the nanomaterials in terms of its occurrence, i.e., as nanoparticles, in food or feed matrix or as food contact material. There is a recommendation of characterization methods in test media for *in-vitro* and *in-vivo* testing of nanomaterials. Sections are also available on solubility of nanomaterials and their degradation or dissolution through enzymatic of chemical processes. Besides, the US National Institute of Standards and Technology (NIST) have also established reference material standards (RMS) for qualifying instrument performance in nanoparticles and nanomaterials measurement. The reference materials are in the form of Au NPs in three sizes of 10, 30, and 60 nm including a report detailing measurement protocols and data of size, size distribution from multiple measurement techniques.

11.3.2 OCCUPATIONAL HEALTH, PUBLIC SAFETY, AND CONSUMER ISSUES

Hazard identification and risk assessment of agricultural inputs such as fertilizers and pesticides on human health can be a herculean task due to the voluminous variety of substances utilized and the accompanying doses as well as geographical and environmental or meteorological factors involved. The assessment of risks and hazards for ENMs is made more challenging as the models and assumptions used for conventional chemicals and materials may not be applicable or appropriate for ENMs. Risk assessment of nanomaterials and nanoparticles require a load of information on the capability to reach the identified sites as well as the effect and responses of the ENMs at that site (Dasgupta et al., 2016).

Several studies showed a low toxic impact of nanomaterials in food and agricultural products (Liu et al., 2016; Dudefoi et al., 2017). There have been some moves to re-evaluate the safety of specific nanomaterials. For instance, TiO_2 has been reported of pre-tumorous damage in the colon of rats over an extended period of 100 days by a group of French researchers. Re-evaluation of this data and recommendations made to various agencies led to an amendment to the French Farm and Food Bill targeting a ban on the import and marketing of TiO_2 as additives in the food product. Interestingly, this also led to follow-up announcements from major food manufacturers such as Mars and Dunkin to remove the particles from their products (He et al., 2019), illustrating the impact of legislation on the application of nanomaterials.

The mechanisms of nanotoxicity have been studied extensively covering ROS and oxidative stress, apoptosis, genotoxicity, and carcinogenic potential (Jain et al., 2018). Numerous studies demonstrate the toxicity of ROS generated by nanomaterials due to its size, surface properties, presence of metal ions and composition. These include studies on oxidative stress-mediated or induced by Nickel nanowires (Hossain and Kleve, 2011), silica NPs (Guo et al., 2015), Cobalt oxide (Chattopadhyay, 2014) and ZnO (De Berardis et al., 2010). ROS has also been identified as critical signaling molecules in cell death, and mechanisms of nanoparticles-induced apoptosis have also been postulated particularly for zinc oxide nanoparticles (ZnO NPs) (Wilhelmi et al., 2013; Wang et al., 2014). Due to their small size and high reactivity, there is a very high probability of nanoparticles interacting with cellular organelles and macromolecules such as DNA, RNA, and proteins inducing changes to genetic materials and mutations as well as altering biochemical pathways. This induction of genotoxicity by nanomaterials through various mechanisms has been studied in recent years (Kumar et al., 2014; Magdolenova et al., 2014; Kansara et al., 2015).

While the risks associated with nanomaterials and nanoparticles are still active areas of research, there are also studies on opportunities to reduce the hazards and toxicity associated with nanomaterials and nanoparticles. Several approaches have been investigated, such as the functionalization of nanoparticles, control of morphology and enhancement of effectiveness. A number of these approaches utilize biomaterials to create eco-friendly green materials from nanomaterials or nanoparticles. It is possible to adsorb biomaterials such as proteins and lipids with functional amide and carboxyl groups onto Au NPs for stabilization purposes (Zhang et al., 2016) and this is further corroborated by another study which showed that proteins, in fact, stabilized Ag NPs (Jain et al., 2011).

Microorganisms and enzymes can also be utilized for these purposes. Thus, the bacteria *Bacillus cereus* has been used on iron oxide nanoparticles as capping and stabilizing agent, the fungi *Aspergillus flavus* on Ag NPs as reducing and capping agent and the yeast *Magnusiomyces ingens* LH-F1 as reducing and stabilizing/capping agent on Au NPs. The enzyme alpha-amylase is utilized as reducing and capping agents in TiO_2 NPs as an antibacterial product. Use of plant extracts has also been reported, such as the use of coffee arabica seed, *Aloe vera* plant and pineapples and oranges fruits, to name a few.

11.3.3 ENVIRONMENTAL CONCERNS

The first concern of nanomaterials and nano-based products, and rightly so, has been on the safety of human beings. There have been several studies, mainly toxicological studies, to address this issue. However, its impact on the environment should not be dismissed lightly. Data in this topic is scant, especially on the environmental fate of nanomaterials and nano-based products, as it does not seem to be the priority focus of producers in agriculture or the food industry. Nanomaterials in the environment can come from natural phenomena such as forest fires, soil erosion, volcanic activities, or human activities such as mining activities, process waste streams and fuel combustions (Smita et al., 2012). As they settle in different environmental conditions, their fate differs depending on the matrices they settle in.

Release of nanomaterials into the air is possible at every step of the process during generation, transportation, utilization, and end-of-life treatments. Possibility of particle size or size distribution changes through photochemical process or condensation which alters settling velocities complicates fate determination models (Hartmann et al., 2014). Nanomaterials enter soil matrices through various products such as fertilizers, pesticides, biosolids, or even sewage water (Batley et al., 2013). Nanomaterials adhere to soil strongly due to their higher surface area and their release from the soil to affect plants or soil microorganisms is dependent on the strength of this adherence due to the soil properties or environmental conditions such as wind or rain run-offs (Jafar and Hamzeh, 2013). Nanomaterials enter the water system through the waste effluent streams or surface run-offs (Batley et al., 2013) and their fate is affected by physical processes such as aggregation or interaction of components of the aquatic environment or biological degradation and it is these changes affect their fate in the environment (Vale et al., 2016).

Carbon-based nanoparticles are nanomaterials that have been shown to be plant growth regulators. Hardy et al. (2018) investigated the effects of long-term exposure to MWCNTs on the growth of three important crops (barley, soybean, and corn) (Hardy et al., 2018). The tested species were cultivated in hydroponics supplemented with 50 μg/mL MWCNTs. After 20 weeks of continuous exposure to the nanomaterials, no significant toxic effects on plant development were observed. Several positive phenotypical changes were recorded, in addition to the enhancement of photosynthesis in MWCNT-exposed crops. Raman spectroscopy with point-by-point mapping proved that the MWCNTs in the hydroponic solution moved into all tested species and were distributed in analyzed organs (leaves, stems, roots, and seeds). Results confirmed the significant potential of CBN in plant agriculture. However, the documented presence of MWCNTs in different organs of all exposed crops highlighted the importance of detailed risk assessment of nanomaterials (Hardy et al., 2018).

11.3.4 TOXICOLOGICAL IMPLICATION OF AGRICULTURE-ASSOCIATED ENMs

The primary reservoir of ENMs in agriculture is the soil. The transformation, transport, and fate of ENMs are controlled by physicochemical properties of soil such as cation exchange capacity, soil composition and pH (Hartmann et al., 2014; Mahdi et al., 2018). In addition, the morphology, size, zeta potential, surface coating and chemical composition of the ENMs are also the contributing factors in understanding the dynamic of ENMS in the soil medium (Marmiroli, 2019). The presence of ENMs in soil may also have toxicological impacts on the terrestrial microorganisms that consequently affect soil integration and fertility. Many studies of soil exposure to various types of nanomaterials (e.g., Cu, Ce, Ag, Ti, Au, GO, Zn) indicate negative effects on the survival and activity of the terrestrial biotas such as the microbiomes (beneficial bacteria), mycorrhiza fungi and invertebrates (Matranga and Corsi, 2012; Mortimer and Holden, 2019; Pullagurala et al., 2019). Consequently, these deplete the nitrogen fixation process by the soil microbiomes, fungi, and invertebrates, which further reduce the soil fertility and affect the plant growth (Karimi and Fard, 2017). Indirectly, this scenario may disrupt the ecosystem and feed chain.

Soil microbiomes such as Actinobacteria, Proteobacteria, Acidobacteria, and Cyanobacteria play roles in waste decomposition, nutrient cycling and plant growth performance contributing to soil fertility (Juan et al., 2017). It was reported that ENMs could lead to toxicity on microbiomes as demonstrated by soil exposure to Ag NP at the concentration of higher than 30 mg/kg (Sillen et al., 2015; Michels et al., 2017). It was identified that

that ionic dissolution of the ENMs may have more impact, whereby the Ag ions were shown to be 20–48 times more toxic than Ag NP on denitrifier *Pseudomonas stutzeri*, the nitrogen fixer *Azotobacter vinelandii* and nitrifier *Nitrosomonas europaea* in terms of cellular and transcriptional responses (Yang et al., 2013). Nevertheless, variation in responses towards ENMs was demonstrated in different soil environments and species-specific. For instance, it was reported that Gram-positive bacteria, *B. subtilis* was more susceptible to the bactericidal effect of ZnO NP compared to the Gram-negative bacteria, *Pseudomonas aerugunosa* (Azam et al., 2012). Another study indicates that exposure of soil bacteria in two different soil types (a sandy loam (Bet-Dagan) and a sandy clay loam (Yatir)), to metal oxide nanoparticles resulted in higher susceptibility of the bacteria to change when exposed to the nanoparticle in Bet-Dagan soil than in Yatir soil (Frenk et al., 2013). The changes refer to the bacterial hydrolytic activity, oxidative potential, community composition, and size. It was postulated that the resistance of Yatir soil to the nanoparticles may be due to the higher organic matter concentrations, clay fraction of the soil, and higher native community richness and diversity.

The presence of arbuscular mycorrhizal fungi (AMF) protects against pathogens and toxic stresses and promote plant growth and development (Jeffries et al., 2003). AMF is a symbiotic microorganism to most terrestrial plant species which secrete glomalin, a glycoprotein which act as a metal chelator in the rhizosphere, a region of soil near the plant root. Ghasemi et al. (2017) reported that the reduction of ZnO NP availability and uptake by the plant may be due to the secretion of glomalin by AMF (Siani et al., 2017). It was also reported that AMF reduced the toxicity effects of ENMs in plants, although the colonization and diversity of AMF themselves are being inhibited by ENMs (Marmiroli, 2019). This indicates the protective roles of this rhizospheric AMF in preventing nanotoxicity of ENMs in plants and emphasize the importance to include soil-microbial interactions when assessing nanophytotoxicology and risks.

It is evident that the presence of ENMs in soil has toxic effects against invertebrates such as earthworm species through ingestion and internalization of the ENMs into their digestive gut epithelium leading to immune suppression and mortality. ENMs disrupt the pathways related to ribosomal functions, sugar/protein metabolism, energy production and histone activity which led to toxicity in invertebrates (Novo et al., 2015). The bioindicator of ENMs toxicity in invertebrates is the presence of amino acids such as leucine, valine, isoleucine, and sugar (glucose and maltose) (Lankadurai et al., 2015). Hence, this could be one of the assessment tools in monitoring the toxicity effects of ENMs usage in agriculture.

One of the limitations in the use of ENMs in agriculture is the lack of information in the trophic transfer of the ENMs in the terrestrial food chain, this refers to the movement or transfer of ENMS between organisms, including humans (Lowry et al., 2010; Gardea-Torresdey et al., 2014). This relates to the fate and life cycle of the material within the ecosystem. There is evidence that organism that feeds indirectly by eating another organism displayed higher bioaccumulation of ENMs compared to an organism that feeds directly. Majumdar et al. (2016) described the trophic pathway of nanoceria (CeO_2 NP) in comparison to bulk CeO_2 from soil to kidney bean plants (*P. vulgaris*) to Mexican bean beetles (*Epilachna varivestis*) and to spined soldier bug (*Podicus maculiventris*) (Majumdar et al., 2015). It is interesting to note that the translocation of the bulk CeO_2 from root to shoot (first trophic level) was twice higher than that of the nanoceria, whilst in the last trophic level (in Spined soldier bug) showed higher biomagnification factor (BMF) of nanoceria compared to the bulk CeO_2. During the mid-trophic level (Mexican bean beetles), the larvae excreted most of the ingested Ce at about 88–98% whilst the adult beetles retained most of Ce and only excreted about 32–36% Ce. This shows that nano-sized materials have higher chances to retain in the feed chain and affect the final users at most. The age of the organism could also be the determining factor in BMF of the material or analyte. Hence, the information on BMF may also useful in determining bioaccumulation of ENMs and its effect on human health, the final consumer.

Nevertheless, there is still a knowledge gap in understanding the ENMs' transformation in soil, transfer process into the food chain and the effects on human health albeit a number of published studies. The mechanistic effects on biotic responses in molecular (omic) level and particle properties such as quantum dots, graphenes, and complex composite material in consumer product, are significantly lacking (Matranga and Corsi, 2012; Mortimer and Holden, 2019). Currently, a number of studies focus on identifying biomarkers in ecotoxicology including ecotoxicogenomics-based endpoints and whole-cell-array library to assess and predict the effects of chemical stressors on aquatic species and ecosystems (Gou et al., 2010; Connon et al., 2012). Further, the mechanistic effect of the chemical stressor could be determined, for instance, the pathway of DNA damage, genotoxicity, and identify possible mode of action (MOA). These data can be used in the regulatory and risk assessment framework as an alternative toxicity assessment endpoint (Gou et al., 2010). However, due to the multifaceted variation in the behavior and mechanism of ENMs on the agriculture ecosystem, case-by-case studies according to type of ENMs is required (Marmiroli, 2019).

11.3.5 SAFETY ASSESSMENT AND PREVENTIVE MEASURES IN THE USE OF AGRI-BASED ENMs

To date, a better understanding of the ENMs' properties, applications, safety aspect and specific needs of nanomaterials lead to more interest and initiatives for legislation by policymakers and industrial stakeholders (Opinions, 2007, 2009). Latest progress in ENMs application and impacts are analyzed and monitored by policymakers and also supported by relevant authorities, for instance, the USA National Nanotechnology Initiative (http://www.nano.gov/you/environmental-health-safety) (Rasmussen et al., 2019). Despite having a list of risk assessment frameworks specific for ENMs, these frameworks differ in terms of their aims, applicability domain, basic assumption, and alignment different regulations (Oomen et al., 2018). Thus, they lack the specific decision points and associated methods needed for decision making that are required for actual application and on current scientific knowledge.

11.3.5.1 TOXICITY RISK ASSESSMENT ON THE EXPOSURE AND HANDLING OF THE AGRICULTURE-RELATED NANOMATERIALS

Continuous and timely safety assessment on the potential environmental (particularly to soil and aquatic microorganism) and human health impacts of nanotechnology will enable regulators to assess potential risks, provide the industry with information needed to develop safer nanomaterials, and improve public trust of nanotechnology. This information is important to prevent hazardous effect on occupational health during production and handling of the nanomaterial, which to date is quite limited.

Toxicity testing is essential in human hazard and risk assessment, whereby toxicity can be investigated *in silico*, *in vitro* or *in vivo* (Gupta and Xie, 2018). Human exposure to toxic chemicals typically occurs through inhalation, ingestion, injection/insertion, and skin absorption (De Matteis, 2017). Nanotoxicology aims to determine the relationship between exposure route and dosage of ENMs and to assess the role played by its physico-chemical properties through the data obtained from *in vitro* and *in vivo* studies (Oberdörster, 2010). *In vitro* investigations are very useful to enrich clinical and epidemiological studies, which are pertinent for the manufacturers of NM-based commercial products (Stone et al., 2009). Chronic toxicity studies, biodistribution, and toxicokinetic in organs require *in vivo* assessment (Fröhlich and Salar-Behzadi, 2014). Additionally, computational methods are important tools in the nanotoxicology field, which allow the conduct of *in vivo* studies that have both ethical and experimental constraints (Aillon et al., 2009). *In silico* toxicology can be used to estimate the toxicity of NMs and strengthen the standard *in vitro* test (Raies and Bajic, 2016). This information shall build up the current database on the use of nanomaterials and strategies to protect the whole ecosystem and living organisms.

One caveat in the testing procedure of toxicity testing on ENMs is the lack of information on the interaction between ENMs and test system, chemically, and biologically. For instance, the interaction between ENMs and the culture medium, or biological fluids, is one of the critical parameters to consider in toxicology studies (Jiang et al., 2009). This interaction could lead to NP aggregation and dispersion, that influence cell uptake which could cause toxicity (Powers et al., 2006). Hence, several investigations focused on the preparation of stable NPs dispersion prior to testing to ensure the reliability of the data and avoid false-positive outcomes (Buford et al., 2007; Sager et al., 2007).

The OECD Working Party on Manufactured Nanomaterials (WPMN), an expert group on ecotoxicology and environmental fate of nanomaterials (NMs), identified OECD test guidelines (TGs) for chemicals which were generally applicable for the testing of NM, except for TG 105 (water solubility) and 106 (adsorption-desorption) (Table 11.1) (Kühnel and Nicke, 2014). It was proposed that sample preparation, dispersion, analysis, dosimetry, and characterization when condcting OECD chemical TG on nanomaterials shall be considered and analyzed prior to environmental fate and behavior test. This is because these parameters are the determining factors in the efficacy of the environmental fate and behavior test.

Currently, the number of approved TGs has significantly addressed the physical-chemical properties, effects on biotic systems, environmental fate and behavior, and health effects of nanomaterials (Rasmussen et al., 2019). Relevant OECD TGs were adapted to fit in most nanomaterial applications, and several new TGs for nano-relevant properties were developed. In addition to TGs listed in Table 11.1, three TGs specifically applicable to

TABLE 11.1 Specific Test Guidelines (TGs) and Guideline Documents (GDs) Recommended Through the OECD Working Party on Manufactured Nanomaterials (WPMN) an Expert Meeting on Ecotoxicology and Environmental Fate of ENMs*

	Compartment	OECD TG/ GD	Generally Applicable	Recommendation
Guidance Documents		New GD	–	GD on physical-chemical characterization of ENM should be developed
		New GD	–	ENM detection method is soil and sediment
		GSPD	–	Inclusion of decision tree (three tiers: (i) stock/stem suspension preparation; (ii) preparation of exposure solution; and (iii) conducting the test)
TG on Ecotoxicology	Water	201	Yes	No specific amendments, but several critical points identified to be considered within GD/GSPD
	Water	202	Yes	No specific amendments, but several critical points identified to be considered within GD/GSPD
	Water	211	Yes	No specific amendments, but several critical points identified to be considered within GD/GSPD
	Soil and sediment	222	Yes	Dry and wet spiking allowed
	Water, soil, and sediment	225	Yes	Dry and wet spiking allowed
TG on Fate Behavior	Water	105	No	Development of new TG specifying dissolution behavior of NM necessary
	Soil and sediment	106	No	Development of a new TG necessary, specifying the application of NM, shaking time, liquid to solid ratio, type, and concentration of electrolytes, and number concentration of ENM
	Water	305	Yes	As BCF is considered not applicable for water-phase, dietary spiking should be favored, nanospecific guidance

TABLE 11.1 *(Continued)*

	Compartment	OECD TG/ GD	Generally Applicable	Recommendation
	Soil and sediment	312	Yes	Wet spiking should be favored Specific guidance regarding the type of soil, the pre-assessment of dissolution, dispersibility, and application method
	Soil and sediment	315	Yes	Wet spiking should be favored Specific guidance on the application of ENM to the test GD for data interpretation
	Soil and sediment	317	Yes	Wet and dry spiking Specific guidance on the application of ENM to the test GD for data interpretation
		New TG	–	NM dispersibility and dispersion stability
		New TG	–	NM biodegradation

[*] *Source:* Reproduced with permission from: Kühnel and Nickel (2014); Copyright © Elsevier (2014).

manufactured nanomaterials on mammalian toxicology, have been adopted which are TG318 "Dispersion Stability of Nanomaterials in Simulated Environmental Media," and adaptation of TG412 and TG413 on subacute inhalation toxicity: 28-day study/90-day study. Related guideline document, GD39 on inhalation toxicity testing has also been revised. These TGs are also complemented by OECD harmonized template for data collection which is implemented in the International Uniform Chemical Information Database (IUCLID) (Iuclid, 2000).

This initiative is supported by an establishment of a database platform consisting of the latest available safety assessment publication and tools which are free and publicly accessible. One of the latest public database platforms for nanomaterial is the 'NANoREG Toolbox' (Figure 11.3), an inventory with unique metadata set in Excel® format, developed by NANoREG (Jantunen et al., 2018). NANoREG focused on safety assessment under the EU regulation on registration, evaluation, authorization, and restriction of chemicals (REACH) and the instruments listed in the Toolbox are relevant and useful for nanomaterial safety assessments worldwide. The Toolbox is a ready-to-use inventory that covers ove 500 current tools for all tasks in nanomaterial safety assessment, accessible globally. This inventory is in accordance to the NanoREG Framework, assembling relevant tools to implement the regulatory provision under REACH and safe-by-design (SbD), risk prioritization assessment (nanoRA) and life cycle assessment (LCA) for nanomaterials (Jantunen et al., 2018). The toolbox enables knowledge transfer between regulatory, stakeholder, researchers, industries, and users for safer handling of agriculture-related nanomaterials or ENMS.

Apart from the aforementioned initiatives, the EFSA has developed a stepwise framework for nano-related hazard identification and characterization in food/feed. The purpose of the framework is to assess potential risks due to applications of nanoscience and nanotechnologies in the food and feed chain (Hardy et al., 2018). Guidance on the assessment is on: (i) the physicochemical characterization requirements of ENMs usage as food additives, enzymes, food contact materials, feed additives and pesticides, and (ii) testing approaches to identify and characterize hazards due to the nano-related properties including information from *in vitro* genotoxicity, absorption, distribution, metabolism, and excretion and repeated dose 90-day oral toxicity studies in rodents. The pre-requisite of the framework is to determine whether a material display the characteristic of a nanomaterial can qualify for the nano-specific testing by determining the rate of degradation of the nanomaterial to non-nanomaterial. If a nanomaterial fully dissolved in a fast manner, then it may be subjected to non-nanomaterial assessment, hence, avoid unnecessary testing (nano-specific testing).

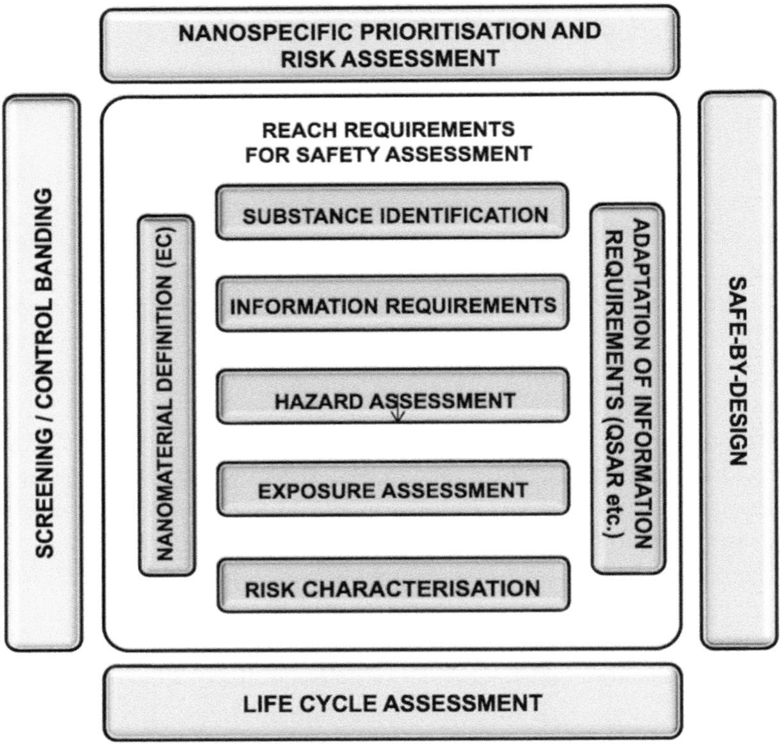

FIGURE 11.3 NANoREG Toolbox structure: The two Excel® workbooks are organized into 11 worksheets (WS). These WSs provide an overview of the tools useful for implementing the two Parts (I and II) of the NANoREG framework. The blue rectangles refer to Part I, which addresses the regulatory requirements on safety assessment (mainly in REACH) and discusses their implementation for NMs. The green rectangles refer to Part II, which proposes three forward-looking strategies that may facilitate or accelerate the implementation of those requirements. The yellow rectangle refers to tools with other purposes related to NM safety assessment.

Source: Reprinted from Jantunen et al. (2018); Creative Commons Attribution 4.0 International.

11.3.5.2 THE MITIGATION AND PREVENTIVE MEASURES

Nanotechnology improves the efficiency of water, light, and agrochemicals in agriculture such as the nano-enabled solar distillation and crystallization technologies convert high-salinity wastewater into clean water for crop agriculture, enhance water-use efficiency by improving water retention in crops, and functioned as 'smart plants' utilizing nano-sensor that can relay their water needs to farmers (Lowry et al., 2019). However, there are concerns regarding the side effects on safety, health, and environment from this fast-emerging technology. To address increasing concern on the side effects of nanomaterials usage in agriculture, relevant studies and innovation on the existing nanotechnology approaches have been conducted to mitigate and prevent potential environmental and health issues related to the introduction of nanomaterials in agriculture activities.

In farming, ZnO is one of the forms of Zn used in commercial fertilizers commonly used for acidic soils due to deficiency of zinc, that leached out from the plant rooting area (Cakmak, 2008; Alloway, 2009; McBeath and McLaughlin, 2014). However, the occurrence of phytotoxicity in crops grown in elevated Zn-soils was also reported (Chaney, 1993). Phytotoxicity could be any adverse effect on plants caused by specific substances (phytotoxins) or groing conditions which may delay seed germination or inhibits plant growth. Corresponding toxicological effects were measured based on plant growth, chlorophyll production, Zn bioaccumulation, H_2O_2 generation, stress enzyme activity, and lipid peroxidation using different cellular, molecular, and biochemical approaches (Mukherjee et al., 2014). There are more concerns on the use of miniaturized ZnO in fertilizer as it could easily penetrate any biological system inaccessible by the bare ZnO. It was indicated that ZnO nanoparticles produce toxic responses in mammalian cells, bacteria, copepods, crustaceans, fish, and aquatic microorganisms (Ma et al., 2014). Interestingly, the presence of additives in combination with ENMs could prevent the phytotoxic effect. The study showed that iron-doping reduces the toxicity of ZnO NPs towards terrestrial plants, green pea plants (*Pisum sativum* L.). The toxicity level of doped ZnO NPs is less than that of bare ZnO NPs as per zinc uptake, chlorophyll content, and ROS (H_2O_2) production are considered (Mukherjee et al., 2014). Thus, iron doping method could be considered as a safer approach to reduce the toxicity of ZnO NPs in terrestrial plants.

Studies indicate that Ag NPs promote and enhance plant growth, yield, and nutrient use efficiency, but could be phytotoxic (Almutairi and Alharbi, 2015; Yang et al., 2017; Kannaujia et al., 2019; Marmiroli, 2019; Sadak, 2019). Comparison between chemically synthesized Ag NPs, chemically synthesized Ag NPs capped with phytochemical and $AgNO_3$ salt was conducted for phytotoxicity, according to these parameters; growth parameters, ROS production, cytotoxicity assay and silver accumulation in two wheat varieties (HD-2967 and DBW-17) (Kannaujia et al., 2019). The phytochemicals capped B-Ag NPs was shown to be the least toxic and effective in promoting early seedling growth at lower concentration by delivering a potent antioxidant that decreases ROS toxicity. This supports previous findings by Rani et al. (2016) on the effect of synthetic and biosynthesized Ag NPs on growth, physiology, and oxidative stress of water hyacinth: *Eichhornia crassipes* (Mart) Solms (Rani et al., 2016). Moreover, Stevenson et al. (2013) reported freshwater organisms and their environment may impact the potential toxicity of Ag NPs. This finding highlights the effect of citrate-coated Ag NPs on the freshwater algae *Chlamydomonas reinhardtii* in different stages of growth in batch cultures. The extracellular molecules produced by the algal cells themselves mitigate both the nanoparticle-specific and ionic toxicity of Ag NPs (Stevenson et al., 2013). Thus, beside modification of the nanostructure, environmental feedback towards ENMs play an important role in the mitigation of toxicity.

11.4 REGULATIONS AND LEGISLATION APPLICABLE TO ENMs

Regulations and legislation are a fundamental component of all products which relate to either quality or safety. The regulatory environment plays a key role in ensuring the product meets quality standards as well as safety concerns. In addition, the regulatory process has a major impact on the price at which the consumer needs to pay for these products, i.e., the higher the regulatory requirements, the higher the price of the product. There is a trade-off between safety and price and finding the balance between these two extremes can be a complex process.

In the case of nanomaterials, much effort has been made to establish a regulatory framework through the formation of governing bodies or agencies by various governments, but the process of agreements on various aspects has faced various limitations. These limitations include the understanding of nanoscale interactions, functions, toxicities, and the control of its properties at the nanometer scale. Thus, it is difficult to make that first recommendation to consider regulatory intervention as that regulation may not have enough data. Furthermore, each nanomaterial has its unique properties leading to case-specific or study-specific evaluations for regulation and legislation.

11.5 CHALLENGES IN THE IMPLEMENTATION OF THE REGULATION RELATED TO ENMs IN THE AGRICULTURE INDUSTRY

There are communication gaps in the implementation and enforcement of hazard assessment and regulation related to nanotechnology. This is due to the immature safety governance framework for ENMs, a disconnect between human and environment nanosafety, environmental nano-risk and human nano-risk challenge (Lombi et al., 2019). Despite the significant research progress towards understanding ENM hazards and risk, there is still no harmonized basis for ENMs risk governance across different sectors. There is a need for transparency in data sharing and knowledge transfer across multidisciplinary area in nanotechnology as indicated by the One Health framework, which involves the integration between human, environment, and animal health domain (Lebov et al., 2017). This information is significant to facilitate advance nanosafety research, supporting the development and implementation of reliable frameworks for risk assessment and decision-making and to facilitate risk communication with relevant stakeholders such as the industries, regulators, insurance companies, civil society organizations and the general public (Stone et al., 2018).

The European Commission has taken the initiatives to develop a clear understanding of the nanotechnology risks, management practices, and societal perceptions among transdisciplinary risk governance framework to improve the current status quo of immature and underdeveloped cross-sectorial governance (Comission, 2018). Three large international projects which RiskGONE, NanoRIGO, and Gov4Nano have been funded to establish frameworks for nanotechnology risk governance. Besides, good governance of agri-nanotechnologies is critically required. One caveat in the development of the nanosafety framework is a lack in communication between human health risk and environmental health risk comprises of researchers interested in mammalian cells, tissues, and organism versus researchers working in single cells organisms, plant tissues, invertebrates, and fish (Haas, 1992). The lack of communication is due to different challenges faced by both parties.

The advancing application of nanotechnology in the agricultural and industrial area leads to larger incident/accident of ENMs released into the environment. Human and environmental exposure pathways become much more complex when agri-nanotechnologies are added to the scenario. Hence, preventive measures to avoid decremental effect on environment and health are highly required. However, it is extremely challenging in detecting and quantifying ENMs in complex environmental matrices due to the complexity of ENMs transformation in the environment from dissolution to agglomeration and changes in composition and surface properties, even though with the availability of protocols for assessment (Lombi et al., 2019).

Despite the complexities, developments of potential environmental release scenario in predicting the environmental concentrations (PECs) of ENMs in the environment using advanced modeling approaches such as probabilistic material flow analysis (Huggett et al., 2003; Scott-Fordsmand et al., 2017) and quantitative method such as single-particle time-of-flight mass spectroscopy as described in Nanodefine project template for nanomaterial characterization of tier 1 and tier 2 measurement methods (Hodoroaba and Mielke, 2014). The risk assessment of agri-based ENMs shall involve: (a) material characterization, (b) release, fate, and exposure modeling (to obtain environment concentration, PECs), (c) hazard characterization (to derive predicted no-effect concentrations: PNECs), and (d) risk characterization (normally by comparing PECs and PNECs) (Hodoroaba and Mielke, 2014; Scott-Fordsmand et al., 2017).

Prioritization of assessment of ENMs shall be considered according to their occurrences in the environment, usage, or availability of database on its chemical and biological behaviors. For instance, there is a question whether silver sulfite NP which is more reactive in a wide range of environmental condition, should be given a priority for risk assessment compared to pristine metallic silver which is rarely studied on exposure to ambient environment nevertheless commonly tested for ecotoxicity (Pulido-Reyes et al., 2017). Due to its miniaturized size, large surface area to mass ratio and surface characteristic, ENMs can permeate through the bloodstream, across the body membrane barriers, and interact with biological milieu at cellular and molecular levels. This could result in cellular injury and toxicity effects as the ENMs penetrate into the cellular compartment. Oxidative stress has been shown to be an underlying mechanism of possible toxicity of ENMs, causing both immunotoxicity and genotoxicity. ENM risk assessment on human health requires consideration of multiple exposure pathways (inhalation, ingestion, dermal absorption, or injection) which considerably depending on the source of exposure.

To date, the connection between environmental exposure pathway to human exposure is still insufficient whereby their applications are discipline-specific with minimal communication between the assessors. Hence, one unified, interdisciplinary, and ideally transdisciplinary approach is urgently required. As such that embodied in the One Health concept (Lombi et al., 2019), is a unifying framework to ensure human safety and animal health, to reduce disease threats and to ensure a safe food supply through effective and responsible management of natural resources (Lebov et al., 2017). It is endorsed by the World Health Organization (WHO), the FAO of the United Nations on food safety as well as human, animal NMs and environmental health, the World Organization for Animal Health (OIE), and various governments.

Adopting a One Health perspective can help (a) actively recognizing the interconnected nature of social and biological systems and working to incorporate both different disciplines and stakeholders in knowledge building and decision-making processes, (b) navigating the acceptance for agri-nanotechnologies in the public domain, (c) create the conditions for comparative nanotoxicology to progress rapidly, an essential step before a SbD approach is proven effective, and (d) facilitate the development of appropriate regulatory frameworks for nano-agritechnologies (Lombi et al., 2019). Therefore, to overcome the barriers and challenges in agri-nanotechnologies, transdisciplinary knowledge sharing and integration shall be established for: (a) efficient inter-disciplinary communication and cooperation in human and environmental toxicology, (b) harmonized nanometrology and testing procedures with consistent datasets combining both human and environmental toxicology, (c) stimulating comparative toxicological studies and learning, (d) attracting the public and stakeholders in research and innovation, and (e) bridging regulators from different areas such as food, medicine, agrochemicals, veterinary to contribute to the development of a transdisciplinary risk governance framework for nanotechnology.

11.6 CONCLUSION AND TRENDS

In recent years, nanotechnology has been considered as a novel tool to flourish agricultural production through several ways in enhancing nutrient management efficiency, improving plant protection practices and disease detection, improving crop varieties, and monitoring plant growth. Nanotechnology offers generous visions to boost the agricultural sector through unveiling the potentiality of products for advanced applications to increase global crop production volume for the increased world population in coming decades. Currently, ENMs are very efficiently manipulated in the form of nanofertilizers, nanoherbicides, nanopesticides, and plant growth regulators in order to improve crop yield at the same time alleviating toxicity. In fact, nanotechnology possesses huge potentials for diverse applications to maintain sustainable agriculture in years to come.

Several studies on nanoparticle-plant interaction have already been reported. Taking advantage of the hugely positive implications of nano incorporation in agriculture, a risk assessment should be addressed with priority to achieve long-term agricultural sustainability. Therefore, the ultimate target should be the multifarious application of NPs in agriculture, allowing a risk-free environment for the availability of soil microorganisms as well as macro and micro-nutrients, improving plant traits against environmental stresses and diseases, improving the delivery of agriculture chemicals, and mitigation of different abiotic stresses. Interestingly, nanoparticles-based formulations have been proved to be superior over bioformulations in addressing all these issues. Consequently, present-day agriculture is embracing the innovative approach of nanotechnology for combating the global challenge of crop production, food security, climate change, and sustainability. Besides, nanotechnology-based applications are also relevant in various aspects of food science, e.g., food processing, food packaging, food nutrition, and food contact materials. However, the concentration-dependent behavior of ENMs with the living system is a major concern that must be addressed prior to their widespread application to minimize their harmful impacts (Arif et al., 2016; Mishra et al., 2017).

Comprehensive investigations are further needed to ensure the safety of nano-related products to develop improved alternatives to traditional formulations. Nanotechnology has immense potentialities particularly, in the field of postharvest processing of agricultural waste materials through producing nanocellulose, nanocomposites, and biochar. Immediate attention should be focused on developing novel nanocomposites capable of carrying active agents like nutrients, pesticides, and fertilizers. Other accomplishments yet to be achieved include climate-oriented farming systems, breeding new drought-resistant crop species, and restoration of infertile soil in order to establish sustainability in global agricultural practices. Additionally, environmental and health risk assessments and the principles of green chemistry need to be incorporated into nanotechnological applications at the source. Thus, despite noticeable advancements in nanotechnology, many issues are still to be resolved, to produce a significant impact on agriculture.

ACKNOWLEDGMENTS

Financial support from the University of Malaya Research Grant, UMRG (RP045A-17AET and IIRG003B-19FNW) are acknowledged.

KEYWORDS

- **crop improvement**
- **genetic modification**
- **nano-based kits**
- **nanoemulsions**
- **nanoliposomes**
- **nanomaterial safety**
- **nanosensors**
- **pathogen identification**
- **plant diseases**
- **polymeric nanoparticles**
- **quantum dots**
- **seed germination**
- **toxicity risk assessment**
- **weed control**

REFERENCES

Abd-Elsalam, K. A., (2013). Nanoplatforms for plant pathogenic fungi management. *Fungal Genom. Biol., 2*, e107.

Agrawal, S., & Rathore, P., (2014). Nanotechnology pros and cons to agriculture: A review. *Int. J. Curr. Microbiol. App. Sci., 3*(3), 43–55.

Ahmed, F., Arshi, N., Kumar, S., Gill, S. S., Gill, R., Tuteja, N., & Koo, B. H., (2013). Nanobiotechnology: Scope and potential for crop improvement. In: *Crop Improvement Under Adverse Conditions* (pp. 245–269). Springer, Berlin.

Aillon, K. L., Xie, Y., El-Gendy, N., Berkland, C. J., & Forrest, M. L., (2009). Effects of nanomaterial physicochemical properties on *in vivo* toxicity. *Advanced Drug Delivery Reviews, 61*(6), 457–466. https://doi.org/10.1016/j.addr.2009.03.010.

Alharby, H. F., Hakeem, K. R., & Qureshi, M. I., (2019). Weed control through herbicide-loaded nanoparticles. In: *Nanomaterials and Plant Potential* (pp. 507–527). Springer, Berlin.

Alloway, B. J., (2009). Soil factors are associated with zinc deficiency in crops and humans. *Environmental Geochemistry and Health, 31*(5), 537–548.

Almutairi, Z., & Alharbi, A., (2015). Effect of silver nanoparticles on seed germination of crop plants. *Journal of Advances in Agriculture, 4*(1), 280–285. https://doi.org/10.24297/jaa.v4i1.4295.

Amna, T., Hassan, M. S., Yousef, A., Mishra, A., Barakat, N. A., Khil, M. S., & Kim, H. Y., (2013). Inactivation of foodborne pathogens by NiO/TiO_2 composite nanofibers: A novel biomaterial system. *Food and Bioprocess Technology, 6*(4), 988–996.

Arif, N., Yadav, V., Singh, S., Singh, S., Mishra, R., Sharma, S., et al., (2016). Current trends of engineered nanoparticles (ENPs) in sustainable agriculture: An overview. *J. Environ. Anal. Toxicol., 6*(5).

Arora, N. K., (2018). Agricultural sustainability and food security. *Environmental Sustainability, 1*, 217–219. https://doi.org/10.1007/s42398-018-00032-2.

Azam, A., Ahmed, A. S., Oves, M., Khan, M. S., Habib, S. S., & Memic, A., (2012). Antimicrobial activity of metal oxide nanoparticles against gram-positive and gram-negative bacteria: A comparative study. *Int. J. Nanomedicine, 7*, 6003–6009.

Baker, S., Volova, T., Prudnikova, S. V., Satish, S., & Prasad, N., (2017). Nanoagroparticles emerging trends and future prospects in modern agriculture system. *Environmental Toxicology and Pharmacology, 53*, 10–17.

Batley, G. E., Kirby, J. K., & McLaughlin, M. J., (2013). Fate and risks of nanomaterials in aquatic and terrestrial environments. *Accounts of Chemical Research, 46*(3), 854–862. doi: 10.1021/ar2003368.

Berekaa, M. M., (2015). Nanotechnology in food industry; advances in food processing, packaging and food safety. *Int. J. Curr. Microbiol. App. Sci., 4*(5), 345–357.

Bhagat, D., Samanta, S. K., & Bhattacharya, S., (2013). Efficient management of fruit pests by pheromone nanogels. *Scientific Reports, 3*, 1294.

Bottero, J. Y., Wiesner, M. R., Labille, J., Auffan, M., Vidal, V., & Santaella, C., (2017). Life cycle of nanoparticles in the environment: Nanomaterials and agriculture application-life cycle through their direct application and indirect application in bio-sludge-amended soils. In: Axelos, M. A., & Van De, V. M. H., (eds.), *Nanotechnology in Agriculture and Food Science* (pp. 333–346). Wiley-VCH, Weinheim.

Brayner, R., Ferrari-Iliou, R., Brivois, N., Djediat, S., Benedetti, M. F., & Fiévet, F., (2006). Toxicological impact studies based on *Escherichia coli* bacteria in ultrafine ZnO nanoparticles colloidal medium. *Nano Letters, 6*(4), 866–870.

Brody, A. L., Bugusu, B., Han, J. H., Sand, C. K., & Mchugh, T. H., (2008). Innovative food packaging solutions. *Journal of Food Science, 73*(8), 107–116.

Buford, M. C., Hamilton, R. F., & Holian, A., (2007). A comparison of dispersing media for various engineered carbon nanoparticles. *Particle and Fibre Toxicology, 4*(1), 6. doi: 10.1186/1743-8977-4-6.

Busolo, M. A., Fernandez, P., Ocio, M. J., & Lagaron, J. M., (2010). Novel silver-based nanoclay as an antimicrobial in polylactic acid food packaging coatings. *Food Additives and Contaminants, 27*(11), 1617–1626.

Cakmak, I., (2008). Enrichment of cereal grains with zinc: Agronomic or genetic biofortification? *Plant and Soil, 302*(1, 2), 1–17.

Campos, E. V. R., De Oliveira, J. L., Da Silva, C. M. G., Pascoli, M., Pasquoto, T., Lima, R., et al., (2015). Polymeric and solid lipid nanoparticles for sustained release of carbendazim and tebuconazole in agricultural applications. *Scientific Reports, 5*, 13809.

Carbone, M., Donia, D. T., Sabbatella, G., & Antiochia, R., (2016). Silver nanoparticles in polymeric matrices for fresh food packaging. *Journal of King Saud University-Science, 28*(4), 273–279.

Carpita, N., Sabularse, D., Montezinos, D., & Delmer, D. P., (1979). Determination of the pore size of cell walls of living plant cells. *Science, 205*(4411), 1144–1147.

Chaney, R. L., (1993). Zinc phytotoxicity. In: Robson, A. D., (ed.), *Zinc in Soils and Plants* (pp. 135–150). Springer, Dordrecht.

Chattopadhyay, S., Dash, S. K., Kar, M. S., et al., (2014). Chitosan-modified cobalt oxide nanoparticles stimulate TNF-α-mediated apoptosis in human leukemic cells. *J. Biol. Inorg. Chem., 19*, 399. doi: 10.1007/s00775-013-1085-2.

Chawengkijwanich, C., & Hayata, Y., (2008). Development of TiO_2 powder-coated food packaging film and its ability to inactivate *Escherichia coli in vitro* and in actual tests. *International Journal of Food Microbiology, 123*(3), 288–292.

Chen, H., & Yada, R., (2011). Nanotechnologies in agriculture: New tools for sustainable development. *Trends in Food Science and Technology, 22*(11), 585–594.

Chinnamuthu, C., & Boopathi, P. M., (2009). Nanotechnology and agroecosystem. *Madras Agricultural Journal, 96*(1/6), 17–31.

Commission, E., (2018). *Risk Governance of Nanotechnology (RIA).* https://ec.europa.eu/info/funding-tenders/opportunities/portal/screen/opportunities/topic-details/nmbp-13-2018 (accessed on 8 July 2021).

Connon, R. E., Geist, J., & Werner, I., (2012). Effect-based tools for monitoring and predicting the ecotoxicological effects of chemicals in the aquatic environment. *Sensors, 12*(9), 12741–12771.

Corredor, E., Testillano, P. S., Coronado, M. J., González-Melendi, P., Fernández-Pacheco, R., Marquina, C., et al., (2009). Nanoparticle penetration and transport in living pumpkin plants: *In situ* subcellular identification. *BMC Plant Biology, 9*(1), 45.

Cruz-Romero, M., Murphy, T., Morris, M., Cummins, E., & Kerry, J., (2013). Antimicrobial activity of chitosan, organic acids and nano-sized solubilisates for potential use in smart antimicrobially-active packaging for potential food applications. *Food Control, 34*(2), 393–397.

Cursino, L., Li, Y., Zaini, P. A., De La Fuente, L., Hoch, H. C., & Burr, T. J., (2009). Twitching motility and biofilm formation are associated with tonB1 in *Xylella fastidiosa. FEMS Microbiology Letters, 299*(2), 193–199.

Cutter, C. N., Willett, J., & Siragusa, G., (2001). Improved antimicrobial activity of nisin-incorporated polymer films by formulation change and addition of food-grade chelator. *Letters in Applied Microbiology, 33*(4), 325–328.

Da Silva, L. C., Oliva, M. A., Azevedo, A. A., & De Araujo, J. M., (2006). Responses of restinga plant species to pollution from an iron pelletization factory. *Water, Air, and Soil Pollution, 175*(1–4), 241–256.

Dasgupta, N., Ranjan, S., Chakraborty, A. R., Ramalingam, C., Shanker, R., & Kumar, A., (2016). Nanoagriculture and water quality management. In: Ranjan, S., Dasgupta, N., & Lichtfouse, E., (eds.), *Nanoscience in Food and Agriculture* (pp. 1–42). Springer International Publishing, Switzerland.

Dasgupta, N., Ranjan, S., Mundekkad, D., Ramalingam, C., Shanker, R., & Kumar, A., (2015). Nanotechnology in agro-food: From field to plate. *Food Research International, 69*, 381–400.

De Berardis, B., Civitelli, G., Condello, M., Lista, P., Pozzi, R., Arancia, G., & Meschini, S., (2010). Exposure to ZnO nanoparticles induces oxidative stress and cytotoxicity in human colon carcinoma cells. *Toxicology and Applied Pharmacology, 246*(3), 116–127.

De Matteis, V., (2017). Exposure to inorganic nanoparticles: Routes of entry, immune response, biodistribution and *in vitro/in vivo* toxicity evaluation. *Toxics, 5*(4), 29.

Debnath, N., Das, S., Seth, D., Chandra, R., Bhattacharya, S. C., & Goswami, A., (2011). Entomotoxic effect of silica nanoparticles against *Sitophilus oryzae* (L.). *Journal of Pest Science, 84*(1), 99–105.

Dekkers, S., Krystek, P., Peters, R. J., Lankveld, D. P., Bokkers, B. G., Hoeven-Arentzen, P. H. V., et al., (2011). Presence and risks of nanosilica in food products. *Nanotoxicology, 5*(3), 393–405.

Dhingra, R., Naidu, S., Upreti, G., & Sawhney, R., (2010). Sustainable nanotechnology: Through green methods and life-cycle thinking. *Sustainability, 2*(10), 3323–3338.

Ditta, A., (2012). How helpful is nanotechnology in agriculture? *Advances in Natural Sciences: Nanoscience and Nanotechnology, 3*(3), 033002.

Dixit, R., Malaviya, D., Pandiyan, K., Singh, U., Sahu, A., & Shukla, R., (2015). Bioremediation of heavy metals from soil and aquatic environment: An overview of principles and criteria of fundamental processes. *Sustainability, 7*(2), 2189–2212.

Dudefoi, W., Moniz, K., Allen-Vercoe, E., Ropers, M. H., & Walker, V. K., (2017). Impact of food-grade and nano-TiO_2 particles on a human intestinal community. *Food and Chemical Toxicology, 106*, 242–249.

EFSA, (2016). *Annual Report of the EFSA Scientific Network of Risk Assessment of Nanotechnologies in Food and Feed for 2015*. https://efsa.onlinelibrary.wiley.com/doi/epdf/10.2903/sp.efsa.2016.EN-939 (accessed on 8 July 2021).

Fernández-Baldo, M. A., Messina, G. A., Sanz, M. I., & Raba, J., (2009). Screen-printed immunosensor modified with carbon nanotubes in a continuous-flow system for the *Botrytis cinerea* determination in apple tissues. *Talanta, 79*(3), 681–686.

Frederiksen, H. K., Kristensen, H. G., & Pedersen, M., (2003). Solid lipid microparticle formulations of the pyrethroid gamma-cyhalothrin-incompatibility of the lipid and the pyrethroid and biological properties of the formulations. *Journal of Controlled Release, 86*(2, 3), 243–252.

Frenk, S., Ben-Moshe, T., Dror, I., Berkowitz, B., & Minz, D., (2013). Effect of metal oxide nanoparticles on microbial community structure and function in two different soil types. *PLoS One, 8*(12), e84441.

Fröhlich, E., & Salar-Behzadi, S., (2014). Toxicological assessment of inhaled nanoparticles: Role of *in vivo*, *ex vivo*, *in vitro*, and *in silico* studies. *International Journal of Molecular Sciences, 15*(3), 4795–4822.

Fujino, T., & Itoh, T., (1998). Changes in pectin structure during epidermal cell elongation in pea (*Pisum sativum*) and its implications for cell wall architecture. *Plant and Cell Physiology, 39*(12), 1315–1323.

Gajbhiye, M., Kesharwani, J., Ingle, A., Gade, A., & Rai, M., (2009). Fungus-mediated synthesis of silver nanoparticles and their activity against pathogenic fungi in combination with fluconazole. *Nanomedicine: Nanotechnology, Biology and Medicine, 5*(4), 382–386.

Gardea-Torresdey, J. L., Rico, C. M., & White, J. C., (2014). Trophic transfer, transformation, and impact of engineered nanomaterials in terrestrial environments. *Environmental Science and Technology, 48*(5), 2526–2540.

Gelvin, S., (2003). *Agrobacterium*-mediated plant transformation: The "gene-jockeying" tool: The biology behind. *Microbiol. Mol. Biol. Rev., 67*, 16–37.

Ghormade, V., Deshpande, M. V., & Paknikar, K. M., (2011). Perspectives for nano-biotechnology enabled protection and nutrition of plants. *Biotechnology Advances, 29*(6), 792–803.

González-Gálvez, D., Janer, G., Vilar, G., Vílchez, A., & Vázquez-Campos, S., (2017). The life cycle of engineered nanoparticles. In: Tran, L., Bañares, M. A., & Rallo, R., (eds.), *Modeling the Toxicity of Nanoparticles* (pp. 41–69). Springer International Publishing, Switzerland.

González-Melendi, P., Fernández-Pacheco, R., Coronado, M. J., Corredor, E., Testillano, P., Risueño, M. C., et al., (2007). Nanoparticles as smart treatment-delivery systems in plants: Assessment of different techniques of microscopy for their visualization in plant tissues. *Annals of Botany, 101*(1), 187–195.

Gopinath, K., Gowri, S., Karthika, V., & Arumugam, A., (2014). Green synthesis of gold nanoparticles from fruit extract of *Terminalia arjuna*, for the enhanced seed germination activity of *Gloriosa superba. Journal of Nanostructure in Chemistry, 4*(3), 115.

Gou, N., Onnis-Hayden, A., & Gu, A. Z., (2010). Mechanistic toxicity assessment of nanomaterials by whole-cell-array stress genes expression analysis. *Environmental Science and Technology, 44*(15), 5964–5970.

Gruère, G., Narrod, C., & Abbott, L., (2011). *Agricultural, Food, and Water Nanotechnologies for the Poor*. International Food Policy Research Institute, Washington, DC.

Guan, H., Chi, D., Yu, J., & Li, H., (2010). Dynamics of residues from a novel nano-imidacloprid formulation in soybean fields. *Crop Protection, 29*(9), 942–946.

Guo, C., Xia, Y., Niu, P., Jiang, L., Duan, J., Yu, Y., et al., (2015). Silica nanoparticles induce oxidative stress, inflammation, and endothelial dysfunction *in vitro* via activation of the MAPK/Nrf2 pathway and nuclear factor-κB signaling. *International Journal of Nanomedicine, 10*, 1463.

Gupta, R., & Xie, H., (2018). Nanoparticles in daily life: Applications, toxicity and regulations. *Journal of Environmental Pathology, Toxicology and Oncology, 37*(3).

Haas, P. M., (1992). Epistemic communities and international policy coordination. *International Organization, 46*(1), 1–35.

Hafeez, A., Razzaq, A., Mahmood, T., & Jhanzab, H. M., (2015). Potential of copper nanoparticles to increase growth and yield of wheat. *J. Nanosci. Adv. Technol., 1*(1), 6–11.

Halder, D., Mitra, A., Bag, S., Raychaudhuri, U., & Chakraborty, R., (2011). Study on gelatin-silver nanoparticle composite towards the development of bio-based antimicrobial film. *Journal of Nanoscience and Nanotechnology, 11*(12), 10374–10378.

Hardy, A., Benford, D., Halldorsson, T., Jeger, M. J., Knutsen, H. K., More, S., et al., (2018). Guidance on risk assessment of the application of nanoscience and nanotechnologies in the food and feed chain: Part 1, human and animal health. *EFSA Journal, 16*(7).

Hartmann, N. I. B., Skjolding, L. M., Hansen, S. F., Baun, A., Kjølholt, J., & Gottschalk, F., (2014). *Environmental Fate and Behavior of Nanomaterials: New Knowledge on Important Transformation Processes*. Danish Environmental Protection Agency, Denmark.

He, L., Liu, Y., Mustapha, A., & Lin, M., (2011). Antifungal activity of zinc oxide nanoparticles against *Botrytis cinerea* and *Penicillium expansum. Microbiological Research, 166*(3), 207–215.

He, X., Deng, H., & Hwang, H. M., (2019). The current application of nanotechnology in food and agriculture. *Journal of Food and Drug Analysis, 27*(1), 1–21. https://doi.org/10.1016/j.jfda.2018.12.002.

Heredia, A., Guillen, R., Jimenez, A., & Fernandez-Bolanos, J., (1993). Plant cell wall structure. *Revista Espanola de Ciencia y Tecnologia de Alimentos, 33*(2), 113–131.

Hodoroaba, V. D., & Mielke, J., (2014). *Templates for Nanomaterial Characterization of Tier 1 and Tier 2 Measurement Methods*. Nanodefine technical report D7.2, nanodefine consortium, Wageningen.

Hossain, M. Z., & Kleve, M. G., (2011). Nickel nanowires induced and reactive oxygen species-mediated apoptosis in human pancreatic adenocarcinoma cells. *International Journal of Nanomedicine, 6*, 1475.

Huggett, D. B., Cook, J. C., Ericson, J. F., & Williams, R. T., (2003). A theoretical model for utilizing mammalian pharmacology and safety data to prioritize potential impacts of human pharmaceuticals to fish. *Human and Ecological Risk Assessment, 9*(7), 1789–1799.

Iuclid, F., (2000). *International Uniform Chemical Information Database*. European Commission ISPRA CD ROM.

Jafar, G., & Hamzeh, G., (2013). Ecotoxicity of nanomaterials in soil. *Ann. Biol. Res., 4*(1), 86–92.

Jain, A., Ranjan, S., Dasgupta, N., & Ramalingam, C., (2018). Nanomaterials in food and agriculture: An overview on their safety concerns and regulatory issues. *Critical Reviews in Food Science and Nutrition, 58*(2), 297–317.

Jain, N., Bhargava, A., Majumdar, S., Tarafdar, J. C., & Panwar, J., (2011). Extracellular biosynthesis and characterization of silver nanoparticles using *Aspergillus flavus* NJP08: A mechanism perspective. *Nanoscale, 3*(2), 635–641.

Jantunen, A. P. K., Gottardo, S., Rasmussen, K., & Crutzen, H. P., (2018). An inventory of ready-to-use and publicly available tools for the safety assessment of nanomaterials. *NanoImpact, 12*, 18–28.

Jeffries, P., Gianinazzi, S., Perotto, S., Turnau, K., & Barea, J. M., (2003). The contribution of arbuscular mycorrhizal fungi in sustainable maintenance of plant health and soil fertility. *Biology and Fertility of Soils, 37*(1), 1–16.

Jia, G., Wang, H., Yan, L., Wang, X., Pei, R., Yan, T., et al., (2005). Cytotoxicity of carbon nanomaterials: Single-wall nanotube, multi-wall nanotube, and fullerene. *Environmental Science and Technology, 39*(5), 1378–1383.

Jiang, J., Oberdörster, G., & Biswas, P., (2009). Characterization of size, surface charge, and agglomeration state of nanoparticle dispersions for toxicological studies. *Journal of Nanoparticle Research, 11*(1), 77–89.

Jin, T., & He, Y., (2011). Antibacterial activities of magnesium oxide (MgO) nanoparticles against foodborne pathogens. *Journal of Nanoparticle Research, 13*(12), 6877–6885.

Jo, Y. K., Kim, B. H., & Jung, G., (2009). Antifungal activity of silver ions and nanoparticles on phytopathogenic fungi. *Plant Disease, 93*(10), 1037–1043.

Johnston, J. H., Grindrod, J. E., Dodds, M., & Schimitschek, K., (2008). Composite nano-structured calcium silicate phase change materials for thermal buffering in food packaging. *Current Applied Physics, 8*(3, 4), 508–511.

Joseph, T., & Morrison, M., (2006). Nanotechnology in agriculture and food. *Nanoforum Report, 2*(2), 3.

Juan, W., Kunhui, S. H. U., Zhang, L. I., & Youbin, S. I., (2017). Effects of silver nanoparticles on soil microbial communities and bacterial nitrification in suburban vegetable soils. *Pedosphere, 27*(3), 482–490.

Kah, M., (2015). Nanopesticides and nanofertilizers: Emerging contaminants or opportunities for risk mitigation? *Frontiers in Chemistry, 3*, 64.

Kah, M., Beulke, S., Tiede, K., & Hofmann, T., (2013). Nanopesticides: State of knowledge, environmental fate, and exposure modeling. *Critical Reviews in Environmental Science and Technology, 43*(16), 1823–1867.

Kandasamy, S., & Prema, R. S., (2015). Methods of synthesis of nanoparticles and its applications. *J. Chem. Pharm. Res., 7*, 278–285.

Kanjana, D., (2015). Potential applications of nanotechnology in major agriculture divisions: A review. *International Journal of Agriculture, Environment and Biotechnology, 8*(3), 699.

Kannaujia, R., Srivastava, C. M., Prasad, V., Singh, B. N., & Pandey, V., (2019). *Phyllanthus emblica* fruit extract stabilized biogenic silver nanoparticles as a growth promoter of wheat varieties by reducing ROS toxicity. *Plant Physiology and Biochemistry, 142*, 460–471.

Kansara, K., Patel, P., Shah, D., Shukla, R. K., Singh, S., Kumar, A., & Dhawan, A., (2015). TiO_2 nanoparticles induce DNA double-strand breaks and cell cycle arrest in human alveolar cells. *Environmental and Molecular Mutagenesis, 56*(2), 204–217.

Kaphle, A., Navya, P., Umapathi, A., & Daima, H. K., (2018). Nanomaterials for agriculture, food and environment: Applications, toxicity and regulation. *Environmental Chemistry Letters, 16*(1), 43–58.

Karimi, E., & Fard, E. M., (2017). Nanomaterial effects on soil microorganisms. In: Ghorbanpour, M., Khanuja, M., & Varma, A., (eds.), *Nanoscience and Plant-Soil Systems* (pp. 137–200). Springer International Publishing, Switzerland.

Kashyap, P. L., Xiang, X., & Heiden, P., (2015). Chitosan nanoparticle based delivery systems for sustainable agriculture. *International Journal of Biological Macromolecules, 77*, 36–51.

Keller, A. A., McFerran, S., Lazareva, A., & Suh, S., (2013). Global life cycle releases of engineered nanomaterials. *Journal of Nanoparticle Research, 15*(6), 1692.

Khatem, R., Bakthi, A., & Hermosín, M., (2016). *Comparison of the Systemic Nanoherbicide Imazamox-LDH Obtained by Direct Synthesis and Reconstruction: Preliminary Results.* http://www.setcor.org/conferences/Nanotech-France-2016/conference-program/14 (accessed on 8 July 2021).

Khodakovskaya, M., Dervishi, E., Mahmood, M., Xu, Y., Li, Z., Watanabe, F., & Biris, A. S., (2009). Carbon nanotubes are able to penetrate plant seed coat and dramatically affect seed germination and plant growth. *ACS Nano, 3*(10), 3221–3227.

Khot, L. R., Sankaran, S., Maja, J. M., Ehsani, R., & Schuster, E. W., (2012). Applications of nanomaterials in agricultural production and crop protection: A review. *Crop Protection, 35*, 64–70.

Kingsley, J. D., Ranjan, S., Dasgupta, N., & Saha, P., (2013). Nanotechnology for tissue engineering: Need, techniques and applications. *Journal of Pharmacy Research, 7*(2), 200–204.

Kiser, M., Westerhoff, P., Benn, T., Wang, Y., Perez-Rivera, J., & Hristovski, K., (2009). Titanium nanomaterial removal and release from wastewater treatment plants. *Environmental Science and Technology, 43*(17), 6757–6763.

Knox, J., (1995). The extracellular matrix in higher plants. 4. Developmentally regulated proteoglycans and glycoproteins of the plant cell surface. *The FASEB Journal, 9*(11), 1004–1012.

Kole, C., Kole, P., Randunu, K. M., Choudhary, P., Podila, R., Ke, P. C., et al., (2013). Nanobiotechnology can boost crop production and quality: First evidence from increased plant biomass, fruit yield and phytomedicine content in bitter melon (*Momordica charantia*). *BMC Biotechnology, 13*(1), 37.

Kuhnel, D., & Nickel, C., (2014). The OECD expert meeting on ecotoxicology and environmental fate - towards the development of improved OECD guidelines for the testing of nanomaterials. *Science of the Total Environment, 472*, 347–353.

Kumar, A., Najafzadeh, M., Jacob, B. K., Dhawan, A., & Anderson, D., (2014). Zinc oxide nanoparticles affect the expression of p53, Ras p21 and JNKs: An *ex vivo*/*in vitro* exposure study in respiratory disease patients. *Mutagenesis, 30*(2), 237–245.

Laborie, M., (2009). Bacterial cellulose and its polymeric nanocomposites. In: Lucia, L., & Rojas, O. R., (eds.), *The Nanoscience and Technology of Renewable Biomaterials* (pp. 231–271). Wiley, Chichester.

Lankadurai, B. P., Nagato, E. G., Simpson, A. J., & Simpson, M. J., (2015). Analysis of *Eisenia fetida* earthworm responses to sub-lethal C60 nanoparticle exposure using ^{1}H-NMR based metabolomics. *Ecotoxicology and Environmental Safety, 120*, 48–58. https://doi.org/10.1016/j.ecoenv.2015.05.020.

Lebov, J., Grieger, K., Womack, D., Zaccaro, D., Whitehead, N., Kowalcyk, B., & MacDonald, P. D. M., (2017). A framework for one health research. *One Health, 3*, 44–50.

Leo, B. F., Chen, S., Kyo, Y., Herpoldt, K. L., Terrill, N. J., Dunlop, I. E., et al., (2013). The stability of silver nanoparticles in a model of pulmonary surfactant. *Environmental Science and Technology, 47*(19), 11232–11240. doi: 10.1021/es403377p.

Leo, B. F., Sarah, F., Daniel, G. C., Ioannis, T., Pakatip, R., Angela, G., David, M., et al., (2019). Label-free TOF-SIMS imaging of sulfur producing enzymes inside microglia cells following exposure to silver nanowires. *ACS Analytical Chemistry.* doi: 10.1021/acs.analchem.9b01704.

Lin, D., & Xing, B., (2007). Phytotoxicity of nanoparticles: Inhibition of seed germination and root growth. *Environmental Pollution, 150*(2), 243–250.

Liu, F., Wen, L. X., Li, Z. Z., Yu, W., Sun, H. Y., & Chen, J. F., (2006). Porous hollow silica nanoparticles as controlled delivery system for water-soluble pesticide. *Materials Research Bulletin, 41*(12), 2268–2275.

Liu, J., Wang, F. H., Wang, L. L., Xiao, S. Y., Tong, C. Y., Tang, D. Y., & Liu, X. M., (2008). Preparation of fluorescence starch-nanoparticle and its application as plant transgenic vehicle. *Journal of Central South University of Technology, 15*(6), 768–773.

Liu, R., & Lal, R., (2015). Potentials of engineered nanoparticles as fertilizers for increasing agronomic productions. *Science of the Total Environment, 514*, 131–139.

Liu, R., Zhang, H., & Lal, R., (2016). Effects of stabilized nanoparticles of copper, zinc, manganese, and iron oxides in low concentrations on lettuce (*Lactuca sativa*) seed germination: Nanotoxicants or nanonutrients? *Water, Air, and Soil Pollution, 227*(1), 42.

Liu, X., Zhang, F., Zhang, S., He, X., Fang, R., Feng, Z., & Wang, Y. J., (2005). Effects of nano-ferric oxide on the growth and nutrients absorption of peanut. *Plant Nutr. Fert. Sci., 11*, 14–18.

Liu, Y., Tong, Z., & Prudhomme, R. K., (2008). Stabilized polymeric nanoparticles for controlled and efficient release of bifenthrin. *Pest Management Science: Formerly Pesticide Science, 64*(8), 808–812.

Lombi, E., Donner, E., Dusinska, M., & Wickson, F., (2019). A one health approach to managing the applications and implications of nanotechnologies in agriculture. *Nature Nanotechnology, 14*(6), 523.

Lowry, G. V., Avellan, A., & Gilbertson, L. M., (2019). Opportunities and challenges for nanotechnology in the agri-tech revolution. *Nature Nanotechnology, 14*(6), 517.

Lowry, G. V., Hotze, E. M., Bernhardt, E. S., Dionysiou, D. D., Pedersen, J. A., Wiesner, M. R., & Xing, B., (2010). Environmental occurrences, behavior, fate, and ecological effects of nanomaterials: An introduction to the special series. *Journal of Environmental Quality, 39*(6), 1867–1874.

Ma, X., Geisler-Lee, J., Deng, Y., & Kolmakov, A., (2014). Corrigendum to "interactions between engineered nanoparticles (ENPs) and plants: Phytotoxicity, uptake and accumulation." *Science of the Total Environment, 481*, 635.

Ma, Y., Kuang, L., He, X., Bai, W., Ding, Y., Zhang, Z., et al., (2010). Effects of rare earth oxide nanoparticles on root elongation of plants. *Chemosphere, 78*(3), 273–279.

Maarouf, S., Tazi, B., & Guenoun, F., (2016). Synthesis and characterization of new composite membranes based on polyvinylpyrrolidone, polyvinyl alcohol, sulfosuccinic acid, phosphomolybdic acid, and silica. *J. Chem. Pharmaceut. Res., 8*, 387–395.

Magdolenova, Z., Collins, A., Kumar, A., Dhawan, A., Stone, V., & Dusinska, M., (2014). Mechanisms of genotoxicity. A review of *in vitro* and *in vivo* studies with engineered nanoparticles. *Nanotoxicology, 8*(3), 233–278.

Mahdi, K. N. M., Peters, R., Van, D. P. M., Ritsema, C., & Geissen, V., (2018). Tracking the transport of silver nanoparticles in soil: A saturated column experiment. *Water, Air, and Soil Pollution, 229*(10), 334.

Mahmoodzadeh, H., Nabavi, M., & Kashefi, H., (2013). Effect of nanoscale titanium dioxide particles on the germination and growth of canola (*Brassica napus*). *Journal of Ornamental Plants, 3*(1), 25–32.

Majumdar, S., Almeida, I. C., Arigi, E. A., Choi, H., VerBerkmoes, N. C., Trujillo-Reyes, J., et al., (2015). Environmental effects of nanoceria on seed production of common bean (*Phaseolus vulgaris*): A proteomic analysis. *Environmental Science and Technology, 49*(22), 13283–13293.

Marchiol, L., (2018). Nanotechnology in agriculture: New opportunities and perspectives. In: Celic, O., (ed.), *New Visions in Plant Science*. doi: 10.5772/intechopen.74425.

Marmiroli, N., White, J., & Song, J., (2019). *Exposure to Engineered Nanomaterials in the Environment*. Elsevier, Amsterdam.

Matranga, V., & Corsi, I., (2012). Toxic effects of engineered nanoparticles in the marine environment: Model organisms and molecular approaches. *Marine Environmental Research, 76*, 32–40.

Matsumura, Y., Yoshikata, K., Kunisaki, S. I., & Tsuchido, T., (2003). Mode of bactericidal action of silver zeolite and its comparison with that of silver nitrate. *Appl. Environ. Microbiol., 69*(7), 4278–4281.

Mauriello, G., De Luca, E., La Storia, A., Villani, F., & Ercolini, D., (2005). Antimicrobial activity of a nisin-activated plastic film for food packaging. *Letters in Applied Microbiology, 41*(6), 464–469.

McBeath, T. M., & McLaughlin, M. J., (2014). Efficacy of zinc oxides as fertilizers. *Plant and Soil, 374*(1/2), 843–855.

Mewis, I., & Ulrichs, C., (2001). Effects of diatomaceous earth on water content of *Sitophilus granarius* (L.) (Col., Curculionidae) and its possible use in stored product protection. *Journal of Applied Entomology - Zeitschrift fur Angewandte Entomologie, 125*(6), 351–360.

Michels, C., Perazzoli, S., & Soares, H. M., (2017). Inhibition of an enriched culture of ammonia-oxidizing bacteria by two different nanoparticles: Silver and magnetite. *Science of the Total Environment, 586*, 995–1002.

Mishra, S., Keswani, C., Abhilash, P., Fraceto, L. F., & Singh, H. B., (2017). Integrated approach of agri-nanotechnology: Challenges and future trends. *Frontiers in Plant Science, 8*, 471.

Mishra, V., Mishra, R. K., Dikshit, A., & Pandey, A. C., (2014). Interactions of nanoparticles with plants: An emerging perspective in the agriculture industry. In: Ahmad, P., & Rasool, S., (eds.), *Emerging Technologies and Management of Crop Stress Tolerance* (pp. 159–180). Elsevier, Amsterdam.

Misra, A. N., Misra, M., & Singh, R., (2013). Nanotechnology in agriculture and food industry. *International Journal of Pure and Applied Sciences and Technology, 16*(2), 1.

Misra, P., Shukla, P. K., Pramanik, K., Gautam, S., & Kole, C., (2016). Nanotechnology for crop improvement. In: Kole, C., Sakthi, K. D., & Khodakovskaya, M. V., (eds.), *Plant Nanotechnology* (pp. 219–256). Springer, Cham.

Mitrano, D. M., Motellier, S., Clavaguera, S., & Nowack, B., (2015). Review of nanomaterial aging and transformations through the life cycle of nano-enhanced products. *Environment International, 77*, 132–147.

Mohammed, F. A., Balaji, K., Girilal, M., Kalaichelvan, P., & Venkatesan, R., (2009). Mycobased synthesis of silver nanoparticles and their incorporation into sodium alginate films for vegetable and fruit preservation. *Journal of Agricultural and Food Chemistry, 57*(14), 6246–6252.

Mondal, A., Basu, R., Das, S., & Nandy, P., (2011). Beneficial role of carbon nanotubes on mustard plant growth: An agricultural prospect. *Journal of Nanoparticle Research, 13*(10), 4519.

Monica, R. C., & Cremonini, R., (2009). Nanoparticles and higher plants. *Caryologia, 62*(2), 161–165.

Morones, J. R., Elechiguerra, J. L., Camacho, A., Holt, K., Kouri, J. B., Ramírez, J. T., & Yacaman, M. J., (2005). The bactericidal effect of silver nanoparticles. *Nanotechnology, 16*(10), 2346.

Mortimer, M., & Holden, P. A., (2019). Fate of engineered nanomaterials in natural environments and impacts on ecosystems. In: Marmiroli, N., White, J., & Song, J., (eds.), *Exposure to Engineered Nanomaterials in the Environment* (pp. 61–103). Elsevier, Amsterdam.

Mufamadi, M., & Sekhejane, P., (2017). Nanomaterial-based biosensors in agriculture application and accessibility in rural smallholding farms: Food security. In: Prasad, R., Kumar, M., & Kumar, V., (eds.), *Nanotechnology* (pp. 263–278). Springer, Singapore.

Mukherjee, A., Pokhrel, S., Bandyopadhyay, S., Mädler, L., Peralta-Videa, J. R., & Gardea-Torresdey, J. L., (2014). A soil mediated Phyto-toxicological study of iron-doped zinc oxide nanoparticles (Fe@ ZnO) in green peas (*Pisum sativum* L.). *Chemical Engineering Journal, 258*, 394–401.

Muniandy, S. T., Appaturi, J. N., Thong, K. L., Lai, C. W., Ibrahim, F., & Leo, B. F., (2019b). A reduced graphene oxide-titanium dioxide nanocomposite based electrochemical aptasensor for rapid and sensitive detection of *Salmonella enterica. Bioelectrochemistry, 127*, 136–144.

Muniandy, S., Teh, S. J., Thong, K. L., Thiha, A., Dinshaw, I. J., Lai, C. W., & Leo, B. F., (2019a). Carbon nanomaterial-based electrochemical biosensors for foodborne bacterial detection. *Critical Reviews in Analytical Chemistry, 49*(6), 1–24.

Nair, R., Varghese, S. H., Nair, B. G., Maekawa, T., Yoshida, Y., & Kumar, D. S., (2010). Nanoparticulate material delivery to plants. *Plant Science, 179*(3), 154–163.

Novo, M., Lahive, E., Díez-Ortiz, M., Matzke, M., Morgan, A. J., Spurgeon, D. J., et al., (2015). Different routes, same pathways: Molecular mechanisms under silver ion and nanoparticle exposures in the soil sentinel *Eisenia fetida*. *Environmental Pollution, 205*, 385–393.

Oberdörster, G., (2010). Safety assessment for nanotechnology and nanomedicine: Concepts of nanotoxicology. *Journal of Internal Medicine, 267*(1), 89–105.

Oerke, E. C., (2006). Crop losses to pests. *The Journal of Agricultural Science, 144*(1), 31–43.

Oomen, A. G., Steinhäuser, K. G., Bleeker, E. A. J., Van, B. F., Sips, A., Dekkers, S., et al., (2018). Risk assessment frameworks for nanomaterials: Scope, link to regulations, applicability, and outline for future directions in view of needed increase in efficiency. *NanoImpact, 9*, 1–13.

Opinions, S., (2007). *Scientific Committee on Emerging and Newly Identified Health Risks (SCENIHR) Opinion on the Appropriateness of the Risk Assessment Methodology in Accordance with the Technical Guidance Documents for New and Existing Substances for Assessing the Risks of Middle Materials*. Adopted by the SCENIHR at the 19th Plenary Meeting.

Opinions, S., (2009). *Scientific Committee on Emerging and Newly Identified Health Risks (SCENIHR) Opinion: Risk Assessment of Products of Nanotechnologies*. Adopted by the SCENIHR During the 28th Plenary Meeting.

Pal, S., Tak, Y. K., & Song, J. M., (2007). Does the antibacterial activity of silver nanoparticles depend on the shape of the nanoparticle? A study of the gram-negative bacterium *Escherichia coli*. *Appl. Environ. Microbiol., 73*(6), 1712–1720.

Panáček, A., Kolář, M., Večeřová, R., Prucek, R., Soukupova, J., Kryštof, V., et al., (2009). Antifungal activity of silver nanoparticles against *Candida* spp. *Biomaterials, 30*(31), 6333–6340.

Pandey, G., (2018). Challenges and future prospects of agri-nanotechnology for sustainable agriculture in India. *Environmental Technology and Innovation, 11*, 299–307.

Parisi, C., Vigani, M., & Rodríguez-Cerezo, E., (2015). Agricultural nanotechnologies: What are the current possibilities? *Nano Today, 10*(2), 124–127.

Pavel, A., & Creanga, D. E., (2005). Chromosomal aberrations in plants under magnetic fluid influence. *Journal of Magnetism and Magnetic Materials, 289*, 469–472.

Pérez-Masiá, R., López-Rubio, A., Fabra, M. J., & Lagaron, J. M., (2014). Use of electrohydrodynamic processing to develop nanostructured materials for the preservation of the cold chain. *Innovative Food Science and Emerging Technologies, 26*, 415–423.

Peters, R. J., Bouwmeester, H., Gottardo, S., Amenta, V., Arena, M., Brandhoff, P., et al., (2016). Nanomaterials for products and application in agriculture, feed and food. *Trends in Food Science and Technology, 54*, 155–164.

Pillai, S. K., Ray, S. S., Scriba, M., Ojijo, V., & Hato, M. J., (2013). Morphological and thermal properties of photodegradable biocomposite films. *Journal of Applied Polymer Science, 129*(1), 362–370.

Popov, V., Hinkov, I., Diankov, S., Karsheva, M., & Handzhiyski, Y., (2015). Ultrasound-assisted green synthesis of silver nanoparticles and their incorporation in antibacterial cellulose packaging. *Green Processing and Synthesis, 4*(2), 125–131.

Powers, K. W., Brown, S. C., Krishna, V. B., Wasdo, S. C., Moudgil, B. M., & Roberts, S. M., (2006). Research strategies for safety evaluation of nanomaterials. Part VI. Characterization of nanoscale particles for toxicological evaluation. *Toxicological Sciences, 90*(2), 296–303.

Prasad, R., Bhattacharyya, A., & Nguyen, Q. D., (2017). Nanotechnology in sustainable agriculture: Recent developments, challenges, and perspectives. *Frontiers in Microbiology, 8*, 1014.

Prasad, R., Kumar, V., & Prasad, K. S., (2014). Nanotechnology in sustainable agriculture: Present concerns and future aspects. *African Journal of Biotechnology, 13*(6), 705–713.

Prasad, R., Pandey, R., & Barman, I., (2016). Engineering tailored nanoparticles with microbes: Quo vadis? *Wiley Interdisciplinary Reviews: Nanomedicine and Nanobiotechnology, 8*(2), 316–330.

Pulido-Reyes, G., Leganes, F., Fernández-Piñas, F., & Rosal, R., (2017). Bio-nano interface and environment: A critical review. *Environmental Toxicology and Chemistry, 36*(12), 3181–3193.

Pullagurala, V. L. R., Adisa, I. O., Rawat, S., White, J. C., Zuverza-Mena, N., Hernandez-Viezcas, J. A., et al., (2019). Fate of engineered nanomaterials in agro-environments and impacts on agroecosystems. In: Marmiroli, N., White, J., & Song, J., (eds.), *Exposure to Engineered Nanomaterials in the Environment* (pp. 105–142). Elsevier, Amsterdam.

Racuciu, M., & Creanga, D., (2007). Cytogenetic changes induced by aqueous ferrofluids in agricultural plants. *Journal of Magnetism and Magnetic Materials, 311*(1), 288–290.

Racuciu, M., & Creanga, D., (2007). TMA-OH coated magnetic nanoparticles internalized in vegetal tissue. *Romanian Journal of Physics, 52*(3, 4), 395.

Rai, M., & Ingle, A., (2012). Role of nanotechnology in agriculture with special reference to management of insect pests. *Applied Microbiology and Biotechnology, 94*(2), 287–293.

Raies, A. B., & Bajic, V. B., (2016). *In silico* toxicology: Computational methods for the prediction of chemical toxicity. *Wiley Interdisciplinary Reviews: Computational Molecular Science, 6*(2), 147–172.

Ramachandraiah, K., Han, S. G., & Chin, K. B., (2015). Nanotechnology in meat processing and packaging: Potential applications: A review. *Asian-Australasian Journal of Animal Sciences, 28*(2), 290.

Rani, P. U., Yasur, J., Loke, K. S., & Dutta, D., (2016). Effect of synthetic and biosynthesized silver nanoparticles on growth, physiology and oxidative stress of water hyacinth: *Eichhornia crassipes* (Mart.) Solms. *Acta Physiologiae Plantarum, 38*(2), 58.

Rasmussen, K., Rauscher, H., Kearns, P., González, M., & Sintes, J. R., (2019). Developing OECD test guidelines for regulatory testing of nanomaterials to ensure mutual acceptance of test data. *Regulatory Toxicology and Pharmacology, 104*, 74–83.

Ravishankar, R. V., & Jamuna, B. A., (2011). Nanoparticles and their potential application as antimicrobials. In: Méndez-Vilas, A., (ed.), *Communicating Current Research and Technological Advances* (pp. 197–209). Formatex, Spain.

Roy, A., Singh, S. K., Bajpai, J., & Bajpai, A. K., (2014). Controlled pesticide release from biodegradable polymers. *Central European Journal of Chemistry, 12*(4), 453–469.

Rungraeng, N., Cho, Y. C., Yoon, S. H., & Jun, S., (2012). Carbon nanotube-polytetrafluoroethylene nanocomposite coating for milk fouling reduction in plate heat exchanger. *Journal of Food Engineering, 111*(2), 218–224.

Sadak, M. S., (2019). Impact of silver nanoparticles on plant growth, some biochemical aspects, and yield of fenugreek plant (*Trigonella foenum-graecum*). *Bulletin of the National Research Centre, 43*(1), 38.

Sager, T. M., Porter, D. W., Robinson, V. A., Lindsley, W. G., Schwegler-Berry, D. E., & Castranova, V., (2007). Improved method to disperse nanoparticles for *in vitro* and *in vivo* investigation of toxicity. *Nanotoxicology, 1*(2), 118–129.

Sanchez-Garcia, M. D., Lopez-Rubio, A., & Lagaron, J. M., (2010). Natural micro and nanobiocomposites with enhanced barrier properties and novel functionalities for food biopackaging applications. *Trends in Food Science and Technology, 21*(11), 528–536.

Sarkar, B., Bhattacharjee, S., Daware, A., Tribedi, P., Krishnani, K., & Minhas, P., (2015). Selenium nanoparticles for stress-resilient fish and livestock. *Nanoscale Research Letters, 10*(1), 371.

Sasson, Y., Levy-Ruso, G., Toledano, O., & Ishaaya, I., (2007). Nanosuspensions: Emerging novel agrochemical formulations. In: Ishaaya, I., Nauen, R., & Horowitz, A. R., (eds.), *Insecticides Design Using Advanced Technologies* (pp. 1–39). Springer-Verlag, Berlin Heidelberg.

Scott-Fordsmand, J. J., Peijnenburg, W. J. G. M., Semenzin, E., Nowack, B., Hunt, N., Hristozov, D., et al., (2017). Environmental risk assessment strategy for nanomaterials. *International Journal of Environmental Research and Public Health, 14*(10), 1251. doi: 10.3390/ijerph14101251.

Sharon, M., Choudhary, A. K., & Kumar, R., (2010). Nanotechnology in agricultural diseases and food safety. *Journal of Phytology, 2*(4), 83–92.

Shatkin, J. A., & Kim, B., (2015). Cellulose nanomaterials: Life cycle risk assessment, and environmental health and safety roadmap. *Environmental Science: Nano, 2*(5), 477–499.

Siani, N. G., Fallah, S., Pokhrel, L. R., & Rostamnejadi, A., (2017). Natural amelioration of zinc oxide nanoparticle toxicity in fenugreek (*Trigonella foenum-gracum*) by arbuscular mycorrhizal (*Glomus intraradices*) secretion of glomalin. *Plant Physiology and Biochemistry, 112*, 227–238.

Sillen, W. M. A., Thijs, S., Abbamondi, G. R., Janssen, J., Weyens, N., White, J. C., & Vangronsveld, J., (2015). Effects of silver nanoparticles on soil microorganisms and maize biomass are linked in the rhizosphere. *Soil Biology and Biochemistry, 91*, 14–22.

Singh, S., Singh, B. K., Yadav, S., & Gupta, A., (2015). Applications of nanotechnology in agriculture and their role in disease management. *Res. J. Nanosci. Nanotechnol., 5*(1), 1–5.

Singh, T., Shukla, S., Kumar, P., Wahla, V., Bajpai, V. K., & Rather, I. A., (2017). Application of nanotechnology in food science: Perception and overview. *Frontiers in Microbiology, 8*, 1501.

Sivamani, E., DeLong, R. K., & Qu, R., (2009). Protamine-mediated DNA coating remarkably improves bombardment transformation efficiency in plant cells. *Plant Cell Reports, 28*(2), 213–221.

Smita, S., Gupta, S. K., Bartonova, A., Dusinska, M., Gutleb, A. C., & Rahman, Q., (2012). Nanoparticles in the environment: Assessment using the causal diagram approach. *Environmental Health, 11*(1), S13.

Srivastava, A. K., Dev, A., & Karmakar, S., (2018). Nanosensors and nanobiosensors in food and agriculture. *Environmental Chemistry Letters, 16*(1), 161–182.

Stadler, T., Buteler, M., & Weaver, D. K., (2010). Novel use of nanostructured alumina as an insecticide. *Pest Management Science: Formerly Pesticide Science, 66*(6), 577–579.

Stevenson, L. M., Dickson, H., Klanjscek, T., Keller, A. A., McCauley, E., & Nisbet, R. M., (2013). Environmental feedbacks and engineered nanoparticles: Mitigation of silver nanoparticle toxicity to *Chlamydomonas reinhardtii* by algal-produced organic compounds. *PLoS One, 8*(9), e74456.

Stone, V., Führ, M., Feindt, P. H., Bouwmeester, H., Linkov, I., Sabella, S., et al., (2018). The essential elements of a risk governance framework for current and future nanotechnologies. *Risk Analysis, 38*(7), 1321–1331.

Stone, V., Johnston, H., & Schins, R. P. F., (2009). Development of *in vitro* systems for nanotoxicology: Methodological considerations. *Critical Reviews in Toxicology, 39*(7), 613–626.

Suman, R. P., Jain, V. K., & Varma, A., (2010). Role of nanomaterials in symbiotic fungus growth enhancement. *Current Science, 99*(9), 1189–1191.

Tankhiwale, R., & Bajpai, S., (2012). Preparation, characterization and antibacterial applications of ZnO-nanoparticles coated polyethylene films for food packaging. *Colloids and Surfaces B: Biointerfaces, 90*, 16–20.

Thirumurugan, A., Ramachandran, S., & Shiamala, G. A., (2013). Combined effect of bacteriocin with gold nanoparticles against food spoiling bacteria: An approach for food packaging material preparation. *International Food Research Journal, 20*(4).

Torney, F., Trewyn, B. G., Lin, V. S. Y., & Wang, K., (2007). Mesoporous silica nanoparticles deliver DNA and chemicals into plants. *Nature Nanotechnology, 2*(5), 295.

Tsuji, K., (2001). Microencapsulation of pesticides and their improved handling safety. *Journal of Microencapsulation, 18*(2), 137–147.

Vale, G., Mehennaoui, K., Cambier, S., Libralato, G., Jomini, S., & Domingos, R. F., (2016). Manufactured nanoparticles in the aquatic environment-biochemical responses on freshwater organisms: A critical overview. *Aquatic Toxicology, 170*, 162–174.

Vijayakumar, P. S., Abhilash, O. U., Khan, B. M., & Prasad, B. L., (2010). Nanogold-loaded sharp-edged carbon bullets as plant-gene carriers. *Advanced Functional Materials, 20*(15), 2416–2423.

Wakabayashi, K., & Böger, P., (2004). Phytotoxic sites of action for molecular design of modern herbicides (Part 1): The photosynthetic electron transport system. *Weed Biology and Management, 4*(1), 8–18.

Wang, J., Deng, X., Zhang, F., Chen, D., & Ding, W., (2014). ZnO nanoparticle-induced oxidative stress triggers apoptosis by activating JNK signaling pathway in cultured primary astrocytes. *Nanoscale Research Letters, 9*(1), 117.

Wang, X., Liu, X., & Han, H., (2013). Evaluation of antibacterial effects of carbon nanomaterials against copper-resistant *Ralstonia solanacearum*. *Colloids and Surfaces B: Biointerfaces, 103*, 136–142.

Wani, A., & Shah, M., (2012). A unique and profound effect of MgO and ZnO nanoparticles on some plant pathogenic fungi. *Journal of Applied Pharmaceutical Science, 2*(3), 4.

Weir, A., Westerhoff, P., Fabricius, L., Hristovski, K., & Von, G. N., (2012). Titanium dioxide nanoparticles in food and personal care products. *Environmental Science and Technology, 46*(4), 2242–2250.

Wilhelmi, V., Fischer, U., Weighardt, H., Schulze-Osthoff, K., Nickel, C., Stahlmecke, B., et al., (2013). Zinc oxide nanoparticles induce necrosis and apoptosis in macrophages in a p47phox-and Nrf2-independent manner. *PLoS One, 8*(6), e65704.

Wu, L., & Liu, M., (2008). Preparation and properties of chitosan coated NPK compound fertilizer with controlled-release and water-retention. *Carbohydrate Polymers, 72*(2), 240–247.

Yamaaka, M., Hara, K., & Kudo, J, (2005). Bactericidal action of a silver ion solution on *Escherichia coli*, studied by energy-filtering transmission electron microscopy and proteoic analysis. *Appl. Enviro. Microbiol., 71*(11), 758–7593.

Yan, J., Huang, K., Wang Y., & Liu, S., (2005). Study on ani-pollution nano-preparation of dimethomorph and its performance. *Chinese Science Bulletin, 50*(2), 108–112.

Yang, F. L., Li, X. G., Zhu, F., & Lei, C. L., (2009). Structural characterization of nanoparticles loaded with garlic essential oil and their insecticidal activity against *Tribolium castaneum* (Herbst) (Coleoptera: Tenebrinidae). *Journal of Agricultural and Food Chemistry, 57*(21), 10156–10162.

Yang, J., Cao, W., & Rui, Y., (2017). Interactions between nanoparticles and plants: Phytotoxicity and defense mechanisms. *Journal of Plant Interactions, 12*(1), 158–169.

Yang, W., Dominici, F., Fortunati, E., Kenny, J. M., & Puglia, D., (2015). Melt free radical grafting of glycidyl methacrylate (GMA) onto fully biodegradable poly (lactic) acid films: Effect of cellulose nanocrystals and a masterbatch process. *RSC Advances, 5*(41), 32350–32357.

Yang, Y., Wang, J., Xiu, Z., & Alvarez, P. J. J., (013). Impacts of silver nanoparticles on cellular and transcriptional activity of nitrogen-cycling bacteria. *Environmental Toxicology and Chemistry, 32*(7), 1488–1494.

Yemmireddy V. K., Farrell, G. D., & Hung, Y. C., (2015). Development of titanium dioxide (TiO_2) nanocoatings on food contact surfaces and method to evaluate their durability and photocatalytic bactericidal property. *Journal of Food Science, 80*(8), N1903–N1911.

Yildiz, N., & Pala, A., (2012). Effects of small-diameter silver nanoparticles on microbial load in cow milk. *Journal of Dairy Science, 95*(3), 1119–1127.

Zemke-White, W., Clements, K., & Harris, P., (2000). Acid lysis of macroalgae by marine herbivorous fishes: Effects of acid pH on cell wall porosity. *Journal of Experimental Marine Biology and Ecology, 245*(1), 57–68.

Zhang, X., Qu, Y., Shen, W., Wang, J., Li, H., Zhang, Z., et al., (2016). Biogenic synthesis of gold nanoparticles by yeast *Magnusiomyces ingens* LH-F1 for catalytic reduction of nitrophenols. *Colloids and Surfaces A: Physicochemical and Engineering Aspects, 497*, 280–285.

Zheng, L., Hong, F., Lu, S., & Liu, C., (2005). Effect of nano-TiO_2 on strength of naturally aged seeds and growth of spinach. *Biological Trace Element Research, 104*(1), 83–91.

Zhu, H., Han, J., Xiao, J. Q., & Jin, Y., (2008). Uptake, translocation, and accumulation of manufactured iron oxide nanoparticles by pumpkin plants. *Journal of Environmental Monitoring, 10*(6), 713–717.

CHAPTER 12

SUSTAINABLE EXPLOITATION OF AGRICULTURAL, FORESTRY, AND FOOD RESIDUES FOR GREEN NANOTECHNOLOGY APPLICATIONS

LUCIANO PAULINO SILVA,[1,2,3] ARIANE PANDOLFO SILVEIRA,[1,2] CÍNTHIA CAETANO BONATTO,[1,4] EDUARDO FERNANDES BARBOSA,[5] KELLIANE ALMEIDA MEDEIROS,[6] LÍVIA CRISTINA DE SOUZA VIOL,[7] TATIANE MELO PEREIRA,[1,3] THAÍS RIBEIRO SANTIAGO,[8] VERA LÚCIA PERUSSI POLEZ,[7] and VICTORIA BAGGI MENDONÇA LAURIA[1,2]

[1]*Laboratório de Nanobiotecnologia (LNANO), Embrapa Recursos Genéticos e Biotecnologia, Pq. Est. Biol. Final W5 Norte, Asa Norte, Brasília–70770-917, DF, Brazil*

[2]*Programa de Pós-graduação em Nanociência e Nanobiotecnologia, Universidade de Brasília, Instituto de Ciências Biológicas, Asa Norte, Brasília–70910-900, DF, Brazil*

[3]*Programa de Pós-graduação em Ciências Biológicas (Biologia Molecular), Universidade de Brasília, Instituto de Ciências Biológicas, Asa Norte, Brasília–70910-900, DF, Brazil*

[4]*TecSinapse, Pesquisa Aplicada, São Paulo–04583-110, SP, Brazil*

[5]*Universidade Federal do Oeste da Bahia, Centro das Ciências Biológicas e da Saúde, Campus Reitor Edgard Santos, Rua Bertioga, 892, Morada Nobre I, Barreiras–47810-059, BA, Brazil*

[6]*Hospital das Forças Armadas, Diretoria Técnica de Ensino e Pesquisa, Divisão de Pesquisa, Setor HFA, Sudoeste, Brasília–70673-900, DF, Brazil*

[7]*Laboratório de Prospecção de Compostos Bioativos (LPCB), Embrapa Recursos Genéticos e Biotecnologia, Pq. Est. Biol. Final W5 Norte, Asa Norte, Brasília–70770-917, DF, Brazil*

[8]*Programa de Pós-Graduação em Fitopatologia, Universidade de Brasília, Instituto de Ciências Biológicas, Asa Norte, Brasília–70910-900, DF, Brazil*

12.1 INTRODUCTION

For centuries new technological achievements were completely disconnected from concerns about the environment. However, the last two decades changed the way how people explore the planet, and now a growing number of products and processes are based on the use of renewable resources and greener processes. In fact, this is true even to the highly technical areas, and nanotechnology is an example about how it is possible to develop innovative and sustainable solutions that satisfy customer demands without making negative impacts towards the environment. Currently, green nanotechnology represents a paradigm shift in the way nanotechnologists and their stakeholders deal with some of the most challenging issues related to the field. Indeed, virtually all current advances and outcomes covered by green nanotechnology take support from tangible research, development, and innovation (RD&I) projects with sustainability in focus and commonly using bioresources as raw materials and inputs.

Nowadays, agriculture, agroindustry, forestry, and food industry generate huge amounts of co-products, by-products, and residues. Notably, such materials depict an important business issue to be managed but otherwise can also be considered attractive economically and transformable into materials with high value-added, including nanomaterials. Really, waste valorization represents a major concern in the global economy of the 21st century, and nanotechnology offers the means of developing novel materials based on wastes and residues that exhibit sustainable aspects and innovative properties. Particularly, green nanotechnology solutions emerge from physical, chemical, biological, and engineering sciences as the most relevant example of integrating technological innovation with sustainability principles. Hence, the purpose is to achieve competitive advantages aligned with Sustainable

Development Goals to ensure environmental sustainability, healthy lives, and promote entrepreneurship initiatives.

Thus, areas like biology, biotechnology, biomedicine, chemical sciences, physical sciences, food sciences, and material sciences, and sectors like agriculture and industries can now benefit from green nanotechnology-based developments. Thereupon, these new technologies include nanosystems (nanomaterials and nanodevices) with high potential to break paradigms and solve real-world problems (Figure 12.1). In fact, these greener nanotechnologies can improve and even revolutionize the manner in which several areas and sectors perform their activities by offering new horizons and possibilities. Therefore, this chapter explores (but not exhausts) the myriad of recent advances and trends about RD&I opportunities related to the sustainable exploitation of renewable resources aiming at producing and applying novel nanotechnological solutions.

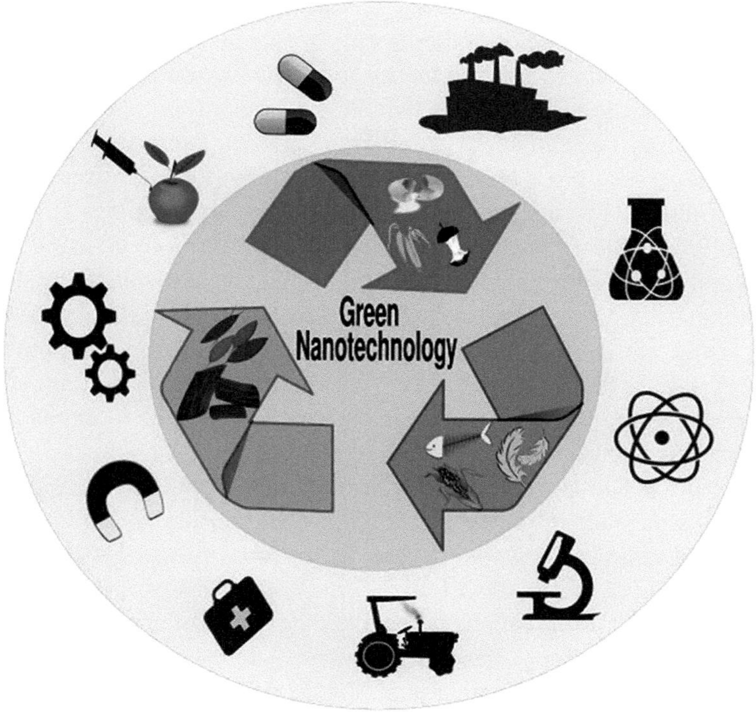

FIGURE 12.1 Green nanotechnology-based solutions from agricultural, forestry, and food residues.

12.2 BIOLOGICAL APPLICATIONS OF GREEN NANOTECHNOLOGY-BASED SOLUTIONS

In recent years, professionals from all over the world have been investing in the development of sustainable and eco-friendly nanotechnologies. Moreover, the production of nanomaterials by green routes aims at the rational use of biological resources from biodiversity and the various production chains, and numerous components can be used as precursors and/or essential blocks during the synthesis steps. In addition, green routes result in the production of nanomaterials with biocompatible, biodegradable, and low-cost production characteristics.

Notably, several biological resources, including animals, plants, algae, fungi, bacteria (as well as the diversity of compounds produced by them) and by-products derive from agricultural and food industry processes involving some of these organisms have the potential for use during green nanomaterials synthesis. This is because they act as sources of compounds that can be used as bioreductive agents, stabilizers, surfactants, among others. Particularly, this fact enables the production of a diverse range of nanomaterials, including metallic nanoparticles (MNPs) (e.g., using primary and secondary metabolites as reducing and stabilizing agents), polymeric nanoparticles (PNPs) (e.g., using alginate, carboxymethylcellulose, chitosan, and other biopolymers as structuring biomaterials), lipid nanoparticles (LNPs) (e.g., using oils and fats as essential constituents), liposomes (e.g., using phospholipids and other complex lipids from eggs and cells as synthetic membranes), virosomes (e.g., using viral proteins as self-assembling blocks), emulsions (e.g., using naturally sourced animal and vegetable oils as raw materials), nanocrystals (e.g., using partially hydrolyzed cellulose as nanostructured material), among others; and this is of great importance in expanding their biological applications.

Therefore, nanoparticles and other nanomaterials have been the subject of significant studies due to the extensive possibility of different compositions and characteristics according to the final properties of each nanostructure (Figure 12.2). MNPs from green synthesis processes are usually produced using biological resources in the presence of metal salts (e.g., gold, silver, copper, iron, etc.), resulting in nanosystems with different compositions, shapes, and sizes (Kumar et al., 2018). Hence, the use of MNPs is due to the broad inhibitory spectrum towards bacteria (Gram-positive and Gram-negative), fungi (uni- and multicellular), and viruses; biosensors and nanobiosensors development; and drug delivery systems. On the other hand, PNPs are nanostructured systems composed of polymers which have dimensions within the nanometric scale and used in encapsulation (nanocapsules) or entrapment (nanospheres) processes of numerous active principles (Crucho and Barros, 2017). PNPs are more commonly used for the delivery of hydrophilic components. Otherwise, LNPs are among the nanomaterials that can be used for the delivery of hydrophobic active ingredients, and which can be classified into solid lipid nanoparticles (SLNPs) and carrier lipid nanoparticles (CLNPs). SLNPs are LNPs that are in the solid phase at room temperature, and surfactants are used for emulsification (Mistry and Sarker, 2016), whereas CLNPs consist of liquid (oil) and solid (fat) phase lipids that result in LNPs with even more controlled encapsulation/entrapment properties due to the higher degree of crystalline imperfection (Zhang et al., 2013).

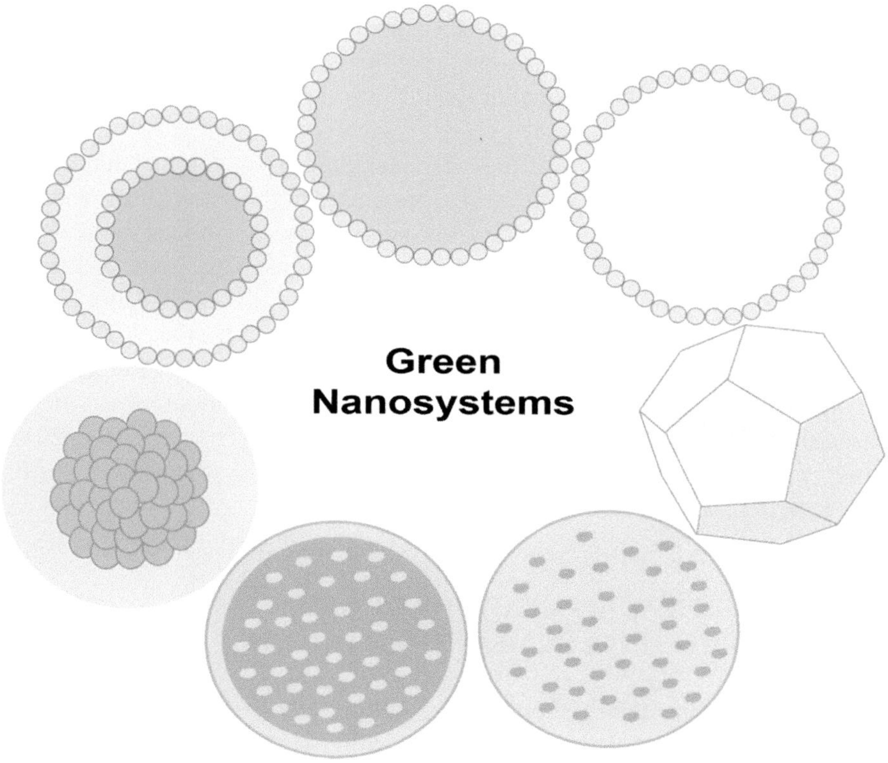

FIGURE 12.2 Stylized representations of some green nanosystem types.

Other types of nanosystems (e.g., emulsions and liposomes) have the potential to carry both hydrophobic and hydrophilic compounds with distinct yields. Then, emulsions are systems composed of two immiscible liquids, one of which is completely dispersed as globules (internal phase, discontinuous, or dispersed) in the other liquid (external phase, continuous, or dispersant) and normally stabilized by surfactants, in varying proportions according to desired characteristics (Tadros, 2013). Already liposomes are biologically inert and biocompatible spherical vesicles consisting of one (unilamellar) or several concentric (multilamellar) lipid bilayers that isolate one or more internal aqueous compartments from the external environment (Pattni et al., 2015).

Still, nanocrystals represent a class of nanomaterials that can be formed by soft hydrolysis of biopolymers, including cellulose, and this allows the production of structures with crystalline domains that show a significant increase in surface area/volume ratio (Albernaz et al., 2015). Another recent, but relevant type of nanomaterial is the carbon dots or carbon quantum dots (C-dots) which are zero-dimensional photoluminescent carbon-based materials exhibiting dimensions smaller than 20 nm (Sun et al., 2006; Sagbas and Sahiner, 2019).

Among biological applications of nanomaterials developed with basis on green nanotechnology approaches, it can be highlighted those focused on nanosystems aiming sustained-release (drugs, hormones, cosmetics, vaccines, fertilizers); nanomaterials for control of pathogens and pests (bacteria, fungi, viruses, parasites, insects) (Khan et al., 2018; Lateef et al., 2018) or showing activity against tumor cells (Al-Sheddi et al., 2018; Khan et al., 2018); nanomaterials exhibiting antioxidant properties (Lateef et al., 2018); nanomaterials showing anticoagulant activity (Lateef et al., 2018); nanosensors for detection of pathogens and diseases (Drygin et al., 2012; Zhao et al., 2014); and development of nanostructured systems aiming effluents remediation (Francis et al., 2017).

12.3 BIOMEDICAL APPLICATIONS OF GREEN NANOTECHNOLOGY-BASED SOLUTIONS

In the green synthesis of nanoparticles, additionally to organic compounds, different types of metals and non-metals, inorganic agents, can be used. In fact, MNPs (e.g., magnetic nanoparticles and noble metal nanoparticles) are among the most used type of nanomaterials aiming biomedical applications. Inside this group, MNPs produced by green synthesis approaches have been extensively studied and proposed to be used in biomedical applications in replacement of the traditional nanotechnology strategies. In recent years, such nanosystems have been particularly useful in diagnostics and therapeutics of different diseases (theranostics). Hence, these biomedical applications are possible because of the potential interaction of metals with microorganisms, eukaryotic cells, and molecules according to the specific physicochemical characteristics (Sharma et al., 2018).

The broad potential of green nanoparticles aiming biomedical applications also happens because their synthesis typically does not involve hazardous chemicals or expensive physical processes, being cheaper, biocompatible, and non-toxic. Thus, the choice of biosource, an organic compound, with known properties in traditional medicine, such as plants, algae, or microorganisms' extracts, can allow the formation and may also enhance green nanoparticles' activity (Sharma et al., 2018). In fact, there are many nanoparticles synthesized using plants, with well-known biomedical applications in the literature. Some of them exhibit the autofluorescence feature typically desired and used in diagnosis as a sensor or a biomarker, and also show selective oxygen release functionality towards the production of reactive oxygen species (ROS), used in therapy, hence leading cell death (Ovais et al., 2016).

Corroborating the potential biomedical application of MNPs produced by green routes, metals are used in traditional medicine for hundreds of years, but they commonly lead to toxicity and accumulation in the body when administered to patients in free forms, but if they are used in green synthesis towards the production of MNPs they typically show less toxicity and have the release of ions controlled according to the medium pH. Thus, considering a tissue microenvironment containing healthy cells which have neutral pH, the metal ions are not released, but in contrast tumor cells and injured tissues, which have acidic pH, would lead to the release of ions causing toxicity and even cell death (Ovais et al., 2016). Still, metal choice happens according to the desired application. For example, in bone reconstruction, titanium is used because it is more resistant (Razavi et al., 2015), while gold is used when catalytic action is sought (Debjani and Pratyusha, 2013) and silver is used for antimicrobial action (Ahmed et al., 2016). Titanium oxide nanoparticles (TiO_2 NPs) have been mainly applied in regenerative medicine (e.g., bone reconstruction, because this nanomaterial presents mechanical characteristics of high mechanical resistance and low corrosion rate. In addition, when TiO_2 NPs are exposed to ultraviolet radiation, they can promote the release of ROS and consequently exhibit a potent antimicrobial effect. The antimicrobial action of TiO_2 NPs produced using *Aspergillus flavus* as a reducer agent was observed in *Escherichia coli* (Rajakumar et al., 2012), and also by TiO_2 NPs synthesized using *Psidium guajava* extract as a reducer, which also showed antibacterial effects against *Staphylococcus aureus* (Santhoshkumar et al., 2014).

Additionally, magnetically-responsive metal oxide nanoparticles are also known for their antitumor activity, killing cancer cells in processes such as magnetic hyperthermia. In such cases, nanoparticles promote local heat when activated by an alternating magnetic field source. This activity can be enhanced by specifically functionalizing the nanoparticles with other molecules such as proteins and antibodies. Ramirez-Núñez et al. (2018) synthesized iron oxide nanoparticles with *Cinnamomum verum* and *Vanilla planifolia* extracts, which showed activity against microglial cells. In addition, Balaji et al. (2017) developed zirconia nanoparticles with *Eucalyptus globulus* extract which showed antioxidant and antitumor activity, mainly against human colon cells. The various nanoparticles activities are determined according to the composition, ratio between surface area and volume, and other physicochemical properties.

A wide spectrum of activities has been revealed in studies with silver nanoparticles (Ag NPs), gold nanoparticles (Au NPs), and zinc oxide nanoparticles (ZnO NPs). The MNPs synthesis using these metal ions has been studied since the early 21st century (Sonker et al., 2017). Au NPs exhibit antitumor, antibacterial, and antioxidant activities, and can also be used as drug carriers and imaging probes (Muthukumar et al., 2016). In the case of Ag NPs, several authors have reported these activities as well as antiangiogenic, antiviral, anti-inflammatory, and antifungal activities (Khatami et al., 2018). Still, ZnO NPs, besides acting as antibacterial, antitumor, antifungal, and drug delivery systems, they also exhibit antidiabetic effects by inducing pancreas insulin release (Bala et al., 2015). In the study of Thatoi et al. (2016), Ag NPs, and ZnO NPs produced with aqueous plant extracts of *Heritiera fomes* and *Sonneratia apetala* presented moderate antioxidant activity. In addition, the Ag NPs presented enhanced activities as antidiabetic and antimicrobial agents; and the ZnO NPs presented potent activity as an anti-inflammatory.

12.4 BIOTECHNOLOGICAL APPLICATIONS OF GREEN NANOTECHNOLOGY-BASED SOLUTIONS

Definitely, green nanotechnology can be used extensively in the development of bio-based technologies. In fact, biotechnology provides important insights and practical applications to several areas such as medicine, nutrition, agriculture, and industry. Indeed, biotechnological applications of nanotechnology include nanomedicines (NMs) and biological interactions, biomolecular detection strategies, structural DNA nanotechnology, biosensors development, and antioxidant activity approaches. Thus, these applications can use innovative tools and technologies to solve challenging issues such as prevention, detection, and treatment of diseases, prevention of environmental contaminants, among others.

A biosensor is a device to detect, transmit, and record the information of a biological analyte such as DNA, RNA, proteins (i.e., enzymes), antibodies, antigens, and other biological components as glucose (El-Ansary and Faddah, 2010; Abu-Salah et al., 2015; Florea et al., 2019). Already nanobiosensors are the biosensors made up of nanomaterials that can detect nanoscale events with very high sensitivity, specificity, and versatility (Abu-Salah et al., 2015). Then, nanobiosensors can be used in the fields such as food analysis, environmental monitoring, medical diagnostics (i.e., identification of diseases and pathogens), among others (Thakur and Ragavan, 2013; Abu-Salah et al., 2015; Chandrasekaran, 2017; Florea et al., 2019). In fact, nanoparticles-based biosensors are potential tools for rapid, specific, and highly sensitive detection of the analyte of interest (Koedrith et al., 2015).

Additionally, nanomaterials with catalytic properties such as gold, copper sulfide, cobalt oxide, copper nanoclusters, and copper oxide nanoparticles (CuO NPs) have been used as peroxidase mimics for the detection of glucose, hydrogen peroxide, cholesterol, sarcosine, melamine, among others (Dutta et al., 2013; Hu et al., 2013; Deng et al., 2014; Rastogi et al., 2015, 2017). Particularly, those nanomaterials synthesized by green nanotechnology approaches using plant extracts, resins, latex, and gums exhibit high potential for biotechnological applications related to the detection of compounds. In fact, tree gum exudates have been used for food applications, environmental remediation, catalytic systems and biosensing (Padil et al., 2018). Rastogi et al. (2017) developed a colorimetric method for glucose nanodetection in biological samples. In their study, palladium nanoparticles were synthesized using kondagogu gum (*Cochlospermum gossypium*) to produce peroxidase mimics and showed a strong affinity to the substrate (hydrogen peroxide) and the chromogen (3,3′,5,5′-tetramethylbenzidine). Thus, these peroxidase-like activities can be used in other applications such as an efficient substitute for peroxidase enzyme (Rastogi et al., 2017). Sumitha et al. (2018) reported the use of durian seeds (*Durio zibethinus*) as renewable sources to produce Ag NPs. In their study, scanning electron microscopy analysis revealed spherical and rod-shaped nanoparticles with 20–72 nm. These nanoparticles exhibited antibacterial activity against *Salmonella typhimurium*, *Staphylococcus haemolyticus*, *S. aureus*, *Bacillus subtilis*, *E. coli,* and *Salmonella typhi* as well as a cytotoxic agent against *Artemia* spp. (Sumitha et al., 2018). Additionally, these Ag NPs also showed catalytic activities (peroxidase-like activity) for the methylene blue degradation and could be used in the field of water treatment, biomedicine, pharmaceuticals, biosensor, among others (Sumitha et al., 2018).

As mentioned previously, another important biotechnological application would be the antioxidant action exhibited by some nanomaterials synthesized using green routes. Technically, antioxidants are substances (molecules or nanomaterials) that at relatively low concentrations inhibit or delay the oxidation of a substrate. Therefore, green nanomaterials are promising in frontiers of research for improved antioxidants such as those organic (i.e., melanin), non-metallic (i.e., selenium), and metallic (i.e., gold, silver, copper, iron, zinc oxide) (Phull et al., 2016; Chung et al., 2017; Liu et al., 2017; Annu et al., 2018; Hamelian et al., 2018; Keshari et al., 2018; Valgimigli et al., 2018; Chen et al., 2019; Rajeswaran et al., 2019; Vinotha et al., 2019). Nanomaterials-based antioxidants showed intrinsic redox activity that can be associated with free radicals trapping and/or with catalase-like and superoxide dismutase-like activities (Valgimigli et al., 2018). In this context, three citrus fruits (*Citrus limon*, *Citrus sinensis*, and *Citrus limetta*) peel extracts were used to produce Ag NPs (Annu et al., 2018). Thus, transmission electron microscopy (TEM) results indicated the presence of spherical-shaped nanoparticles within the range of 9–46 nm. These Ag NPs showed antimicrobial activity against *S. aureus* and *E. coli* as well as anticancer activity against human lung cancer (LC) cell line A549. Moreover, these nanoparticles revealed potent antioxidant activity (Annu et al., 2018). Another example of an eco-friendly method to produce nanoparticles using a bio-waste was obtained from the fruit peel of *Carica papaya* (Kokila et al., 2016). TEM studies revealed the presence of spherical-shaped nanoparticles ranging from 10 to 35 nm, which inhibited the growth of clinically isolated multidrug-resistant human pathogens such as *S. aureus*, *B. subtilis*, *Klebsiella pneumonia*, and *E. coli* (Kokila et al., 2016). Additionally, these nanoparticles also showed potential for scavenging 1,1-diphenyl-2-picrylhydrazyl (DPPH) and 2,2′-Azino-bis (3-ethylbenzothiazoline-6-sulfonic acid) (ABTS) radicals by surface reaction phenomenon in a dose-dependent manner indicating antioxidant activity (Kokila et al., 2016). Thus, green nanotechnology can promote innovation and sustainability for the effective use of bio-waste to produce nanoparticles with different applications in biotechnology and other related fields.

12.5 CHEMICAL SCIENCE APPLICATIONS OF GREEN NANOTECHNOLOGY-BASED SOLUTIONS

Surely, chemical science is the basis of scientific and technological development to obtain new compounds and products for applications in several sectors. Despite its crucial importance, the development of chemicals is often associated with the use of potentially hazardous reagents and solvents, the generation of undesirable emissions, and the disposal of products and by-products that are potentially harmful to the health and environment. In the last few decades, the concept of green chemistry emerged mainly due to its potential to reduce or eliminate these undesirable consequences of the use of "conventional" chemistry (Fleischer and Grunwald, 2008). In this sense, the use of agricultural and food residues, for example, has been a strategy for the substitution of synthetic compounds in several applications (Figure 12.3).

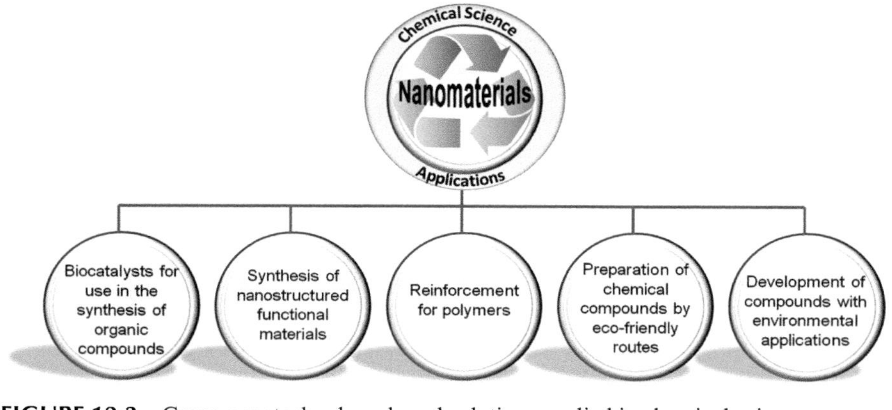

FIGURE 12.3 Green nanotechnology-based solutions applied in chemical science.

Within the 12 principles of green chemistry formulated in 1998, there are many requirements in which nanotechnology has played an important role, especially with regard to applications in chemical science (Fleischer and Grunwald, 2008). In addition to providing new research tools and strategies, nanotechnology allows the acquisition of new materials with peculiar properties and offers new processes for the preparation of nanomaterials (Whitesides, 2005). Among them, the development of nanostructured biocatalysts for use in the synthesis of organic compounds stands out and deserves more mention.

Molaei and Javanshir (2018) prepared a new bio-based nanocomposite (BNC PS/ZnO) as a heterogeneous catalyst using lignocellulosic waste peanut shells (LCWPS) for sustainable synthesis of organic compounds involving multicomponent reactions, for example, pyrimido benzazoles. The LCWPS were prepared *in situ* by hydrothermal synthesis and turned into a value-added product, a new BNC PS/ZnO. That product was used in a sustainable catalytic method for the synthesis of pyrimido [1,2-β] benzazole derivatives in water under microwave irradiation. The time of this reaction was significantly reduced using this catalytic system, and the new system provided a remarkably high atom economy and accelerated the formation of highly functionalized organic compounds. In another study developed by Sudhakar and Soni (2018), it was prepared a catalytic reduction system using palm shell agro-waste derived carbon (Ag NPs-PSAC) as biodegradable support for Ag NPs. The Ag NPs-PSAC prepared was used as an effective catalyst for the reduction of nitrophenols to aminophenols in a short interval of 5 min. In addition to providing a green solution for remediation of nitrophenols avoiding the use of usual organic solvents, the Ag NPs-PSAC catalysts showed also recyclability up to four cycles with no change in the activity.

Likewise, outstanding, and unique magnetic catalysts based on food waste have also been developed by various researchers. For example, Pathak et al. (2019) reported the synthesis of a novel solid magnetically retrievable iron oxide (Fe_3O_4) nanobase catalyst using banana (*Musa acuminata*) peel ash extracts for the Henry reaction. This magnetic nanobase catalyst showed excellent catalytic activity for the synthesis of exclusively β-nitro alcohol using nitroalkanes with different aldehydes via the Henry reaction at room temperature under solvent-free conditions. In addition to contributing to the one-pot green synthesis of β-nitro alcohol at low cost, the catalyst exhibited a strong magnetic response allowing it to be reused for 4 consecutive cycles and could be easily separated from the reaction mixture by simply applying an external magnetic field.

Recently, Nasrollahzadeh et al. (2019) presented an application in the chemical reactions of Cu-based recyclable nanocatalysts using plant extracts. In this work, copper nanoparticles (Cu NPs) such as Cu NPs, CuO NPs, CuS NPs, Cu_2O NPs, Cu/Fe_3O_4 NPs, Cu/ZnO NPs, and $CuAl_2O_4$ NPs among others, were synthesized using plant extracts from leaves, seeds, peels, bark fruits, flowers, and/or roots of plants. Most of them had their catalyst properties investigated in several chemical reactions such as oxidation, reduction, hydration, cyanation, cycloaddition, and coupling reactions. Furthermore, the use of plants for biosynthesis of MNPs aiming catalytic applications has many advantages, including the elimination of chemical reducing compounds, stabilizing agents, and toxic organic solvents; avoidance of the high reaction temperature; and a cost-effective and eco-friendly procedure.

Currently, the synthesis of nanostructured functional materials with peculiar properties from residues is considerable. For example, carbon nanosheets and carbon nanofibers have been prepared from food and agricultural waste for various applications. Kim et al. (2016) prepared nanoporous carbon nanosheets (NP-CNSs) containing numerous redox-active heteroatoms (e.g., oxygen and nitrogen) by pyrolysis of citrus peels in the presence of potassium ions. Finally, the resulting material with unique properties presented high electrochemical performance as a cathode material for sodium-ion storage. Furthermore, the NP-CNSs also had an electrical conductivity approximately 50 times higher than that of reduced graphene oxide (GO). Another study involving the preparation of carbon nanostructured material from agro-waste in simple steps was performed by Nuilek et al. (2018). The authors prepared carbon nanosheets from peanut shell (CNSs) by activation with KOH and heat treatment, and the smallest thickness of the CNSs was less than 50 nm. Additionally, they showed that the chemicals (KOH and H_2SO_4) and thermal activation affect the formation and separation of carbon layers to form nanosheets of graphite, adding value to the products.

Nanofibers as reinforcement materials for polymers were developed using natural fibers from agricultural and food waste. Hence, the favorable mechanical properties of the refolded nanofibers were achieved without the need for the use of coupling agents or chemical modifications on the surface of the polymers. Adeosun et al. (2016) used groundnut shells (GSP) to reinforce polylactic acid (PLA) fibers and the resulting GSP-PLA material showed strength in mechanical properties such as an increase of 286% in hardness, 1,502% in stiffness, 286% in ultimate tensile strength (UTS), 6.8% in rupture energy, and 1,081% in rupture stress. These expressive increments were attributed to the beads formed due to the agglomeration of GSP in the PLA nanofibers. According to the authors, these produced GSP-PLA nanofibers have potential applications in packaging, tissue engineering, and drug delivery. Another researcher group used wood cellulose nanofibers (CNFs) to reinforce an unsaturated polyester (UP) matrix, one of the most commonly used thermoset resins. In brief, the authors prepared nanostructured UP biocomposites at 45 vol% CNFs with elastic modulus and strength about 3 times higher than pure UP, ductility, apparent fracture toughness double, and a substantially higher glass transition temperature (T_g) (from 66°C for pure material to 78°C). The new properties in the modified materials were attributed to the strong synergy of the CNFs-UP observed in the composites (Ansari et al., 2015).

The use of a nanostructuring strategy to obtain chemical compounds by eco-friendly routes has also been reported. In fact, the utilization of waste materials as a precursor for synthesis makes the whole process cheaper, green, and sustainable. Habte et al. (2019) synthesized nano-calcium oxide from waste eggshells by the sol-gel method. The use of waste materials as a precursor for synthesis by the sol-gel technique results in many advantages over other methods for preparing nanoparticles such as being simple, economical, eco-friendly, and require no expensive equipment neither high temperature and pressure. According to the authors, the synthesized nanoparticles can be applied in heavy metal removal from industrial wastewater, paper industry, and as a filler material in automobile industry. Additionally, the synthesis of graphite from the raw material of pyrolyzed bagasse was proposed by Jannatin et al. (2019) as an alternative to commercial graphite. The resulting nanomaterial was used to synthesize GO-nano Fe_3O_4 composite that could be used for protein adsorption, dye removal, heavy metal removal, drug delivery, and sensor development. Firstly, graphite was obtained from raw material of pyrolyzed bagasse and then converted into a GO-Fe_3O_4 bagasse/magnetite composite through oxidation followed by a co-precipitation reaction. Finally, the strategy used by the authors allowed to enhance the economic feasibility of the waste, in particular bagasse, to be the raw material of graphite.

The development of compounds with environmental applications such as the uptake of noxious substances and heavy metals has also been an alternative for the use of agricultural wastes. Valdebenito et al. (2018) produced CNFs-based films from two agricultural waste feedstocks to be used as CO_2 adsorbent materials. Then, the authors demonstrated the good CO_2 adsorption capacity of CNF films, in which the adsorbed concentration of CO_2 was 0.90, 1.27, and 2.11 mmol CO_2/g polymer for CNF films from corn husks, oat hulls, and kraft pulp, respectively. Still, another type of metal absorbent green nanomaterial was presented by Kaliannan et al. (2019). The authors prepared nano-silica (NS) from *Saccharum officinarum* leaves and used as an adsorbent to remove Pb^{2+} and Zn^{2+} from aqueous solutions. The NS was able to absorb 148 mg/g and 137 mg/g of Pb^{2+} and Zn^{2+}, respectively, at room temperature. Thus, the new NS synthesized by an economically viable route can be used as an efficient adsorbent for dyes and heavy metal ions from industrial wastewater and aqueous solutions. In another approach, Abdessemed et al. (2019) produced inexpensive activated carbon support from olive wastes for environmental remediation applications. The authors synthesized ZnO-loaded porous activated carbon (AC-ZnO) from the Algerian olive-waste cakes and utilized it as a photocatalyst for the degradation of ethyl violet dye under UV irradiation. The results of this study indicated that AC-ZnO is a potential catalyst for aquatic pollutant removal.

In this way, the use of agricultural and food residues linked to the development of green nanotechnology-based products and processes has provided advances in the chemical sciences. In addition to the development of new compounds by more environmentally friendly routes, such nano-compounds can be used to reduce pollutants as well as to substitute toxic substances usually used both in the laboratory and industry.

12.6 AGRICULTURAL APPLICATIONS OF GREEN NANOTECHNOLOGY-BASED SOLUTIONS

Currently, agriculture faces wide range of challenges such as productivity stagnation, weather conditions fluctuation, crop reduction due to insect attack and phytopathogens infection, loss of soil fertility, and insufficient or contaminated water and soil that have entailed to severe losses (Pouratashi and Iravani, 2012). Moreover, aiming to increase and sustain productivity, harmful agrochemicals soluble in toxic solvents are constantly commercialized, thereby impacting the agrosystem and allowing the emergence of agent-resistant noxious organisms (Ishaaya et al., 2007). Trying to solve these problems, innovative technologies are essential in agriculture, and thus, nanotechnology has revolutionized the manner in which farmers produce agricultural goods by using nanoparticles and other nanomaterials instead of traditional inputs.

The nanometric size of nanomaterials increases the surface area of contact and consequently enhances the effect and significantly improves their activity (Baker and Satish, 2012). Particularly, those nanomaterials produced by green routes offer the additional benefit of the sustainability aiming applications in agriculture field. Therefore, the desired characteristics of greenly-produced nanoparticles are ideal for application in agriculture since they can be inexpensive, biocompatible, selective, stable, degradable, and allow the controlled release of compounds.

With the growing awareness of society, scientific communities have employed rational, eco-friendly synthesis of nanoparticles using biogenic sources like plants and microorganisms (Baker and Satish, 2012). In fact, biological compounds influence the solubility in water, enhance the colloidal stability, and decrease the harmful environmental impact (Syed et al., 2016). Some biological compounds involved in these green synthesis strategies are flavonoids, phenols, tannins, terpenes, and saponins (Baker et al., 2017). Markedly, nanoparticles used in agriculture are composed mainly of silver, gold, copper, zinc, aluminum, silicon, graphene, and clay or conjugated with carriers as cellulose, chitosan, and hydrogels. In agriculture, these nanoparticles and nanomaterials are applied as fungicides, bactericides, nematicides, seed coating agents, fertilizers, herbicides, pesticides, biosensors, plant growth regulators and hormones, and others (Duhan et al., 2017; Prasad et al., 2017) (Figure 12.4).

Green nanoparticles can cause damage in cell membranes and promote cell permeability, diminishing the activity of membranous enzymes, inhibiting DNA replication, and causing damage to genetic material that consequently results in alteration of transcription and translation, thereby influencing all cellular processes of microorganisms (Rajeshkumar, 2019). Recent studies demonstrated the efficiency of green nanoparticles in suppressing fungi, bacteria, and nematodes growth and detecting phytopathogens.

Ag NPs synthesized using the latex extract of *Euphorbia tirucalli* reduced gall formation, the number of egg masses, eggs per egg mass in roots, and juvenile J2 population of *Meloidogyne incognita* in tomato. Biocompatible iron oxide nanoparticles reduced by 6% the severity in plants infested by bacterial wilt *Ralstonia solanacearum* (Alam et al., 2019). Similarly, Hermida-Montero et al. (2019) studied the antifungal activity of green Cu NPs against *Fusarium solani*, *Fusarium oxysporum*, and *Neofusicoccum* sp. In sum, the authors reported the effects of nanoparticles on cell membranes and alteration in mycelium morphology. The employability of green nanoparticles is also important in controlling multidrug-resistant fungi/bacteria present in the fields. In fact, spherical nanoparticles ranging from 5 to 40 nm inhibited the activity of normal and multidrug-resistant phytopathogenic bacteria *Agrobacterium tumefaciens* (Chowdhury et al., 2014).

Technologies based on green nanoparticles are strategies compatible with integrated management of diseases and also undesirable weeds and insects, reducing the amount and number of agrochemical applications due

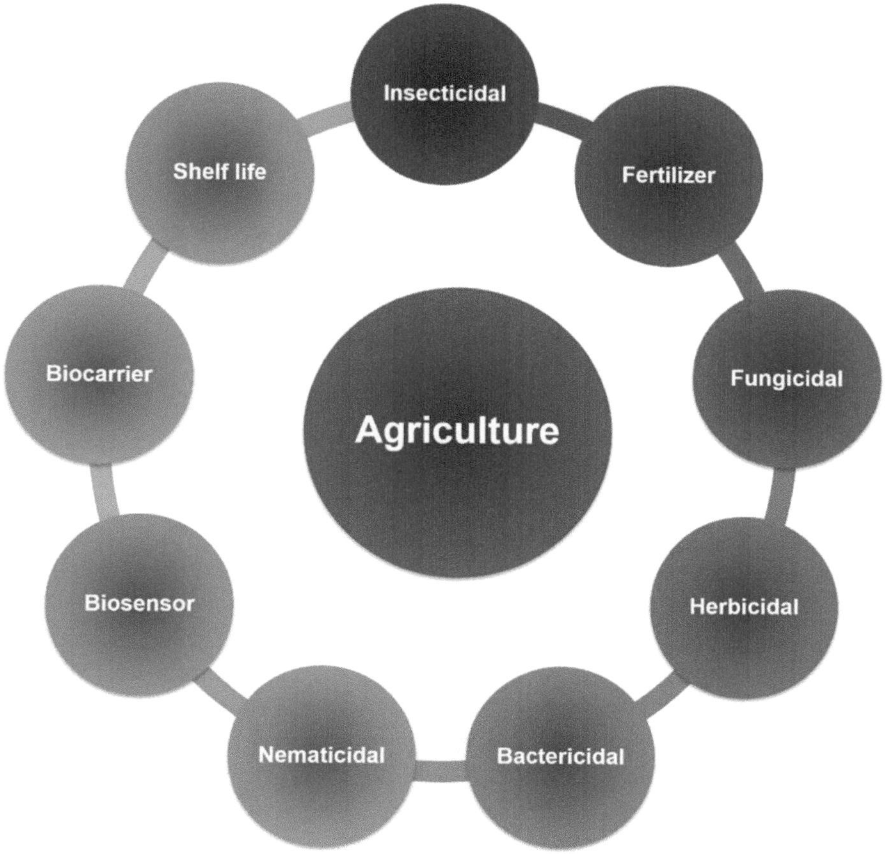

FIGURE 12.4 Green nanotechnology applied in agriculture with their effects and uses.

to association with nanoparticles and regulate the controlled release of the active ingredient in the field. According to Le et al. (2019), chitosan was used to reduce silver ions and stabilize Ag NPs, and afterward synergized with fungicide Antracol (Ag@CS/An) to test the antimicrobial activity against *Phytophthora capsici*. The results showed that Ag@CS/An has a better antifungal performance when compared with each component alone. Also, biosynthesized Ag NPs decreased the amount of fungicides isoprothiolane and azoxystrobin necessary to control *Magnaporthe grisea* (Rabab and El-Shafey, 2013). In addition, nanomaterials may be used as carriers of highly reactive chemicals, especially under flooded conditions. Yu et al.

(2015) synthesized chitosan nanoparticles linked by disulfide bonds to diuron, aiming to obtain the controlled release of the herbicide dependently of the glutathione concentration. Similar results were observed when chitosan was applied to synthesize Cu NPs and the nanoformulation controlled the release of pendimethalin (Itodo et al., 2017).

Other critical aspects in agriculture are the early and correct disease identification and the detection of possible contaminants. A solution is the fabrication of nanobiosensors which are highly sensitive devices capable to detect up to ppb level quickly, accurately, and without the need for trained personnel. The nanomaterials commonly used in nanobiosensors are nanochannels, Au NPs, and carbon nanotubes (CNTs). Au NPs are used to amplify the sensitivity of immunological assays. For example, Zhao et al. (2014) developed an electrochemical enzyme-linked immunoassay (ECEIA) containing Au NPs associated with antibodies labeled with horse-radish peroxidase to detect *Pantoea stewartii* subsp. *stewartii*. Compared with conventional assays, ECEIA was 20 times more sensitive with a detection limit of 7.8×10^3 cfu/mL. Drygin et al. (2012) developed lateral flow immunoassays using antibody-based sandwich format and Au NPs as labels to detect Potato virus X. The specificity of the assay was tested against the major potato viruses (Potato virus Y, Potato virus M, and Potato virus A), obtaining a sensitivity of 2 ng/mL. *Cyamopsis tetragonoloba* extract was used to reduce Ag NPs, which can be used as a nanobiosensor to determine ammonia concentration, depending on the distance of Ag NPs inside a nanocomposite (Pandey et al., 2012). In addition, nanobiosensors can also be used in agriculture to detect mycotoxins, pesticides concentration, and pollutants (Baker et al., 2017).

Several studies demonstrated the potential of green nanoparticles as insecticides (Bhattacharyya et al., 2016). Ag NPs synthesized using *Nerium oleander* presented strong larvicidal activity against malaria vector *Anopheles stephensi* (Suganya et al., 2013). Moreover, in the study conducted by Zahir et al. (2012), it was reported the first pesticide based on Ag NPs syntesized using aqueous leaves extract of *Euphorbia prostrata* which were active against *Sitophilus oryzae* with complete mortality observed after seven days.

Notwithstanding the advances related to studies of green nanotechnology solutions to control pests and diseases, green nanomaterials can also be employed to increase productivity as nano-fertilizer. In fact, some green nanoparticles can enhance plant growth and promote seed germination. More than half of the nitrogen and phosphate applied to soils stay unavailable to the plant due to the incorporation into soil organic matter. So, the use of nanoparticles and nanomaterials can reduce this loss of macronutrients and micronutrients. In such case, the nutrients can be applied individually or together in the same nanosystem. Shebl et al. (2019) synthesized manganese zinc ferrite nanoparticles ($Mn_{0.5}Zn_{0.5}Fe_2O_4$ NPs) using a green chemistry route and applied via foliar in squash plants (*Cucurbita pepo* L). In sum, the use of the nano-fertilizer at lower concentration resulted in the highest vegetative growth when the nanoparticles were synthesized at 180°C compared with nanoparticles prepared at lower microwave holding temperature (140 and 160°C). Nanocarriers (NCs) are also reported enhancing the plant growth and to control the release of nutrients. Joshi et al. (2018) studied the influence of CNTs in germination and growth of wheat. CNTs dilated in 80% the xylem and phloem, enhancing the transport of water and nutrients, consequently increasing the yield in 63%. Additionally, the controlled release of nitrogen was investigated using urea-hydroxyapatite nanoparticles. Kottegoda et al. (2017) observed a rate of 12 times slower of nitrogen compared to pure urea.

Green nanoparticles can also contribute to the last step of the agricultural chain as food processing materials. Recently, nanoparticles offered better texture, high nutrient content, and longer shelf time to agricultural products without to affect taste and appearance (Prasad et al., 2017). Mohammadi et al. (2016) evaluated the effects of nanostructured chitosan loaded with *Zataria multiflora* extract on the shelf-life and antioxidant activities of harvested cucumber. After stored for 21 days, the cucumber treated with chitosan-*Z. multiflora* nanomaterial had the best quality when compared to water treatment.

Despite a lot of information about applications and advantages of green nanotechnology in agriculture, scanty reports are yet available about the green synthesis of products applied in the field until shelf time. Indeed, the development of a specific regulation can be important to increase the exploration of this new area of nanotechnology, since sustainable solutions can benefit agriculture from seeding to post-harvesting.

12.7 FOOD SCIENCE APPLICATIONS OF GREEN NANOTECHNOLOGY-BASED SOLUTIONS

Food waste and loss are a worldwide challenging issue to be considered. According to the Food and Agriculture Organization of the United Nations (FAO, 2017), about 1.3 billion tons of food is wasted or lost along the food supply chains. In addition, fruits, vegetables, roots, and tubers have the highest wastage rates, with percent ranging from 40% to 50% (FAO, 2017).

According to Marchetto et al. (2008), only 58% of fruit and 75% of vegetables are effectively consumed. Besides, part of these foods, such as peels and seeds, is commonly neglected or wasted. Only a small share of agricultural residues and food wasted is harnessed, for example, in organic composting processes. On the other hand, the vast majority is simply discarded and neglected, instead of being used for creating some economic value and mainly mitigate negative impacts on the environment. Undesirably, the uncontrolled decomposition of these elements results in gases emission that contributes to ecosystem climate changes (Hall et al., 2009).

Thus, it is paramount to develop new strategies to reduce food loss and waste rates. Bearing that in mind, the world has witnessed a growing development of sustainable technological innovations that foster green chemistry (Lenardão et al., 2003; Prado, 2003) and these innovations usually include nanotechnological solutions, which are aimed at creating a safe, economical, and eco-friendly nanoscale systems that allow developing new processes and products, which can be applied in different food areas (Figure 12.5). Such green nanotechnology solutions in the food industry often use bioactive compounds from agricultural and food residues for developing nanomaterials (Iravani, 2011; Adelere and Lateef, 2016).

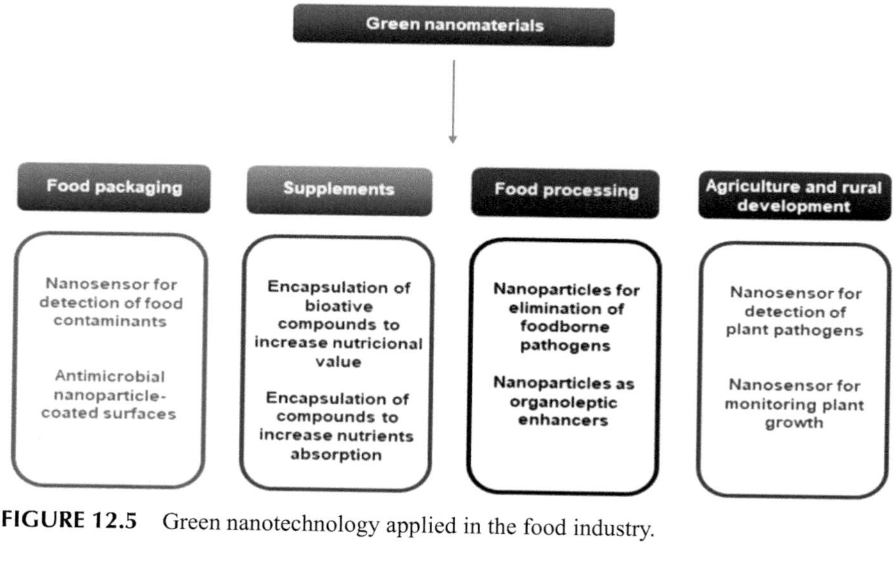

FIGURE 12.5 Green nanotechnology applied in the food industry.

According to Kitts (1994), bioactive compounds are extra nutritional substances, which have numerous chemical and structural characteristics potentially useful. In fact, several studies have shown that these compounds are very beneficial to human health due to their antitumor, antiviral, antimicrobial, anti-inflammatory, antioxidant, and antihypertensive properties (Young and Lowe, 2001; Kris-Etherton et al., 2002; Lu et al., 2012). That is why in recent years the world has seen increasing use of these substances in the pharmaceutical, biomedical, agricultural, and industrial areas. It is worth mentioning these products are usually obtained from parts of plants, such as leaves, stems, shoots of roots, and flowers, as well as from peels and seeds of fruit and vegetables, and they act as reducing and stabilizing agents in the green synthesis of MNPs (Kris-Etherton et al., 2002; Adelere and Lateef, 2016).

A practical example of the use of such residues to foster the sustainable development of agribusiness in the food industry can be found, for instance, in cashew nutshell. So, Ag NPs, and Au NPs successfully synthesized with this bioresource and showed activity against pathogens associated with fish diseases (Velmurugan et al., 2014a). In addition, Au NPs, and Ag NPs produced using *Zingiber officinale* root extract were also used as an antibacterial system against bacterial food pathogens, such as *Staphylococcus* spp. and *Listeria* spp. (Velmurugan et al., 2014b). Furthermore, Patra and Baek (2015) developed a new approach to synthesize Au NPs from watermelon rind, which demonstrated activity against some foodborne pathogens. In addition, food industrial wastes may be used for producing nontoxic silver nanorods. They are considered great materials to keep the quality of raw milk due to their antimicrobial properties (Sivakumar et al., 2013). In sum, these results show that green nanotechnology is a good alternative for extending the shelf life of some food products and these nanomaterials can also be applied in antimicrobial packaging production.

Therefore, another direction that involves green nanotechnology for delay food deterioration is adding active components in edible thin coatings. This technology applies natural compounds with different properties, such as antimicrobial and antioxidant activity, which can be used in food substrates for increasing the shelf life of these products. Thus, nanoemulsions, PNPs, SLNPs, or other nanosystems can be used for developing edible coatings and improving fruit and vegetable conservation (Zambrano-Zaragoza et al., 2018). The application of green nanotechnology in this area is increasing, and it is a good option for preserving the organoleptic characteristics of food. This can be proven due to the good results achieved in different food preservation studies over the years. For instance, chitosan nanoparticles showed excellent fresh-cut apple coatings due to their high antimicrobial activity against bacteria, molds, and yeasts (Pilon et al., 2014). Furthermore, Otoni et al. (2014) developed edible films from methylcellulose and nanoemulsions of essential oils to form antimicrobial and natural films for sliced bread. Lastly, cellulose nanocrystals (CNCs) reinforced chitosan is an interesting and novel coating material due to retarding pears ripening during post-harvest storage (Deng et al., 2017).

In preventing food loses on agricultural and rural fields, *Piper nigrum* leaves and stems are interesting biological resources to synthesize Ag NPs, which presented antibacterial activity againt agricultural plant pathogens such as *Citrobacter freundii* and *Erwinia cacticida* (Paulkumar et al., 2014). Furthermore, it has been noticed the use of oak leaf and fruit for the synthesis of Ag NPs aimed at acting as an effective approach for control of plant-pathogen infection, since the increasing use of some antibiotics can lead to resistant bacteria (Chahardooli et al., 2014).

Other approaches that use green nanotechnology is in monitoring food quality and improve food safety through sensitive and fast response nanosensors. In fact, these systems are used to detect pathogens, pesticides, or other agrochemicals; identify bioactive compounds; elicit possible alterations on food composition; among others. For instance, Zheng et al. (2018) developed carbon dots functionalized with eco-friendly Ag NPs as colorimetric sensors for detection of food contaminants in fruit samples. In addition, green Ag NPs are interesting nanosystems to quickly identify milk adulterants. In the next few years, these green nanoparticles can be used as an excellent detector in other children's food products (Varun et al., 2017). Sun et al. (2017) produced a green surface-enhanced Raman scattering (SERS) film based on polymethyl methacrylate/Ag NPs/graphene, which is a great application for detection and identification of pesticide residues in food products. Furthermore, Xu et al. (2015) used a simple and green method to synthesize highly-fluorescent C-dots, which can be used for synthetic food colorant tartrazine determination in food samples. However, the excessive consumption of this substance can be associated with potentially adverse health effects in humans. So, it is fundamental for developing efficient and rapid measurement techniques to determine the contents of these and other food colorants, which may cause diseases in humans.

Different nanostructures can be useful tools for identifying bioactive compounds and food pathogens. For instance, green Ag NPs and Au NPs are used as low-cost lactose sensors (Bollella et al., 2016). Furthermore, it has been noticed the use of green Au-SnO_2/graphenes single-walled carbon nanotubes (SWCNTs) nanocomposite aimed at acting as a sensitive sensor to determine antioxidant additives in foods (Du et al., 2014). Lastly, high sensitivity nanosensors are being developed to detect and quantify food pathogens, which could be placed into packaging in the future (Ellairaja et al., 2017).

From these numerous examples, it is possible to conclude that the green nanotechnology is present in many food sectors and it is an excellent alternative in high value-added products' development, which could be used for different purposes, with low costs of production and lesser negative impacts to the environment and human health.

12.8 MATERIALS SCIENCE APPLICATIONS OF GREEN NANOTECHNOLOGY-BASED SOLUTIONS

Surface and interface research concerning green nanotechnology-based solutions focus mainly on the electronic and structural properties of functional molecules on nanostructured materials, considering how these unique interaction patterns are able to influence such surfaces behavior (Chi and Lee, 2019). Moreover, notable efforts are devoted to set the nanotechnology to the tracks of a sustainable future (Lu and Ozcan, 2015). For instance, several of these approaches converge to the selection of natural resources as precursors for carbon-based nanomaterials (CBNs) (Titirici et al., 2015), including recycling and/or re-using of by-products from agricultural, forestry, and food residues. In this context, nanocellulose presents itself as an omnipresent green nanomaterial candidate derived from these feedstocks. This biodegradable polymer can be extracted and obtained from lignocellulosic biomass of plant cell walls by several methods, yielding basically nanomaterials of three types: CNCs, CNFs, and bacterial cellulose (BC) (Moon et al., 2011). In common, these nanomaterials show nanometer size in diameter, along with high mechanical strength, stiffness, and surface area (Phanthong et al., 2018). Thus, such properties make this material compatible to so plenty of applications including surface modifications, nanocomposites production, films, and foams formation, photonic systems, and medical devices (Abitbol et al., 2016).

Nowadays, according to Rol et al. (2019), since 2008, cellulose-based nanomaterials reached the status of a key European bio-economy priority, possibly due its widespread renewable sources occurrence and its astonishing chemical versatility. Surface hydroxyl groups spread on the relatively large surface area provide the abundance of nanocellulose sites available for chemical modification (Abitbol et al., 2016), enabling applications in medicine, paper industry, and electronics (Rol et al., 2019).

Considering the vast amount of published reviews related to nanostructured cellulose applied to materials science applications (Luo et al., 2018; Motaung and Liganiso, 2018; QuanLing et al., 2018; Shak et al., 2018; Tayeb et al., 2018; Thomas et al., 2018; Varghese et al., 2018; Bacakova et al., 2019; Bhat et al., 2019; Kian et al., 2019; Panchal et al., 2019; Sharma et al., 2019; Wang, 2019), the versatility of applications for nanocellulose-based materials in materials science is very clear. Furthermore, it is also relevant to highlight the Motaung and Liganiso (2018) approach, which evidenced the most prominent nanocellulose renewable sources from agro-industrial wastes, such as: sugarcane bagasse, maize stalks, rice husk (Albernaz et al., 2015), and sorghum. That study also showed the potential of lignin and lignocellulosic fibers (mainly from flax, sisal, hemp, jute, cotton, oil palm, silk, and pineapple leaves) as a reinforcement for polymer composites (Jawaid and Abdul-Khalil, 2011; Faruk et al., 2012; Monteiro et al., 2012; Parida et al., 2012; Gurunathan et al., 2015; Yan et al., 2016). Against this backdrop, it is possible to contemplate as future perspectives for sustainable exploitation of agricultural, forestry, and food residues the development of new nanomaterials from biomass, by using innovative treatments, extractions, and chemical modifications strategies, as well as new composite formulations, enabling to optimize its usage concomitantly to minimizing preventable wastage.

12.9 PHYSICAL SCIENCES APPLICATIONS OF GREEN NANOTECHNOLOGY-BASED SOLUTIONS

Nanophotonics, nanoelectronics, nanobionics, and nanofluidics are among some branches that contribute to green nanotechnology, which include physical principles. These principles, acting synergistically with properties of biological materials, enabled the development of a series of innovative devices.

In regard to the nanophotonics-branch that studies the behavior of light at the nanoscale, Huang et al. (2016) developed a natural substrate/SERS chip using Taro leaves (*Colocasia esculenta*) coated with Ag NPs. The plant was chosen due to its hydrophobicity caused by its micro/nanostructured morphology composed of micropapillary structures and secondary crossed nanoplates in its leaves. Such morphology allowed the formation of Ag NPs aggregate after deposition of them on the leaf, causing hotspots and consequently increasing the sensitivity of the SERS signal due to the excitation of the metallic nanostructures caused by the localized surface plasmon resonance (LSPR); besides reducing the limit of biochemical detection of possible analytes of interest, that is, enabling the detection of these even in small quantities (e.g., a drop).

Guidetti et al. (2016) produced flexible films of CNCs, commercially acquired, co-assembled with zwitterionic surfactants (dipolar ions), which presented iridescence (also known as goniochromism) and helical chiral molecular geometry, which exhibits different optical responses to different left and right circular polarizations (Šmidlehner et al., 2017). The films could be adjusted according to photonic and mechanical properties such as optical response, tensile strength, and stiffness, controlling the amount of surfactant and pH of the suspension. CNCs, like other organic nanomaterials, demonstrate diamagnetic properties-usually accompanied by some optical anisotropy (Frka-Petesic et al., 2015).

CNC films were also obtained via acid hydrolysis of the Whatman cellulose filter paper (Frka-Petesic et al., 2017) in order to control their optical properties and the color of the film from the orientation of the CNCs magnetic domains using neodymium magnets. The results indicated that the magnetic fields could be used to control the self-assembly process of the films and that the orientation of the helix had a direct impact on the visual appearance of the CNCs.

Yet about the control of the optical properties of nanostructures, Toivonen et al. (2018) prepared transparent, semi-transparent, and white membranes of pure CNFs from birch (tree of the genus *Betula*) pulp, in order to manipulate light transport. The dispersions produced membranes that showed variations in visual appearance (turbidity) according to the medium size and fibrils thickness distribution, being smaller for the transparent membrane and bigger for the white membrane. For the semitransparent and white membranes, an anomalous light behavior was observed due to the anisotropy of the orientations and polydispersity of fibrils and pores observed in disordered photonic nanostructures; differently of the transparent membrane that has fine fibrils and structural homogeneity. Additionally, the authors suggested that the CNFs produced were useful for the development of sustainable and biocompatible materials.

Xue et al. (2015) applied CNFs, derivative from wood flour, in the creation of a composite paper. By dispersing zinc selenide quantum dots in CNFs film, it was feasible to create a transparent and flexible device, with photoluminescent properties, good thermal stability, and low coefficient of thermal expansion; being prospective in the development of functional papers.

CNF films were also used in nanoelectronics-responsible branch about the study of nanotechnology applied to electronic components-as reported by Syrový et al. (2019), as humidity sensors. The films were prepared from sugarcane bagasse, with or without polyethylene glycol (PEG), followed by printing carbon-based interdigitated electrodes (IDEs) using conductive ink. Both sensors presented reproducible responses and potential uses in smart agriculture, among them, in the monitoring of soil, agricultural products, packaging, and wound dressings; however, those films produced with PEG-coated were more suitable for the printed IDEs.

Hsieh et al. (2013) generated flexible electrical devices using cellulose pulp (microsized) and nanopaper (CNFs) as substrates containing metallic conductive lines constituted by gold sputtering, Ag NPs, and silver metallo-organic decomposition, in order to compare the thermal durability of the substrates and the electrical resistance of the lines. Both substrates were thermally stable (structurally and dimensionally) throughout the evaluation process. On the other hand, the conductivity was directly affected by the width of the substrates and the composition of the metallic lines evaluated, being higher for the nanopaper when compared to pulp (for all lines) and higher for the metallic line composed of Ag NPs, showing as a promising material in the roll-to-roll manufacturing process (Park et al., 2016).

Zyoud et al. (2013), used curcumin as a sensitizing agent for TiO_2 NPs aiming to act in the photodegradation of methyl orange (mutagenic compound and commonly applied in wastewater treatment titrations), towards the water purification via solar light. The sensitization process occurs, in this case, when inserting impurities (curcumin)-which absorb visible photons-into a crystalline network of a semiconductor (TiO_2 NPs)-which absorbs only ultraviolet radiation-improving its photovoltaic properties with regard to the treatment of water using sunlight. To further enhance the device photodegradation capacity, the nanosystem was supported on activated carbon, forming a solid compound.

Stavrinidou et al. (2015) developed electronic circuits implanting variations of conducting polymers conjugated in garden rose (*Rosa floribunda*). After being absorbed by the xylem (vascular system) of the flower, the polymers were able to form "wires" along with its structure and the latter proceeded to conduct electricity. In a subsequent stage, the polymers were conjugated to CNFs, and the resulting material was infiltrated (under vacuum) into a garden rose leaf. The results indicated the production of an organic electronic device with electrochromic properties that are able to change color reversibly when applying electric charges/discharges; open up the way for a series of more sustainable technological applications. Consequently, the findings of this research permeated between nanoelectronics and nanobionics.

The nanobionics, more specifically the plant nanobionics, is a branch of the bioengineering that contributes with the nanotechnology regarding the efficiency of the plants, amplifying the skills of the plant tissue and/or offering new functions. Giraldo et al. (2014) demonstrated that SWCNTs enhanced the photosynthetic action of chloroplasts isolated from spinach leaves (*Spinacia oleracea* L.), due to the formation of a semiconductor exciton (Grosso and Parravicini, 2014), promoting strong optical absorption and possibly transferring/transporting energy to the photosynthesis mechanism.

Sensors for the detection of nitroaromatic (explosives) were produced by Wong et al. (2017). SWCNTs coated with polyvinyl alcohol (P-SWCNTs) and bombolitin II (B-SWCNTs) were infiltrated in *Spinacia oleracea* L. leaves for picric acid (commonly used in explosives) detection. Picric acid was administered on the root and the leaf surface of the functionalized plant. The sensors were evaluated by near-infrared fluorescence, coupled to a charge-coupled device camera and a Raspberry Pi (a smartphone-sized computer type), which can be used as a portable electronic device. The results indicated a higher sensitivity for B-SWCNTs due to an effect known as corona phase recognition, where the B-SWCNTs conjugate becomes selecive for the target analyte (picric acid) promoting the fluorescence. Thus, sensors like this-also called nanobionic plants-can be useful in real-time monitoring and remote sensing in agricultural, industrial, and safety zones.

Kwak et al. (2017) designed nanoparticles with surface charge and controlled sizes to mark specific areas of watercress (*Nasturtium officinale*) and control their distributions inside and outside the plant mesophyll, in order to generate visible light emission. *Nasturtium officinale* was chosen due to its expressive production of adenosine triphosphate (ATP), which gives it potential energy storage, capable of converting chemical energy into light energy. Thus, the nanoparticles were inserted into the plant using a technique (pressurized bath infusion of nanoparticles) described by a nanofluidic mathematical model. Finally, the results allowed the creation of an *in vitro* kinetic model to describe the interaction between the nanoparticles and the biochemical route of the plant, in terms of the conversion of ATP to photons, which can be calculated. In addition, bioluminescence could be modulated to "on" and "off." Thus, the study enabled the development of eco-friendly luminescent sources.

Li et al. (2019) produced a nanofluidic-branch that studies the behavior and manipulation of fluids at the nanoscale-flexible membrane composed by CNFs, aiming the ionic transport. After extracting the lignin and hemicellulose from the CNFs, it was possible to obtain numerous open and aligned nanochannels along the fibers. The hydrophilicity and alignment of the CNFs made the membrane plenty effective in fluids transport. Additionally, the nanochannels could be adjusted in relation to the diameter, surface charge density, and geometry; influencing the ionic conductivity and composing a natural and high-performance device.

The ion transport was also evaluated by Wang et al. (2018) when producing a nanofluidic film constituted of BC, aiming at the creation of a sensor of relative humidity during the process of breathing. It is important to mention that BCs, opposite CNFs, are devoid of lignin and hemicellulose, dispensing the extracting step to obtain the nanochannels. The sensor showed ultra-sensitivity and high conductivity and can be applied in both *in vitro* and *in vivo* analyzes as a wearable device.

Thus, the mentioned examples among the approached branches-including functional papers, naturals substrates and semiconductors, flexible films, membranes, and electrical devices, modulation of size, charge, and optical properties of nanosystems, and light and energy transport control-may suggest ideas for improvement and/or development of future devices supported by green nanotechnology taking into account physical principles.

12.10 INDUSTRIAL APPLICATIONS OF GREEN NANOTECHNOLOGY-BASED SOLUTIONS

The emerging search for materials that have each time lesser impact on nature makes eco-friendly processes increasingly the focus of industrial solutions from a wide variety of markets and applications. Industry's interest in nanotechnology comes from the ability to change material properties by making processes that are typically time-consuming, costly, or inaccurate to become faster, cheaper, and more accurately delivering progressively innovative solutions. With this, some industries launch research and investment in search of characteristics and applications of green nanomaterials.

Among the areas that are essential for the industrial application of green nanotechnology is a field that involves catalyzed reactions for, medical, and chemical industries, mainly in products and processes that are only viable through a catalyst (Cadierno et al., 2008). The most sought-after catalysts are heterogeneous ones that are easily removed from the reaction medium, but with a macro size of the support, there is a difficulty of a smaller contact surface between catalyst and solution reactants. So, through nanotechnology, it is possible to increase the contact surface by making nanostructured catalysts (Somorjai et al., 2009). Currently, it has already been possible to develop nanowires, nanoplates, and nanotubes of palladium using green synthesis through vitamin B1, to be applied as supports for catalysts (Polshettiwar and Varma, 2010). In these cases, green nanotechnology not only helps with reducing the size but also with enhanced shape, morphology, and composition, facilitating numerous industries that need this step (Nadagouda and Varma, 2006).

One of the areas that most apply nanotechnology and consequently green nanotechnology is in the medical field. Nanomaterials-based drug carriers are an important field of research and are being widely used because they commonly can cross cell membranes and deliver the drug in specific locations. For example, functionalized graphene nano-sheets gelatin produced by a green method generated good compatibility, physiological stability, and high drug loading capacity (Liu et al., 2011). Other development produced pH-sensitive green NCs that can make selective and sustained delivery of doxorubicin (Ma et al., 2015). Still, another area in medicine that can benefit from green nanotechnology is that of diagnosis, and thus nanobiosensors tend to make diagnosis faster and even more accurate. In addition, there are hybrids with therapeutic and diagnostic capacities (imaging and therapeutics) for some types of cancer (Bahmani et al., 2013). In addition, Ag NPs produced in large scale through green synthesis using plants (Baghizadeh et al., 2015; Santiago et al., 2019) or food waste (Bonatto and Silva, 2014) can be used in various areas of medicine as an anti-inflammatory, antifungal, and antibacterial agents (Sathishkumar et al., 2009), and also as diagnostic and biosensor components.

The construction industry also already uses nanotechnology as an ally. In such a case, residues from styrofoam industry can be used to produce a lightweight concrete using nanotechnology that improves absorption of vibration and noise (Setyowati, 2014). There is also the treatment of materials used in construction such as wood, plaster, and mineral fiber wih Ag NPs to inhibit fungal growth and it was realized that higher concentrations of nano-metal inhibit the growth of fungi on materials (Huang et al., 2015).

In the electronics industry, Langmuir-Blodgett layers coated or mixed with Au NPs that can be synthesized by green routes were successfully constructed and exhibited hysteresis when voltage scan was reversed (Paul et al., 2003). In addition, there is also nanoprinting of circuits, showing that concepts of nanotechnology were introduced in the construction of a supercomputer (Green et al., 2007). An electrode material was constructed that is suitable and promising for the manufacture of supercapacitors, using ZnO NPs decorating GOs that were synthesized by green routes, which maintained an original capacitance after 1,000 cycles, improving conducting network (Haldorai et al., 2014).

There is also a part of the packaging and plastics industry that has been investing the use of green nanotechnology, particularly aiming the development of food packaging that increases shelf time and asepsis of food products (Bumbudsanpharoke and Ko, 2015). There are packages for rice that were coated with films of Ag NPs that confer antimicrobial activity against *Aspergillus flavus*, and that during storage for 35 days it was not possible to identify migration of silver to rice (Li et al., 2017). Also, containers of food loading have been treated with Ag NPs of varying morphology and size, but in this case, it was noticed migration of Ag NPs to the stored materials (Echegoyen and Nerín, 2013). Even if there is a need for improvements in the incorporation of nanomaterials in this type of industry, it can lead to increase the shelf life of food and reduce losses generated by contamination by microorganisms.

The textile industry has already managed to incorporate green nanotechnology-based solutions into its market since there are numerous types of textile fibers functionalized with ZnO nanostructures (Shateri-Khalilabad and Yazdanshenas, 2013), Ag NPs (Hebeish et al., 2011; Velmurugan et al., 2014c), and chitosan nanoparticles (Hebeish et al., 2013). These nanomaterials have the capacity to provide ultraviolet protection, antimicrobial activity, and thermal stability, respectively, offering new solutions with unprecedented properties.

It can be noticed that nanotechnology can enter and is in most diverse areas and branches of industries, always improving and/or offering a new quality and solution to the products. Similarly, green nanotechnology may be increasingly linked to industry, as knowledge about how green nanomaterials can help in the industry is being dissipated. There are also standardized safety tests so that products not only improve but also offer consumers safety with sustainability.

12.11 CONCLUSIONS AND PERSPECTIVES

The rational exploitation of bioresources represents the current paradigm to achieve sustainable development. Probably, the generation of co-products, by-products, residues, and wastes and to identify valuable uses and relevant applications from them through technology-based approaches are among the most challenging issues for evaluation when managing business related to agriculture, agroindustry, bioindustry, and forestry. Currently, there are a growing number of people engaged with this subject, and more initiatives are expected in the next few years, including those based on advanced technologies. Recent trends strongly suggest that sustainable solutions are gradually becoming common and even standard of mainstream nanotechnology. In this sense, nanotechnology's branch that employs green chemistry principles, the so-called green nanotechnology, emerges as a powerful, innovative, and sustainable way to deliver technology-driven products and processes with high value-added. Certainly, green nanotechnology will contribute to re-defining and re-invigorating many fields and industries. In sum, progress from present state-of-the-art towards real application cases and future developments will invoke creative minds to think beyond the current needs of specific companies and truly change the way by which researchers and stakeholders provide and use products that satisfy desired features and also meet sustainability criteria.

KEYWORDS

- **agricultural by-products**
- **agricultural co-products**
- **green chemistry**
- **green nanotechnology**
- **lipid nanoparticles**
- **metallic nanoparticles**
- **nanoagriculture**
- **nanobiotechnology**
- **nanomaterials**
- **nanomedicine**
- **nanoparticles**
- **nanosensors**
- **nanosystems**
- **polymeric nanoparticles**
- **sustainability**

REFERENCES

Abdessemed, A., Rasalingam, S., Abdessemed, S., Djebbar, K. E. Z., & Koodali, R., (2019). Impregnation of ZnO onto a vegetal activated carbon from Algerian olive waste: A sustainable photocatalyst for degradation of ethyl violet dye. *Int. J. Photoenergy., 2019*, 1–13. doi: https://doi.org/10.1155/2019/4714107.

Abitbol, T., Rivkin, A., & Cao, Y., (2016). Nanocellulose, a tiny fiber with huge applications. *Curr. Opin. Biotech., 39*(2016), 76–88. doi: https://doi.org/10.1016/j.copbio.2016.01.002.

Abu-Salah, K. M., Zourob, M. M., Mouffouk, F., Alrokayan, S. A., Alaameru, M. A., & Ansari, A. A., (2015). DNA-based nanobiosensors as an emerging platform for detection of disease. *Sensors, 15*(6), 14539–14568. doi: https://doi.org/10.3390/s150614539.

Adelere, I. A., & Lateef, A., (2016). A novel approach to the green synthesis of metallic nanoparticles: The use of agro-wastes, enzymes, and pigments. *Nanotechnol. Rev., 6*(5), 567–587. doi: https://doi.org/10.1515/ntrev-2016-0024.

Adeosun, S., Taiwo, O., Akpan, E., Gbenebor, O., Gbagba, S., & Olaleya, S., (2016). Mechanical characteristics of groundnut shell particle reinforced polylactide nanofiber. *Revista Matéria, 21*, 482–491. doi: https://doi.org/10.1590/S1517-707620160002.0045.

Ahmed, S., Ahmad, M., Swami, B. L., & Ikram, S., (2016). A review on plants extract mediated synthesis of silver nanoparticles for antimicrobial applications: A green expertise. *J. Adv. Res., 7*, 17–28. doi: https://doi.org/10.1016/j.jare.2015.02.007.

Alam, T., Khan, R. A. A., Ali, A., Sher, H., Ullah, H., & Ali, M., (2019). Biogenic synthesis of iron oxide nanoparticles via *Skimmia laureola* and their antibacterial efficacy against bacterial wilt pathogen *Ralstonia solanacearum*. *Mat. Sci. Eng., 98*, 101–108. doi: https://doi.org/10.1016/j.msec.2018.12.117.

Albernaz, V. L., Joanitti, G. A., Lopes, C. A. P., & Silva, L. P., (2015). Cellulose nanocrystals obtained from rice by-products and their binding potential to metallic ions. *J. Nanomater., 2015*, 1–8. doi: https://doi.org/10.1155/2015/357384.

Al-Sheddi, E. S., Farshori, N. N., Al-Oqail, M. M., Al-Massarani, S. M., Saquib, Q., Wahab, R., Musarrat, J., et al., (2018). Anticancer potential of green synthesized silver nanoparticles using extract of *Nepeta deflersiana* against human cervical cancer cells (HeLA). *Bioinorg. Chem. Appl., 2018*, 9390784. doi: https://doi.org/10.1155/2018/9390784.

Annu, Ahmed, S. A., Kaur, G., Sharma, P., Singh, S., & Ikram, S., (2018). Fruit waste (peel) as bio-reductant to synthesize silver nanoparticles with antimicrobial, antioxidant and cytotoxic activities. *J. Appl. Biomed., 16*(3), 221–231. doi: https://doi.org/10.1016/j.jab. 2018.02.002.

Ansari, F., Skrifvars, M., & Berglund, L., (2015). Nanostructured biocomposites based on unsaturated polyester resin and a cellulose nanofiber network. *Compos. Sci. Technol., 117*, 298–306. doi: https://doi.org/10.1016/j.compscitech.2015.07.004.

Bacakova, L., Pajorova, J., & Bacakova, M., (2019). Versatile application of nanocellulose: From industry to skin tissue engineering and wound healing. *Nanomaterials-Basel, 9*(2), 164. doi: https://doi.org/10.3390/nano9020164.

Baghizadeh, A., Ranjbar, S., Gupta, V. K., Asif, M., Pourseyedi, S., Karimi, M. J., & Mohammadinejad, R., (2015). Green synthesis of silver nanoparticles using seed extract of *Calendula officinalis* in liquid phase. *J. Mol. Liq., 207*, 159–163. doi: https://doi.org/10.1016/j.molliq.2015.03.029.

Bahmani, B., Bacon, D., & Anvari, B., (2013). Erythrocyte-derived photo-theranostic agents: Hybrid nano-vesicles containing indocyanine green for near-infrared imaging and therapeutic applications. *Sci. Rep., 3*(2180), 1–7. doi: https://doi.org/10.1038/srep02180.

Baker, S., & Satish, S., (2012). Endophytes: Toward a vision in synthesis of nanoparticles for future therapeutic agents. *Int. J. Bio-Inorg. Hybrid. Nanomat., 1*(2), 67–77.

Baker, S., Volova, T., Prudnikova, S. V., Satish, S., & Prasad, N. M. N., (2017). Nanoagroparticles emerging trends and future prospect in modern agriculture system. *Environ. Toxicol. Phar., 53*, 10–17.

Bala, N., Saha, A., Chakraborty, M., Maiti, M., Das, S., Basu, R., & Nandy, P., (2015). Green synthesis of zinc oxide nanoparticles using *Hibiscus sabdariffa* leaf extract: Effect of temperature on synthesis, anti-bacterial activity and anti-diabetic activity. *RSC Adv., 5*, 4993–5003. doi: https://doi.org/10.1039/C4RA12784F.

Balaji, S., Mandal, B. K., Ranjan, S., Dasgupta, N., & Chidambaram, R., (2017). Nano-zirconia-evaluation of its antioxidant and anticancer activity. *J. Photochem. Photobiol. B, 170*, 125–133. doi: https://doi.org/10.1016/j.jphotobiol.2017.04.004.

Bhat, A. H., Khan, I., Usmani, M. A., Reddicherla, U., & Al-Kindy, S. M. Z., (2019). Cellulose an ageless renewable green nanomaterial for medical applications: An overview of ionic liquids in extraction, separation and dissolution of cellulose. *Int. J. Biol. Macromol., 129*, 750–777. doi: https://doi.org/10.1016/j.ijbiomac.2018.12.190.

Bhattacharyya, A., Prasad, R., Buhroo, A. A., Duraisamy, P., Yousuf, I., Umadevi, M., Bindhu, M. R., et al., (2016). One-pot fabrication and characterization of silver nanoparticles using *Solanum lycopersicum*: An eco-friendly and potent control tool against rose aphid, *Macrosiphum rosae*. *J. Nanosci., 2016*, 467941. doi: https://doi.org/10.1155/2016/4679410.

Bollella, P., Schulz, C., Favero, G., Mazzei, F., Ludwig, R., Gordon, L., & Antiochia, R., (2016). Green synthesis and characterization of gold and silver nanoparticles and their application for development of a third-generation lactose biosensor. *Electroanal., 29*(1), 77–86. doi: https://doi.org/10.1002/elan.201600476.

Bonatto, C. C., & Silva, L. P., (2014). Higher temperatures speed up the growth and control the size and optoelectrical properties of silver nanoparticles greenly synthesized by cashew nutshells. *Ind. Crop. Prod., 58*, 46–54. doi: https://doi.org/10.1016/j.indcrop.2014.04.007.

Bumbudsanpharoke, N., & Ko, S., (2015). Nano-food packaging: An overview of market, migration research, and safety regulations. *J. Food. Sci., 80*(5), R910–R923. doi: https://doi.org/10.1111/1750-3841.12861.

Cadierno, V., Francos, J., & Gimeno, J., (2008). Selective ruthenium-catalyzed hydration of nitriles to amides in pure aqueous medium under neutral conditions. *Chem. Eur. J., 14*(22), 6601–6605. doi: https://doi.org/10.1002/chem.200800847.

Chahardooli, M., Khodadadi, E., & Khodadadi, E., (2014). Green synthesis of silver nanoparticles using oak leaf and fruit extracts (*Quercus*) and its antibacterial activity against plant pathogenic bacteria. *Int. J. Biosci., 4*(3), 97–103. doi: https://doi.org/10.12692/ijb/4.3.97-103.

Chandrasekaran, A. R., (2017). DNA nanobiosensors: An outlook on signal readout strategies. *J. Nanomater., 2017*, 2820619. doi: https://doi.org/10.1155/2017/2820619.

Chen, W., Yue, L., Jiang, Q., & Xia, W., (2019). Effect of chitosan with different molecular weight on the stability, antioxidant and anticancer activities of well-dispersed selenium nanoparticles. *IET Nanobiotechnol., 13*(1), 30–35. doi: https://doi.org/10.1049/iet-nbt.2018.505.

Chi, L., & Lee, S. T., (2019). Research on functional nanomaterials, interfaces, and applications at Soochow University. *ACS Nano, 13*(3), 2667–2671. doi: https://doi.org/10.1021/acsnano.9b01960.

Chowdhury, S., Basu, A., & Kundu, S., (2014). Green synthesis of protein capped silver nanoparticles from phytopathogenic fungus *Macrophomina phaseolina* (Tassi) gold with antimicrobial properties against multidrug-resistant bacteria. *Nanoscale Res. Lett., 9*(1), 365. doi: https://doi.org/10.1186/1556-276X-9-365.

Chung, I. M., Rahuman, A., Marimuthu, S., Kirthi, A. V., Anbarasan, K., Padmini, P., & Rajakumar, G., (2017). Green synthesis of copper nanoparticles using *Eclipta prostrata* leaves extract and their antioxidant and cytotoxic activities. *Exp. Ther. Med., 14*(1), 18–24. doi: https://doi.org/10.3892/etm.2017.4466.

Crucho, C. I. C., & Barros, M. T., (2017). Polymeric nanoparticles: A study on the preparation variables and characterization methods. *Mater. Sci. Eng. C, 80*, 771–784. doi: https://doi.org/10.1016/j.msec.2017.06.004.

Debjani, N., & Pratyusha, B., (2013). Green nanotechnology: A new hope for medical biology. *Environ. Toxicol. Phar., 36*(3), 997–1014. doi: https://doi.org/10.1016/j.etap.2013.09.002.

Deng, H. H., Li, G. W., Hong, L., Liu, A. L., Chen, W., Lin, X. H., & Xia, X. H., (2014). Colorimetric sensor based on dual-functional gold nanoparticles: Analyte-recognition and peroxidase-like activity. *Food Chem., 15*(147), 257–261. doi: https://doi.org/10.1016/j.foodchem.2013.09.151.

Deng, Z., Jung, J., Simonsen, J., Wang, Y., & Zhao, Y., (2017). Cellulose nanocrystal reinforced chitosan coatings for improving the storability of postharvest pears under both ambient and cold storages. *J. Food. Sci., 82*(2), 453–462. doi: https://doi.org/10.1111/1750-3841.13601.

Drygin, Y. F., Blintsov, A. N., Grigorenko, V. G., Andreeva, I. P., Osipov, A. P., Varitzev, Y. A., Uskov, A. I., et al., (2012). Highly sensitive field test lateral flow immunodiagnostics of PVX infection. *Appl. Microbiol. Biot., 93*(1), 179–189.

Du, Y., Gao, X., Ye, X., Zheng, Z., Feng, Q., Wang, C., & Wu, K., (2014). Composition and architecture-engineered Au–SnO_2/GNs-SWCNTs nanocomposites as ultrasensitive and robust electrochemical sensor for antioxidant additives in foods. *Sensor. Actuat. B: Chem., 203*, 926–934. doi: https://doi.org/https://doi.org/10.1016/j.snb.2014.06.094.

Duhan, J. S., Kumar, R., Kumar, N., Kaur, N., Nehra, K., & Duhan, S., (2017). Nanotechnology: The new perspective in precision agriculture. *Biotech. Rep., 15*, 11–23. doi: https://doi.org/10.1016/j.btre.2017.03.002.

Dutta, A. K., Das, S., Samanta, S., Samanta, P. K., Adhikary, B., & Biswas, P., (2013). CuS nanoparticles as a mimic peroxidase for colorimetric estimation of human blood glucose level. *Talanta, 107*, 361–367. doi: https://doi.org/10.1016/j.talanta.2013.01.032.

Echegoyen, Y., & Nerín, C., (2013). Nanoparticle release from nano-silver antimicrobial food containers. *Food Chem. Toxicol., 62*, 16–22. doi: https://doi.org/10.1016/j.fct.2013.08.014.

El-Ansary, A., & Faddah, L. M., (2010), Nanoparticles as biochemical sensor. *Nanotechnol. Sci. Appl., 3*, 65–76. doi: https://doi.org/10.2147/NSA.S8199.

Ellairaja, S., Krithiga, N., Ponmariappan, S., & Vasantha, V. S., (2017). Novel pyrimidine tagged silver nanoparticle based fluorescent immunoassay for the detection of *Pseudomonas aeruginosa*. *J. Agr. Food Chem., 65*(8), 1802–1812. doi: https://doi.org/10.1021/acs.jafc.6b04790.

FAO, (2007). *The FAO Register*. http://www.fao.org/brasil/noticias/detail-events/en/c/1053836/ (accessed on 8 July 2021).

Faruk, O., Bledzki, A. K., Fink, H. P., & Sain, M., (2012). Biocomposites reinforced with natural fibers: 2000–2010. *Prog. Polym. Sci., 37*(11), 1552–1596. doi: https://doi.org/10.1016/j.progpolymsci.2012.04.003.

Fleischer, T., & Grunwald, A., (2008). Making nanotechnology developments sustainable. A role for technology assessment? *J. Clean. Prod., 16*(8, 9), 889–898. doi: https://doi.org/10.1016/j.jclepro.2007.04.018.

Florea, A., Melinte, G., Simon, I., & Cristea, C., (2019). Electrochemical biosensors as potential diagnostic devices for autoimmune diseases. *Biosensors, 9*(1), 38. doi: https://doi.org/10.3390/bios9010038.

Francis, S., Joseph, S., Koshy, E. P., & Mathew, B., (2017). Green synthesis and characterization of gold and silver nanoparticles using *Mussaenda glabrata* leaf extract and their environmental applications to dye degradation. *Environ. Sci. Pollut. R., 24*(21), 17347–17357. doi: https://doi.org/10.1007/s11356-017-9329-2.

Frka-Petesic, B., Guidetti, G., Kamita, G., & Vignolini, S., (2017). Controlling the photonic properties of cholesteric cellulose nanocrystal films with magnets. *Adv. Mater., 29*(32), 1–7. doi: https://doi.org/10.1002/adma.201701469.

Frka-Petesic, B., Sugiyama, J., Kimura, S.,Chanzy, H., & Maret, G., (2015). Negative diamagnetic anisotropy and birefringence of cellulose nanocrystals. *Macromolecules, 48*(24), 8844–8857. doi: https://doi.org/10.1021/acs.macromol.5b02201.

Giraldo, J. P., Landry, M. P., Faltermeier, S. M., McNicholas, T. P., Iverson, N. M., Boghssian, A. A., Reuel, N. F., et al., (2014). Plant nanobionics approach to augment photosynthesis and biochemical sensing. *Nat. Mater., 13*, 400–408. doi: https://doi.org/10.1038/NMAT3890.

Green, J. E., Choi, J. W., Boukai, A., Bunimovich, Y., Johnston-Halperin, E., DeIonno, E., Luo, Y., et al., (2007). A 160-kilobit molecular electronic memory patterned at 1011 bits per square centimetre. *Nature, 445*(25), 414–417. doi: https://doi.org/10.1038/nature05462.

Grosso, G., Parravicini, G., (2014). Excitons, plasmons, and dielectric screening in crystals. In: (ed) *Solid State Physics* (2nd edn., pp. 287–331). Elsevier, Amsterdam. doi: https://doi.org/10.1016/B978-0-12-385030-0.00007-4.

Guidetti, G., Atifi, S., Vignolini, S., & Hamad, W. Y., (2016). Flexible photonic cellulose nanocrystal films. *Adv. Mater., 28*(45), 10042–10047. doi: https://doi.org/10.1002/adma.20160338.

Gurunathan, T., Mohanty, S., & Nayak, S. K., (2015). A review of the recent developments in biocomposites based on natural fibres and their application perspectives. *Compos Part. A-Appl. S., 77*, 1–25. doi: https://doi.org/10.1016/j.compositesa.2015.06.007.

Habte, L., Shiferaw, N., Mulatu, D., Thenepalli, T., Chilakala, R., & Ahn, J. W., (2019). Synthesis of nano-calcium oxide from waste eggshell by sol-gel method. *Sustainability-Basel, 11*, 1–10. doi: https://doi.org/10.3390/su11113196.

Haldorai, Y., Voit, W., & Shim, J. J., (2014). Nano ZnO@reduced graphene oxide composite for high performance supercapacitor: Green synthesis in supercritical fluid. *Electrochim. Acta, 120*, 65–72. doi: https://doi.org/10.1016/j.electacta.2013.12.063.

Hall, K. D., Guo, J., Dore, M., & Chow, C. C., (2009). The progressive increase of food waste in America and its environmental impact. *PLoS One, 11*(4), e7940. doi: https://doi.org/10.1371/journal.pone.0007940.

Hamelian, M., Varmira, K., & Veisi, H., (2018). Green synthesis and characterization of gold nanoparticles using thyme and survey cytotoxic effect, antibacterial and antioxidant potential. *J. Photoch. Photobio. B, 184*, 71–79. doi: https://doi.org/10.1016/j.jphotobiol.2018.05.016.

Hebeish, A., El-Shafei, A., Sharaf, S., & Zaghloul, S., (2011). Novel precursors for green synthesis and application of silver nanoparticles in the realm of cotton finishing. *Carbohyd. Polym., 84*(1), 605–613. doi: https://doi.org/10.1016/j.carbpol.2010.12.032.

Hebeish, A., Sharaf, S., & Farouk, A., (2013). Utilization of chitosan nanoparticles as a green finish in multifunctionalization of cotton textile. *Int. J. Biol. Macromol., 60*, 10–17. doi: https://doi.org/10.1016/j.ijbiomac.2013.04.078.

Hermida-Montero, L. A., Pariona, N., Mtz-Enriquez, A. I., Carrión, G., Paraguay-Delgado, F., & Rosas-Saito, G., (2019). Aqueous-phase synthesis of nanoparticles of copper/copper oxides and their antifungal effect against *Fusarium oxysporum*. *J. Hazard. Mater., 380*, 120850. doi: https://doi.org/10.1016/j.jhazmat.2019.120850.

Hsieh, M. C., Kim, C., Nogi, M., & Suganuma, K., (2013). Electrically conductive lines on cellulose nanopaper for flexible electrical devices. *Nanoscale, 5*, 9289–9295. doi: https://doi.org/10.1039/c3nr01951a.

Hu, L., Yuan, Y., Zhang, L., Xhao, J., Majeed, S., & Xu, G., (2013). Copper nanoclusters as peroxidase mimetics and their applications to H_2O_2 and glucose detection. *Anal. Chim. Acta., 762*, 83–86. doi: https://doi.org/10.1016/j.ijbiomac.2013.04.078.

Huang, H. L., Lin, C. C., & Hsu, K., (2015). Comparison of resistance improvement to fungal growth on green and conventional building materials by nano-metal impregnation. *Build. Environ., 93*, 119–127. doi: https://doi.org/10.1016/j.buildenv.2015.06.016.

Huang, J., Zhang, Y., Zhao, Y., Zhang, X. L., Sun, M. L., & Zhang, W., (2016). Superhydrophobic SERS chip based on Ag coated natural taro-leaf. *Nanoscale, 8*, 11487–11493. doi: https://doi.org/10.1039/c6nr03285k.

Iravani, S., (2011). Green synthesis of metal nanoparticles using plants. *Green. Chem., 13*(10), 2638–2650. doi: https://doi.org/10.1039/C1GC15386B.

Ishaaya, I., Nauen, R., & Horowitz, A. R., (2007). *Insecticides Design Using Advanced Technologies* (pp. 1–314). Springer, Berlin Heidelberg. doi: https://doi.org/10.1007/978-3-540-46907-0.

Itodo, H. U., Nnamonu, L. A., & Wuana, R. A., (2017). Green synthesis of copper chitosan nanoparticles for controlled release of pendimethalin. *Asian J. Chem. Sci., 2*(3), 1–10. doi: https://doi.org/10.9734/AJOCS/2017/32937.

Jannatin, M., Supriyanto, G., Abdulloh, Ibrahim, W. A. W., & Rukman, N. K., (2019). Graphene oxide from bagasse/magnetite composite: Preparation and characterization. *IOP C Ser. Earth Env., 217*, 1–7. doi: https://doi.org/10.1088/1755-1315/217/1/012007.

Jawaid, M., & Abdul-Khalil, H. P. S., (2011). Cellulosic/synthetic fibre reinforced polymer hybrid composites: A review. *Carbohyd. Polym., 86*(1), 1–8. doi: https://doi.org/10.1016/j.carbpol.2011.04.043.

Joshi, A., Kaur, S., Dharamvir, K., Nayyar, H., & Verma, G., (2018). Multi-walled carbon nanotubes applied through seed-priming influence early germination, root hair, growth and yield of bread wheat (*Triticum aestivum* L.). *J. Sci. Food. Agr., 98*(8), 3148–3160. doi: https://doi.org/10.1002/jsfa.8818.

Kaliannan, D., Palaninaicker, S., Palanivel, V., Mahadeo, M. A., Ravindra, B. N., & Jae-Jin, S., (2019). A novel approach to preparation of nano-adsorbent from agricultural wastes (*Saccharum officinarum* leaves) and its environmental application. *Environ. Sci. Pollut. R., 26*, 5305–5314. doi: https://doi.org/10.1007/s11356-018-3734-z.

Keshari, A. K., Srivastava, R., Singh, P., Yadav, V. B., & Nath, G., (2018). Antioxidant and antibacterial activity of silver nanoparticles synthesized by *Cestrum nocturnum*. *J. Ayurveda. Integer. Med., 11*(1), 1–8. doi: https://doi.org/10.1016/j.jaim.2017.11.003.

Khan, S. A., Kanwal, S., Rizwan, K., & Shahid, S., (2018). Enhanced antimicrobial, antioxidant, *in vivo* antitumor and *in vitro* anticancer effects against breast cancer cell line by green synthesized un-doped SnO_2 and Co-doped SnO_2 nanoparticles from *Clerodendrum inerme*. *Microb. Pathogenesis, 125*, 366–384. doi: https://doi.org/10.1016/j.micpath.2018.09.041.

Khatami, M., Sharifi, I., Nobre, M. A. L., Zafarnia, N., & Aflatoonian, M. R., (2018). Waste-grass-mediated green synthesis of silver nanoparticles and evaluation of their anticancer, antifungal and antibacterial activity. *Green. Chem. Lett. Rev., 11*(2), 125–134. doi: https://doi.org/10.1080/17518253.2018.1444797.

Kian, L. K., Saba, N., Jawaid, M., & Sultan, M. T. H., (2019). A review on processing techniques of bast fibers nanocellulose and its polylactic acid (PLA) nanocomposites. *Int. J. Biol. Macromol., 121*, 1314–1328. doi: https://doi.org/10.1016/j.ijbiomac.2018.09.040.

Kim, N. R., Yun, Y. S., Song, M. Y., Hong, S. J., Kang, M., Leal, C., Park, Y. W., & Jin, H. J., (2016). Citrus peel-derived, nanoporous carbon nanosheets containing redox-active heteroatoms for sodium-ion storage. *ACS Appl. Mater. Inter., 8*, 3175–3181. doi: https://doi.org/10.1021/acsami.5b10657.

Kitts, D. D., (1994). Bioactive substances in food: Identification and potential uses. *Can. J. Physiol. Pharm., 72*(4), 423–434. doi: https://doi.org/10.1139/y94–062.

Koedrith, P., Thasiphu, T., Weon, J. I., Boonprasert, R., Tuitemwong, K., & Tuitemwong, P., (2015). Recent trends in rapid environmental monitoring of pathogens and toxicants: Potential of nanoparticle-based biosensor and applications. *Sci. World. J., 2015*, 510982. doi: https://doi.org/10.1155/2015/510982.

Kokila, T., Ramesh, P. S., & Geetha, D., (2016). Biosynthesis of Ag NPs using *Carica papaya* peel extract and evaluation of its antioxidant and antimicrobial activities. *Ecotox. Environ. Saf., 134*(2), 467–473. doi: https://doi.org/10.1016/j.ecoenv.2016.03.021.

Kottegoda, N., Sandaruwan, C., Priyadarshana, G., Siriwardhana, A., Rathnayake, U. A., Arachchige, D. M. B., Kumarasinghe, A. R., et al., (2017). Urea-hydroxyapatite nanohybrids for slow release of nitrogen. *ACS Nano, 11*(2), 1214–1221. doi: https://doi.org/10.1021/acsnano.6b07781.

Kris-Etherton, P. M., Hecker, K. D., Bonanome, A., Coval, S. M., Binkoski, A. E., Hilpert, F. F., Griel, A. E., & Etherton, T. D., (2002). Bioactive compounds in foods: Their role in the prevention of cardiovascular disease and cancer. *Am. J. Med., 113*(9), 71–88. doi: https://doi.org/10.1016/s0002-9343(01)00995-0.

Kumar, H. K., Venkatesh, N., Bhowmik, H., & Kuila, A., (2018). Metallic nanoparticle: A review. *Biomed. J. Sci. Tech. Res., 4*(2), 3765–3775. doi: https://doi.org/10.26717/BJSTR.2018.04.001011.

Kwak, S. Y., Giraldo, J. P., Wong, M. W., Komar, V. B., Lew, T. T. S., Eli, J., Weidman, M. C., et al., (2017). A nanobionic light-emitting plant. *Nano Lett., 17*, 7951–7961. doi: https://doi.org/10.1021/acs.nanolett.7b04369.

Lateef, A., Bolaji, F., Oladejo, S. M., Akinola, P. O., Beukes, L. S., & Bosco, G. K. E., (2018). Characterization, antimicrobial, antioxidant, and anticoagulant activities of silver nanoparticles synthesized from *Petiveria alliacea* L. leaf extract. *Prep. Biochem. Biotech., 48*(7), 646–652. doi: https://doi.org/10.1080/10826068.2018.1479864.

Le, V. T., Bach, L. G., Pham, T. T., Le, N. T. T., Ngoc, U. T. P. N., Tran, D. H. N., & Nguyen, D. H., (2019). Synthesis and antifungal activity of chitosan-silver nanocomposite synergize fungicide against *Phytophthora capsici*. *J. Macromol. Sci. A, 56*(6), 522–528. doi: https://doi.org/10.1080/10601325.2019.1586439.

Lenardão, E. J., Freitag, R. A., Dabdoub, M. J., Batista, A. C. F., & Silveira, C. C., (2003). "Green chemistry"-Os 12 princípios da química verde e sua inserção nas atividades de ensino e pesquisa. *Quim. Nova, 26*(1), 123–129. doi: https://doi.org/10.1590/S0100-40422003000100020.

Li, L., Zhao, C., Zhang, Y., Yao, J., Yang, W., Hu, Q., Wang, C., & Cao, C., (2017). Effect of stable antimicrobial nano-silver packaging on inhibiting mildew and in storage of rice. *Food Chem., 215*, 477–482. doi: https://doi.org/10.1016/j.foodchem.2016.08.013.

Li, T., Li, S., Kong, W., Chen, C., Hitz, E., Jia, C., Zhang, X., et al., (2019). A nanofluidic ion regulation membrane with aligned cellulose nanofibers. *Sci. Adv., 5*(2), eaau4238. doi: https://doi.org/10.1126/sciadv.aau4238.

Liu, K., Zhang, J., Cheng, F. F., Zheng, T. T., Wang, C., & Zhu, J. J., (2011). Green and facile synthesis of highly biocompatible graphene nanosheets and its application for cellular imaging and drug delivery. *J. Mater. Chem., 21*, 12034–12040. doi: https://doi.org/10.1039/C1JM10749F.

Liu, Y., Ai, K., Ji, X., Askhatova, D., Du, R., Lu, L., & Shi, J., (2017). Comprehensive insights into the multi-antioxidative mechanisms of melanin nanoparticles and their application to protect the brain from injury in ischemic stroke. *J. Am. Chem. Soc., 139*(2), 856–862. doi: https://doi.org/10.1021/jacs.6b11013.

Lu, J. J., Dang, Y. Y., Huang, M., Xu, W. S., Chen, X. P., & Wang, Y. T., (2012). Anti-cancer properties of terpenoids isolated from *Rhizoma curcumae*: A review. *J. Ethnopharmacol., 143*(2), 406–411. doi: 10.1016/j.jep.2012.07.009.

Lu, Y., & Ozcan, S., (2015). Green nanomaterials: On track for a sustainable future. *Nanotoday, 10*(4), 417–420. doi: https://doi.org/10.1016/j.nantod.2015.04.010.

Luo, H., Li, J. J., & Zhou, F. S., (2018). Advances in hard tissue engineering materials-nanocellulose-based composites. *PBM Nanocellulose-Based Composite, 3*(4), 62–76.

Ma, V., Zhang, B., Liu, J., Zhang, P., Li, Z., & Luan, Y., (2015). Green fabricated reduced graphene oxide: Evaluation of its application as nano-carrier for pH-sensitive drug delivery. *Int. J. Pharm., 496*(2), 984–992. doi: https://doi.org/10.1016/j.ijpharm.2015.10.081.

Marchetto, A. M. P., Ataide, H. H., Masson, M. L. F., Pelizer, L. H., Pereira, C. H. C., & Sendão, M. C., (2008). Avaliação das partes desperdiçadas de alimentos no setor de hortifruti visando seu reaproveitamento. *Rev. Simbio-Logias, 1*(2), 1–14.

Mistry, K. R., & Sarker, D. K., (2016). SLNs can serve as the new brachytherapy seed: Determining influence of surfactants on particle size of solid lipid microparticles and development of hydrophobized copper nanoparticles for potential insertion. *J. Chem. Eng. Process. Technol., 7*(3), 1–9. doi: https://doi.org/10.4172/2157-7048.1000302.

Mohammadi, A., Hashemi, M., & Hosseini, S. M., (2016). Postharvest treatment of nanochitosan-based coating loaded with *Zataria multiflora* essential oil improves antioxidant activity and extends shelf-life of cucumber. *Innov. Food Sci. Emerg. Tech., 33*, 580–588. doi: https://doi.org/10.1016/j.ifset.2015.10.015.

Molaei, S., & Javanshir, S., (2018). Preparation, characterization and use of new lignocellulose-based bio nanocomposite as a heterogeneous catalyst for sustainable synthesis of pyrimido benzazoles. *Green Chem. Lett. Rev., 11*, 275–285. doi: https://doi.org/10.1080/17518253.2018.1484180.

Monteiro, S. N., Calado, V., Rodriguez, R. J. S., & Margem, F. M., (2012). Thermogravimetric behavior of natural fibers reinforced polymer composites: An overview. *Mater. Sci. Eng. A, 557*, 17–28. doi: https://doi.org/10.1016/j.msea.2012.05.109.

Moon, R. J., Martini, A., Nairn, J., Simonsen, J., & Youngblood, J., (2011). Cellulose nanomaterials review: Structure, properties and nanocomposites. *Chem. Soc. Rev., 40*, 3941–3994. doi: https://doi.org/10.1039/C0CS00108B.

Motaung, T. E., & Linganiso, L. Z., (2018). Critical review on agrowaste cellulose applications for biopolymers. *Int. J. Plast. Technol., 22*(2), 185–216. doi: https://doi.org/10.1007/s12588-018-9219-6.

Muthukumar, T., Sudhakumari, Sambandam, B., Aravinthan, A., Sastry, T. P., & Kim, J. H., (2016). Green synthesis of gold nanoparticles and their enhanced synergistic antitumor activity using HepG2 and MCF7 cells and its antibacterial effects. *Process Biochem., 51*(3), 384–391. doi: https://doi.org/10.1016/j.procbio.2015.12.017.

Nadagouda, M. N., & Varma, R. S., (2006). Green and controlled synthesis of gold and platinum nanomaterials using vitamin B2: Density-assisted self-assembly of nanospheres, wires and rods. *Green Chem., 8*, 516–518. doi: https://doi.org/10.1039/B601271J.

Nasrollahzadeh, M., Ghorbannezhad, F., Issaabadi, Z., & Sajadi, S. M., (2019). Recent developments in the biosynthesis of Cu-based recyclable nanocatalysts using plant extracts and their application in the chemical reactions. *Chem. Rec. 19*, 601–643. doi: https://doi.org/10.1002/tcr.201800069.

Nuilek, K., Simo, A., & Baumli, P., (2018). Influence of KOH on the carbon nanostructure of peanut shell. *Resolution and Discovery, 3*, 29–32. doi: https://doi.org/10.1556/2051.2018.00060.

Otoni, C. G., Pontes, S. F. O., Medeiros, E. A. A., & Soares, N. F. F., (2014). Edible films from methylcellulose and nanoemulsions of clove bud (*Syzygium aromaticum*) and oregano (*Origanum vulgare*) essential oils as shelf-life extenders for sliced bread. *J. Agr. Food. Chem., 62*(22), 5214–5219. doi: https://doi.org/10.1021/jf501055f.

Ovais, M., Khalil, A. T., Raza, A., Khan, M. A., Ahmad, I., Islam, N. U., Muthupandian, S., et al., (2016). Green synthesis of silver nanoparticles via plant extracts: Beginning a new era in cancer theranostics. *Nanomedicine, 11*(23), 3157–3177. doi: https://doi.org/10.2217/nnm-2016-0279.

Padil, V. V. T., Wacławek, S., Černík, M., & Varma, R. S., (2018). Tree gum-based renewable materials: Sustainable applications in nanotechnology, biomedical and environmental fields. *Biotechnol. Adv., 36*(7), 1984–2016. doi: https://doi.org/10.1016/j.biotechadv.2018.08.008.

Panchal, P., Ogunsona, & Mekonnen, T., (2019). Trends in advanced functional material applications of nanocellulose. *Processes, 7*(1), 10. doi: https://doi.org/10.3390/pr7010010.

Pandey, S., Goswami, G. K., & Nanda, K. K., (2012). Green synthesis of biopolymer-silver nanoparticle nanocomposite: An optical sensor for ammonia detection. *Int. J. Biol. Macromol., 51*(4), 583–589. doi: https://doi.org/10.1016/j.ijbiomac.2012.06.033.

Parida, C., Das, S. C., & Dash, S. K., (2012). Mechanical analysis of bio nanocomposite prepared from *Luffa cylindrica. Procedia Chem., 4*, 53–59. doi: https://doi.org/10.1016/j.proche.2012.06.008.

Park, J., Shin, K., & Lee, C., (2016). Roll-to-roll coating technology and its applications: A review. *Int. J. Precis. Eng. Man., 17*(4), 537–550. doi: https://doi.org/10.1007/s12541-016-0067-z.

Pathak, G., Rajkumari, K., & Rokhum, L., (2019). Wealth from waste: *M. acuminata* peel waste-derived magnetic nanoparticles as a solid catalyst for the Henry reaction. *Nanoscale Adv., 1*, 1013–1020. doi: https://doi.org/10.1039/C8NA00321A.

Patra, J. K., & Baek, K., (2015). Novel green synthesis of gold nanoparticles using *Citrullus lanatus* rind and investigation of proteasome inhibitory activity, antibacterial, and antioxidant potential. *Int. J. Nanomed., 10*, 7253–7264. doi: https://doi.org/10.2147/IJN.S95483.

Pattni, B. S., Chupin, V. V., & Torchilin, V. P., (2015). New developments in liposomal drug delivery. *Chem. Rev., 115*(19), 10938–10966. doi: https://doi.org/10.1021/acs.chemrev.5b00046.

Paul, S., Pearson, C., Molloy, A., Cousins, M. A., Green, M., Kolliopoulou, S., Dimitrakis, P., et al., (2003). Langmuir-Blodgett film deposition of metallic nanoparticles and their application to electronic memory structure. *Nano. Lett., 3*(4), 533–536. doi: https://doi.org/10.1021/nl034008t.

Paulkumar, K., Gnanajobitha, G., Vanaja, M., Rajeshkumar, S., Malarkodi, C., Pandian, K., & Annadurai, G., (2014). *Piper nigrum* leaf and stem assisted green synthesis of silver nanoparticles and evaluation of its antibacterial activity against agricultural plant pathogens. *Sci. World. J., 2014*, 829894. doi: https://doi.org/10.1155/2014/829894.

Phanthong, P., Reubroycharoen, P., Hao, X., Xu, G., Abudula, A., & Guan, G., (2018). Nanocellulose: Extraction and application. *Carbon Resour. Convers., 1*(1), 32–43. doi: https://doi.org/10.1016/j.crcon.2018.05.004.

Phull, A. R., Abbas, Q., Ali, A., Raza, H., Kim, S. J., Zia, M., & Haq, I. U., (2016). Antioxidant, cytotoxic and antimicrobial activities of green synthesized silver nanoparticles from crude extract of *Bergenia ciliata. Future J. Pharm. Sci., 2*, 31–36. doi: https://doi.org/10.1016/j.fjps.2016.03.001.

Pilon, L., Spricigo, P. C., Miranda, M., Moura, M. R., Assis, O. B. G., Mattoso, L. H. C., & Ferreira, M. D., (2014). Chitosan nanoparticle coatings reduce microbial growth on fresh-cut apples while not affecting quality attributes. *Int. J. Food. Sci. Tech., 50*(2), 440–448. doi: https://doi.org/10.1111/ijfs.12616.

Polshettiwar, V., & Varma, R. S., (2010). Green chemistry by nano-catalysis. *RSC Adv., 12*, 743–754. doi: https://doi.org/10.1039/B921171C.

Pouratashi, M., & Iravani, H., (2012). Farmers' knowledge of integrated pest management and learning style preferences: Implications for information delivery. *Int. J. Pest. Manage., 58*(4), 347–353 doi: https://doi.org/10.1080/09670874.2012.724468.

Prado, A. G. S., (2003). Química verde, os desafios da química do novo milênio. *Quim. Nova, 26*(5), 738–744. doi: https://doi.org/10.1590/S0100-40422003000500018.

Prasad, R., Bhattacharyya, A., & Nguyen, Q. D., (2017). Nanotechnology in sustainable agriculture: Recent developments, challenges, and perspectives. *Front. Microbiol., 8*(1014), 1–13. doi: https://doi.org/10.3389/fmicb.2017.01014.

Quan Ling, Y. Q., Jun Wei, Y., Zhu Qun, S., Shu Jie, X., & Chuan Xi, X., (2018). Recent progress of nanocellulose-based electroconductive materials and their applications as electronic devices. *J. Forest. Eng., 3*(3), 1–11.

Rabab, M. A. E., & El-Shafey, R. A. S., (2013). Inhibition effects of silver nanoparticles against rice blast disease caused by *Magnaporthe grisea. Egypt J. Agric. Res., 91*, 1271–1283.

Rajakumar, G., Rahuman, A. A., Roopan, S. M., Khanna, V. G., Elango, G., Kamaraj, C., Zahir, A. A., & Velayutham, K., (2012). Fungus-mediated biosynthesis and characterization of TiO_2 nanoparticles and their activity against pathogenic bacteria. *Spec. Trochim. Acta. A, 91*, 23–29. doi: https://doi.org/10.1016/j.saa.2012.01.011.

Rajeshkumar, S., (2019). Antifungal impact of nanoparticles against different plant pathogenic fungi. In: *Nanomaterials in Plants, Algae and Microorganisms* (1st edn., pp. 197–217). Chennai, India. doi: https://doi.org/10.1016/B978-0-12-811488-9.00010-X.

Rajeswaran, S., Thirugnanasambandan, S. S., Dewangan, N. K., Moorthy, R. K., Kandasamy, S., & Vilwanathan, R., (2019). Multifarious pharmacological applications of green routed eco-friendly iron nanoparticles synthesized by *Streptomyces* sp. (SRT12). *Biol. Trace Elem. Res., 194*(1), 273–283. doi: https://doi.org/10.1007/s12011-019-01777-5.

Ramirez-Nuñez, A. L., Jimenez-Garcia, L. F., Goya, G. F., Sanz, B., & Santoyo-Salanzar, J., (2018). *In vitro* magnetic hyperthermia using polyphenol-coated $Fe_3O_4@\gamma Fe_2O_3$ nanoparticles from *Cinnamomum verum* and *Vanilla planifolia*: the concert of green synthesis and therapeutic possibilities. *Nanotechnology, 29*(7), 074001. doi: https://doi.org/10.1088/1361-6528/aaa2c1.

Rastogi, L., Beedu, S. R., & Kora, A. J., (2015). Facile synthesis of palladium nanocatalyst using gum kondagogu (*Cochlospermum gossypium*): A natural biopolymer. *IET Nanobiotechnol., 9*(6), 362–367. doi: https://doi.org/10.1049/iet-nbt.2014.0055.

Rastogi, L., Karunasagar, D., Sashidhar, R. B., & Giri, A., (2017). Peroxidase-like activity of gum kondagogu reduced/stabilized palladium nanoparticles and its analytical application for colorimetric detection of glucose in biological samples. *Sensor. Actuat, B: Chem., 240*, 1182–1188. doi: https://doi.org/10.1016/j.snb.2016.09.066.

Razavi, M., Salahinejad, E., Fahmy, M., Yazdimamaghani, M., Vashaee, D., & Tayebi, L., (2015). Green chemical and biological synthesis of nanoparticles and their biomedical applications. In: Basiuk, A. V., & Basiuk, E. V., (eds.), *Green Processes for Nanotechnology* (1st edn. pp. 207–223). doi: https://doi.org/10.1007/978-3-319-15461-9_7.

Rol, F., Belgacem, M. N., Gandini, A., & Bras, J., (2019). Recent advances in surface-modified cellulose nanofibrils. *Prog. Polym. Sci., 88*, 241–264. doi: https://doi.org/10.1016/j.progpolymsci.2018.09.002.

Sagbas, S., & Sahiner, N., (2019). Carbon dots: Preparation, properties, and application. In: Khan, A., Jawaid, M., Inamuddin, & Asiri, A. M., (eds.), *Nanocarbon and its Composites* (1st edn., pp. 651–676). doi: https://doi.org/10.1016/b978-0-08-102509-3.00022-5.

Santhoshkumar, T., Rahuman, A. A., Jayaseelan, C., Rajakumar, G., Marimuthu, S., Kirthi, A. V., Velayutham, K., et al., (2014). Green synthesis of titanium dioxide nanoparticles using *Psidium guajava* extracts and its antibacterial and antioxidant properties. *Asian Pac. J. Trop. Med., 7*(12), 968–976. doi: https://doi.org/10.1016/S1995-7645(14)60171-1.

Santiago, T. R., Bonatto, C. C., Rossato, M., Lopes, C. A. P., Lopes, C. A., Mizubuti, E. S. G., & Silva, L. P., (2019). Green synthesis of silver nanoparticles using tomato leaves extract and their entrapment in chitosan nanoparticles to control bacterial wilt. *J. Sci. Food Agr., 99*, 4248–4259. doi: https://doi.org/10.1002/jsfa.9656.

Sathishkumar, M., Sneha, K., Won, S. W., Cho, C. W., Kim, S., & Yun, Y. S., (2009). *Cinnamon zeylanicum* bark extract and powder mediated green synthesis of nano-crystalline silver particles and its bactericidal activity. *Colloid. Surface. B, 73*(2), 332–338. doi: https://doi.org/10.1016/j.colsurfb.2009.06.005.

Setyowati, E., (2014). Eco-building material of Styrofoam waste and sugar industry fly-ash based on nano-technology. *Procedia. Environ. Sci., 20*, 245–253. doi: https://doi.org/10.1016/j.proenv.2014.03.031.

Shak, K. P. Y., Pang, Y. L., & Mah, S. K., (2018). Nanocellulose: Recent advances and its prospects in environmental remediation. *Beilstein J. Nanotech., 9*, 2479–2498. doi: https://doi.org/10.3762/bjnano.9.232.

Sharma, A., Thakur, M., Bhattacharya, M., Mandal, T., & Goswami, S., (2019). Commercial application of cellulose nanocomposites: A review. *Biotechnol. Rep., 21*. doi: https://doi.org/10.1016/j.btre.2019.e00316.

Sharma, V. K., Yngard, R. A., & Lin, Y., (2018). Silver nanoparticles: Green synthesis and their antimicrobial activities. *Adv. Colloid. Interface., 145*(1, 2), 83–96. doi: https://doi.org/10.1016/j.cis.2008.09.002.

Shateri-Khalilabad, M., & Yazdanshenas, M. E., (2013). Bifunctionalization of cotton textiles by ZnO nanostructures: Antimicrobial activity and ultraviolet protection. *Text. Res. J., 83*(10), 993–1004. doi: https://doi.org/10.1177/0040517512468812.

Shebl, A., Hassan, A., Salama, D., El-Aziz, M. E. A., & Elwahed, M. A., (2019). Green synthesis of manganese zinc ferrite nanoparticles and their application as nanofertilizers for *Cucurbita pepo* L. *Beilstein Arch., 45*. doi: https://doi.org/10.3762/bxiv.2019.45.v1.

Sivakumar, P., Sivakumar, P., Anbarasu, K., Pandian, K., & Renganathan, S., (2013). Synthesis of silver nanorods from food industrial waste and their application in improving the keeping quality of milk. *Ind. Eng. Chem. Res., 52*(49), 17676–17681. doi: https://doi.org/10.1021/ie4009327.

Šmidlehner, T., Piantanida, I., & Pescitelli, G., (2017). Polarization spectroscopy methods in the determination of interactions of small molecules with nucleic acids-tutorial. *Beilstein J. Org. Chem., 14*, 84–105. doi: https://doi.org/10.3762/bjoc.14.5.

Somorjai, G. A., Frei, H., & Park, J. Y., (2009). Advancing the frontiers in nanocatalysis, biointerfaces, and renewable energy conversion by innovations of surface techniques. *J. Am. Chem. Soc., 131*(46), 16589–16605. doi: https://doi.org/10.1021/ja9061954.

Sonker, A., Richa, Pathak, J., Rajneesh, Kannaujiya, V. K., & Sinha, R. P., (2017). Characterization and *in vitro* antitumor, antibacterial and antifungal activities of green synthesized silver nanoparticles using cell extract of *Nostoc* sp. strain HKAR-2. *Can. J. Biotech., 1*(1), 26–37. doi: https://doi.org/10.24870/cjb.2017-000103.

Stavrinidou, E., Gabrielsson, R., Gomez, E., Crispin, X., Nilsson, O., Simon, D. T., & Berggren, M., (2015). Electronic plants. *Sci. Adv., 1*(10), e1501136. doi: https://doi.org/10.1126/sciadv.1501136.

Sudhakar, P., & Soni, H., (2018). Catalytic reduction of nitrophenols using silver nanoparticles-supported activated carbon derived from agro-waste. *J. Environ. Chem. Eng., 6*, 28–36. doi: https://doi.org/10.1016/j.jece.2017.11.053.

Suganya, A., Murugan, K., Kovendan, K., Kumar, P. M., & Hwang, J. S., (2013). Green synthesis of silver nanoparticles using *Murraya koenigii* leaf extract against *Anopheles stephensi* and *Aedes aegypti*. *Parasitol. Res., 112*(4), 1385–1397.

Sumitha, S., Vasanthi, S., Shalini, S., Chinni, S. V., Gopinath, S. C. B., Anbu, P., Bahari, M. B., et al., (2018). Phyto-mediated photo catalyzed green synthesis of silver nanoparticles using *Durio zibethinus* seed extract: Antimicrobial and cytotoxic activity and photocatalytic applications. *Molecules, 23*(12), 3311. doi: https://doi.org/10.3390/molecules23123311.

Sun, H., Liu, H., & Wu, Y., (2017). A green, reusable SERS film with high sensitivity for *in-situ* detection of thiram in apple juice. *Appl. Surf. Sci., 416*, 704–709. doi: https://doi.org/10.1016/j.apsusc.2017.04.159.

Sun, Y. P., Zhou, B., Lin, Y., Wang, W., Fernando, K. A. S., Pathak, P., Meziani, M. J., et al., (2006). Quantum-sized carbon dots for bright and colorful photoluminescence. *J. Am. Chem. Soc., 128*(24), 7756–7762. doi: https://doi.org/10.1021/ja062677d.

Syed, B., Prasad, M. N., & Satish, S., (2016). Synthesis and characterization of silver nanobactericides produced by *Aneurinibacillus migulanus* 141, a novel endophyte inhabiting *Mimosa pudica* L. *Arab. J. Chem., 2016*, doi: https://doi.org/10.1016/j.arabjc.2016.01.005.

Syrový, T., Maronová, S., Kuberský, P., Ehman, N. V., Vallwjos, M. E., Preti, S., Felissia, F. E., et al., (2019). Wide range humidity sensors printed on biocomposite films of cellulose nanofibril and poly(ethylene glycol). *J. Appl. Polym. Sci., 15*(36), 1–10. doi: https://doi.org/10.1002/APP.47920.

Tadros, T. F., (2013). Emulsion formation, stability, and rheology. In: Tadros, T. F., (ed.), *Emulsion Formation and Stability* (pp. 1–75). Wiley, New York. doi: https://doi.org/10.1002/9783527647941.ch1.

Tayeb, A. H., Amini, E., Ghasemi, S., & Tajvidi, M., (2018). Cellulose nanomaterials-binding properties and applications: A review. *Molecules, 23*(10), 2684. doi: https://doi.org/10.3390/molecules23102684.

Thakur, M. S., & Ragavan, K. V., (2013). Biosensors in food processing. *J. Food. Sci. Technol., 50*(4), 625–641. doi: https://doi.org/10.1007/s13197-012-0783-z.

Thatoi, P., Kerry, R. G., Gouda, S., Das, G., Pramanik, K., Thatoi, H., & Patra, J. K., (2016). Photo-mediated green synthesis of silver and zinc oxide nanoparticles using aqueous extracts of two mangrove plant species, *Heritiera fomes* and *Sonneratia apetala* and investigation of their biomedical applications. *J. Photochem. Photobiol. B, 163*, 311–318. doi: https://doi.org/10.1016/j.jphotobiol.2016.07.029r.

Thomas, B., Raj, M. C., Athira, K. B., Rubiyah, M. H., Joy, J., Moores, A., Drisko, G. L., & Sanchez, C., (2018). Nanocellulose, a versatile green platform: From bio-sources to materials and their applications. *Chem. Rev., 118*(24), 11575–11625. doi: https://doi.org/10.1021/acs.chemrev.7b00627.

Titirici, M. M., White, R. J., Brun, N., Budarin, V. L., Su, D. S., Monte, F., Clark, J. H., & MacLachlan, M. J., (2015). Sustainable carbon nanomaterials. *Chem. Soc. Rev., 44*(1), 250–290.

Toivonen, M. S., Onelli, O. D., Jacucci, G., Lovikka, V., Rojas, O. J., Ikkala, O., & Vignolini, S., (2018). Anomalous-diffusion-assisted brightness in white cellulose nanofibril membranes. *Adv. Mater., 30*(16), 1–7. doi: https://doi.org/10.1002/adma.201704050.

Valdebenito, F., García, R., Cruces, K., Ciudad, G., Chinga-Carrasco, G., & Habibi, Y., (2018). CO_2 adsorption of surface-modified cellulose nanofibril films derived from agricultural wastes. *ACS Sustain Chem. Eng., 6*, 12603–12612. doi: https://doi.org/10.1021/acssuschemeng.8b00771.

Valgimigli, L., Baschieri, A., & Amorati, R., (2018). Antioxidant activity of nanomaterials. *J. Mater. Chem. B, 6*, 2036–2051. doi: https://doi.org/10.1039/c8tb00107c.

Varghese, A. G., Paul, S. A., & Latha, M. S., (2018). Remediation of heavy metals and dyes from wastewater using cellulose-based adsorbents. *Environ. Chem. Lett., 17*(2), 867–877. doi: https://doi.org/10.1007/s10311-018-00843-z.

Varun, S., Daniel, K., & Gorthi, S. S., (2017). Rapid sensing of melamine in milk by interference green synthesis of silver nanoparticles. *Mater. Sci. Eng. C, 74*, 253–258. doi: https://doi.org/10.1016/j.msec.2016.12.011.

Velmurugan, P., Anbalagan, K., Manosathyadevan, M., Lee, K. J., Cho, M., Lee, S. M., Park, J. H., et al., (2014b). Green synthesis of silver and gold nanoparticles using *Zingiber officinale* root extract and antibacterial activity of silver nanoparticles against food pathogens. *Bioproc. Biosyst. Eng., 37*(10), 1935–1943. doi: https://doi.org/10.1007/s00449-014-1169-6.

Velmurugan, P., Cho, M., Lee, S. M., Park, J. H., Bae, S., & Oh, B. T., (2014c). Antimicrobial fabrication of cotton fabric and leather using green-synthesized nanosilver. *Carbohyd. Polym., 106*, 319–325. doi: https://doi.org/10.1016/j.carbpol.2014.02.021.

Velmurugan, P., Iydroose, M., Lee, S. M.,Cho, M., Park, J. H., Balachandar, V., & Oh, B. T., (2014a). Synthesis of silver and gold nanoparticles using cashew nut shell liquid and its antibacterial activity against fish pathogens. *Indian J. Microbiol., 54*(2), 196–202. doi: https://doi.org/10.1007/s12088-013-0437-5.

Vinotha, V., Iswarya, A., Thaya, R., Govindarajan, M., Alharbi, N. S., Kadaikunnan, S., Khaled, J. M., et al., (2019). Synthesis of ZnO nanoparticles using insulin-rich leaf extract: Anti-diabetic, antibiofilm and anti-oxidant properties. *J. Photochem. Photobiol. B, 197*, 111541. doi: https://doi.org/10.1016/j.jphotobiol.2019.111541.

Wang, C., Wang, S., Chen, G., Kong, W., Ping, W., Dai, J., Pastel, G., et al., (2018). Flexible, bio-compatible nanofluidic ion conductor. *Chem. Mater., 30*(21), 7707–7713. doi: https://doi.org/10.1021/acs.chemmater.8b03006.

Wang, D., (2019). A critical review of cellulose-based nanomaterials for water purification in industrial processes. *Cellulose, 26*(2), 687–701. doi: https://doi.org/10.1007/s10570-018-2143-2.

Whitesides, G. M., (2005). Nanoscience, nanotechnology, and chemistry. *Small, 1*(2), 172–179. doi: https://doi.org/10.1002/smll.200400130.

Wong, M. H., Giraldo, J. P., Kwak, S. Y., Koman, V. B., Sinclair, R., Lew, T. T. S., Bisker, G., et al., (2017). Nitroaromatic detection and infrared communication from wild-type plants using plant nanobionics. *Nat. Mater., 16*(2), 264–272. doi: https://doi.org/10.1038/nmat4771.

Xu, H., Yang, X., Li, G., Zhao, C., & Liao, X., (2015). Green synthesis of fluorescent carbon dots for selective detection of tartrazine in food samples. *J. Agr. Food. Chem., 63*(30), 6707–6714. doi: https://doi.org/10.1021/acs.jafc.5b02319.

Xue, J., Song, F., Yin, X. W., Wang, X. I., & Wang, Y. Z., (2015). Let it shine: A transparent and photoluminescent foldable nanocellulose/quantum dot paper. *ACS Appl. Mater. Interf., 19*(7), 10076–10079. doi: https://doi.org/10.1021/acsami.5b02011.

Yan, L., Kasal, B., & Huang, L., (2016). A review of recent research on the use of cellulosic fibres, their fibre fabric reinforced cementitious, geo-polymer and polymer composites in civil engineering. *Compos. Part. B Eng., 92*, 94–132. doi: https://doi.org/10.1016/j.compositesb.2016.02.002.

Young, A. J., & Lowe, G. M., (2001). Antioxidant and prooxidant properties of carotenoids. *Arch. Biochem. Biophys., 385*(1), 20–27. doi: https://doi.org/10.1006/abbi.2000.2149.

Yu, Z., Sun, X., Song, H., Wang, W., Ye, Z., Shi, L., & Ding, K., (2015). Glutathione-responsive carboxymethyl chitosan nanoparticles for controlled release of herbicides. *Mat. Sci. Appl., 6*(6), 591–604. doi: https://doi.org/10.4236/msa.2015.66062.

Zahir, A. A., Bagavan, A., Kamaraj, C., Elango, G., & Rahuman, A., (2012). Efficacy of plant-mediated synthesized silver nanoparticles against *Sitophilus oryzae*. *J. Biopest., 5*, 95–102.

Zambrano-Zaragoza, M. L., Gónzalez-Reza, R., Mendoza-Muñoz, N., Miranda-Linares, V., Bernal-Couoh, T. F., Mendoza-Elvira, S., & Quintanar-Guerrero, D., (2018). Nanosystems in edible coatings: A novel strategy for food preservation. *Int. J. Mol. Sci., 19*(3), 705. doi: https://doi.org/10.3390/ijms19030705.

Zhang, L., Hayes, D. G., Chen, G., & Zhong, Q., (2013). Transparent dispersions of milk-fat-based nanostructured lipid carriers for delivery of β-carotene. *J. Agric. Food Chem., 61*(39), 9435–9443. doi: https://doi.org/10.1021/jf403512c.

Zhao, Y., Liu, L., Dezhao, K., Kuang, H., Wang, L., & Xu, C. H., (2014). Dual amplified electrochemical immunosensor for highly sensitive detection of *Pantoea stewartii* subsp. *stewartii*. *ACS Appl. Mat. Int., 6*(23), 21178–21183. doi: https://doi.org/10.1021/am506104r.

Zheng, M., Wang, C., Wang, Y., Wei, W., Ma, S., Sun, X., & He, J., (2018). Green synthesis of carbon dots functionalized silver nanoparticles for the colorimetric detection of phoxim. *Talanta, 185*, 309–315. doi: https://doi.org/10.1016/j.talanta.2018.03.066.

Zyoud, A., & Hilal, H., (2013). Curcumin-sensitized anatase TiO_2 nanoparticles for photodegradation of methyl orange with solar radiation. In: *1st International Conference and Exhibition on the Applications of Information Technology to Renewable Energy Processes and Systems*. Jordan, doi: https://doi.org/10.1109/IT-DREPS.2013.6588145.

INDEX

H

I

Q

R